INTRODUCTION TO MINERALOGY AND PETROLOGY

INTRODUCTION TO MINERALOGY AND PETROLOGY

S. K. Haldar

Emeritus Scientist, Dept. of Applied Geology & Environmental System Management, Presidency University, Kolkata-700 073, and IMX Resources Limited, Australia. Formerly, Hindustan Zinc Limited, Hindustan Copper Limited, ESSO INC and BIL Infratech Ltd, India.

Josip Tišljar✠

Formerly Professors, Head, Director, President and Dean, Department of Mining, Geology and Petroleum Engineering, University of Zagreb, Croatia.

✠Deceased

ELSEVIER

AMSTERDAM • BOSTON • HEIDELBERG • LONDON • NEW YORK • OXFORD
PARIS • SAN DIEGO • SAN FRANCISCO • SYDNEY • TOKYO

Elsevier
225 Wyman Street, Waltham, MA 02451, USA
The Boulevard, Langford Lane, Kidlington, Oxford, OX5 1GB, UK
Radarweg 29, PO Box 211, 1000 AE Amsterdam, The Netherlands

Library of Congress Cataloging-in-Publication Data
Haldar, S. K., and Tišljar Josip
Introduction to mineralogy and petrology / S.K. Haldar and Josip Tišljar.
pages cm
Includes bibliographical references and index.
ISBN 978-0-12-408133-8
1. Petrology. 2. Mineralogy. I. Title.
QE431.2.H35 2013
552–dc23
2013027380

British Library Cataloguing in Publication Data
A catalogue record for this book is available from the British Library

ISBN: 978-0-12-408133-8

For information on all Elsevier publications
visit our web site at store.elsevier.com

Dedication

"The soul is neither born nor does it die,
nor having been born does it cease to be at any time.
Unborn, eternal, ever-abiding and primeval,
it is not killed though the body is slain". (II-20)

Bhagavat Gita

Dedicated in the memory of Late Professor Josip Tišljar (1941–2009) who devoted his entire academic life for the development of the nation as a whole and for the growth of the students in particular. May his eternal, all-pervading, stable, immovable and primeval soul rests in peace in the eternity of God.

Table of Contents

Preface xiii
List of Acronyms used in this Book xv
About the Author xvii

1. Rocks and Minerals

1.1. Introduction 1
1.2. Importance of Minerals and Rocks to Society 1
1.3. Minerals 5
1.4. Rocks 12
1.5. Mineral Resources 14
Further Reading 37

2. Basic Mineralogy

2.1. Introduction 39
2.2. Internal Structure of Crystals and Their Properties 40
2.2.1. Crystallized and Amorphous Solid Minerals 40
2.2.2. Formation of Minerals 41
2.2.3. Crystal Lattice 42
2.2.4. Crystallographic Axes, the Crystal Planes and Elements of Crystal Symmetry 43
2.2.5. Crystal Systems, Crystal Forms, Single Crystals and Crystal Twinning 45
2.3. Chemical and Physical Properties of Minerals 47
2.3.1. Chemical Properties of Minerals 47
2.3.2. Physical Properties of Minerals 47
2.4. Polymorphism and Isomorphism 50
2.5. Overview of the Main Rock Forming Minerals 51
2.5.1. Autochthonous Elements 52
2.5.2. Sulfides 52
2.5.3. Oxides and Hydroxides 53
2.5.4. Carbonates 55
2.5.5. Halides 57
2.5.6. Sulfates 57
2.5.7. Phosphates 58
2.5.8. Silicates 58
2.5.8.1. Nesosilicates $[SiO_4]^{4-}$ 59
2.5.8.2. Sorosilicates— $[Si_2O_7]^{6-}$ 61
2.5.8.3. Cyclosilicates— $[Si_nO_{3n}]^{2n-}$ 62
2.5.8.4. Inosilicates 63
2.5.8.5. Phyllosilicates— $[Si_{2n}O_{5n}]^{2n-}$ 68
2.5.8.6. Tectosilicates 73
Further Reading 79

3. Basic Petrology

3.1. Introduction 81
3.2. Structure of the Earth 83
3.3. Classification of Rocks 85
3.3.1. Igneous Rocks 85
3.3.2. Sedimentary Rocks 85
3.3.3. Metamorphic Rocks 86
3.4. Origin of Earth and Theory of Plate Tectonics 86
3.4.1. Origin of the Earth 86
3.4.1.1. The Protoplanet Hypothesis 87
3.4.1.2. The Nebular Hypothesis 88
3.4.1.3. Age of the Earth 88
3.4.2. Plate Tectonics 89
Further Reading 91

4. Igneous Rocks

4.1. Origin of Igneous Rocks 94
4.1.1. Properties of Magma and Lava 94
4.1.2. Bowen's Reaction Series 95
4.1.3. Cooling of Magma after Crystallization 96
4.2. Classification of Igneous Rocks 98
4.2.1. Forms of Appearance and Structure of the Intrusive (Plutonic) Igneous Rocks 100
4.2.1.1. Forms of Intrusive (Plutonic) Igneous Rocks 100
4.2.1.2. Textures of Intrusive (Plutonic) Igneous Rocks 101
4.2.1.3. Shapes and Structures of Veins Igneous Rocks 102

4.2.2. Forms of Appearance and Structure of the Extrusive (Volcanic) Igneous Rocks 102
4.2.2.1. Forms of Extrusive Igneous Rocks 102
4.2.2.2. Textures of Extrusive Igneous Rocks 103
4.3. Main Group of Igneous Rocks and Their Composition 104
4.3.1. Mineral Composition of Intrusive Igneous Rocks 104
4.3.1.1. Felsic Intrusive Igneous Rocks 105
4.3.1.2. Intermediate Intrusive Igneous Rocks 108
4.3.1.3. Mafic Intrusive Igneous Rocks 109
4.3.1.4. Ultrabasic and Ultramafic Intrusive Igneous Rocks 113
4.3.2. Mineral Composition of Extrusive Igneous Rocks 116
4.3.2.1. Felsic Extrusive Igneous Rocks 116
4.3.2.2. Intermediate Extrusive Igneous Rocks 117
4.3.2.3. Mafic Extrusive Igneous Rocks 118
4.3.3. Veins Igneous Rocks 120
Further Reading 120

5. Sedimentary Rocks

5.1. Function, Significance, Classification and Transformation 122
5.2. Sedimentary Rock Formation 124
5.2.1. Weathering 124
5.2.1.1. Physical or Mechanical Weathering 124
5.2.1.2. Chemical Weathering 126
5.2.1.3. Biological Weathering 128
5.2.2. Sediment Transport 128
5.2.2.1. Fluvial Processes 128
5.2.2.2. Aeolian Processes 129
5.2.2.3. Glacial Processes 130
5.2.3. Deposition 130
5.2.4. Lithification 132
5.3. Texture and Structure of Sedimentary Rocks 133
5.3.1. Bedding 133
5.3.1.1. External Bedding 134
5.3.1.2. Internal Bedding 134
5.3.1.3. Upper Bedding Plane Structures 137
5.3.1.4. Lower Bedding Plane Structures 140
5.3.1.5. Forms Created by Underwater Slides and with the Destruction of the Layers 141
5.3.2. Packing of Grains 142
5.4. Classification of Sediments and Sedimentary Rocks 144
5.5. Clastic Sediments and Sedimentary Rocks 145
5.5.1. Genesis and Classification of Clastic Sedimentary Rocks 145
5.5.2. Coarse-Grained Sediments—Rudaceous 146
5.5.2.1. Intraformational Breccias and Conglomerates 147
5.5.2.2. Extraformational Breccias 149
5.5.2.3. Extraformation Conglomerates 152
5.5.3. Medium Granular Clastic Sediments—Arenaceous Rocks 153
5.5.3.1. The Composition and Distribution of Sandy Sediments 153
5.5.3.2. Arenite Sandstones or Arenaceous Rocks 155
5.5.3.3. Graywacke or Wackes 157
5.5.3.4. Mixed or Hybrid Sandstones 159
5.5.4. Fine Granular Clastic Sediments—Pelite 159
5.5.4.1. Classification of Pelitic Sediments 159
5.5.4.2. Marlstone 162
5.5.4.3. Organic Matter in the Argillaceous Sediments 163
5.5.5. Diagenesis of Clastic Sediments 164
5.5.5.1. Diagenetic Processes in Sandy Sediments 164
5.5.5.2. Diagenetic Processes in Clayey Sediments 168
5.5.6. Residual Sediments: Laterite, Kaolin, Bauxite and Terra Rossa 171

5.6. Volcaniclastic Rock 174
5.6.1. Definition and Origin of Volcaniclastic Sediments and Rocks 174
5.6.2. Composition of Volcaniclastic Sediments and Rocks 177
5.6.3. Alteration of Tuff 178
5.7. Chemical and Biochemical Sedimentary Rocks 179
5.7.1. Limestone 179
5.7.1.1. Mineral Composition, Physical, Chemical and Biological Conditions for Foundation of Limestone 179
5.7.1.2. The Structural Components of Limestone 182
5.7.1.3. Limestone Classification 189
5.7.1.4. Limestone Diagenesis 195
5.7.2. Dolomites 199
5.7.2.1. The Origin of Dolomite 200
5.7.2.2. Early Diagenetic Dolomites 201
5.7.2.3. Late-Diagenetic Dolomite 201
5.7.3. Evaporites 203
5.7.3.1. Mineral Composition, Origin and Classification of Evaporites Rocks 203
5.7.3.2. Petrology and Diagenesis of Evaporite Sediments 205
5.7.4. Siliceous Sediments and Rocks 207
5.7.4.1. Mineral Composition, Origin and Classification of Silicon Sediments and Sedimentary Rocks 207
5.7.4.2. Siliceous Sediments and Siliceous Rocks of Biogenic Foundation 208
5.7.4.3. Siliceous Sediments and Siliceous Rocks of Diagenesis Origin 210
Further Reading 212

6. Metamorphic Rocks

6.1. Origin and Structure of Metamorphic Rocks 213
6.2. Types of Metamorphism and Classification of Metamorphic Rocks 219
6.3. Rocks of Dynamic Metamorphism 220
6.4. Rocks of Contact Metamorphism 220
6.5. Rocks of Regional Metamorphism 223
6.5.1. Schists of Low-Grade Metamorphism 223
6.5.2. Schists of High-Grade Metamorphism 225
6.6. Rocks of Plutonic Metamorphism 230
Further Reading 232

7. Precipitation Systems of Major Sedimentary Bodies—Collector Rocks of Oil and Gas

7.1. Introduction 233
7.2. Main Forms of Collector Sedimentary Bodies in Clastites 234
7.2.1. Alluvial Fans 234
7.2.2. Deltas 236
7.2.3. Sand Bodies in Coastal Marine Environments (Beaches and Offshore) 239
7.2.4. Debrites 242
7.2.5. Turbidity Fans 245
7.3. Main Forms of Collector Sedimentary Bodies in Carbonate Rocks 247
7.3.1. Carbonate Platforms 247
7.3.1.1. Carbonates of High-Energy Shallows 248
7.3.1.2. Peritidal Carbonates 250
7.3.1.3. Carbonates of Restricted Shoals, Lagoons, and Inner Shelf 253
7.3.1.4. Carbonate Bodies of Reef and Perireef Limestones in Carbonate Platform 254
7.3.2. Carbonate Debrites and Turbidites or Allodapic Limestones 257
7.3.3. Reef and Perireef Bioclastic Limestones Outside the Carbonate Platforms 258
Further Reading 260

8. Mineral Deposits: Host Rocks and Origin

8.1. Definition 261
8.1.1. Mineral 261
8.1.2. Rock 262
8.1.3. Mineral and Rock Deposit 262
8.2. Classification of Minerals and Mineral Deposits 263
8.2.1. Mineral Classification System Based on Chemical Composition 263
8.2.2. Geographic Distribution 263
8.2.3. Depth of Occurrence 263
8.2.4. Mode of Occurrence 265
8.2.5. Nature of Mineralization 265
8.2.6. Structural Control 265

8.3. Host Rocks 265
8.4. Genetic Model 266
8.4.1. Magmatic 271
8.4.2. Sedimentary 272
8.4.3. Metamorphic 274
8.4.4. Volcanogenic Massive Sulfide and Volcanic-Hosted Massive Sulfide 275
8.4.5. Black Smokers Pipe Type 276
8.4.6. SEDEX/Stratiform 276
8.4.7. Mississippi Valley Type 277
8.4.8. Manto–Chimney-Replacement 277
8.4.9. Irish 278
8.4.10. Pennine 278
8.4.11. Alpine/Bleiberg 278
8.4.12. Skarn 278
8.4.13. Residual 279
8.4.14. Placer 279
Further Reading 279

9. Resource Assessment and Economic Parameters

9.1. Definition 282
9.2. Parameters 282
9.2.1. Cutoff 282
9.2.2. Minimum Width 283
9.2.3. Ore 283
9.2.4. Ore Deposit 283
9.3. Estimation Procedure 284
9.3.1. Small and Medium Size 285
9.3.2. Large and Deep Seated 285
9.3.2.1. Cross-Section 285
9.3.2.2. Long Vertical Section 287
9.3.2.3. Level Plan 287
9.3.3. Statistical Method 288
9.3.4. Geostatistical Method 288
9.3.5. Petroleum (Oil and Gas) 290
9.3.5.1. Analogy Base 290
9.3.5.2. Volumetric Estimate 291
9.3.5.3. Performance Analysis 292
9.4. Resource Classification 292
9.4.1. Metallic/Nonmetallic Minerals 293
9.4.1.1. Conventional/Traditional Classification System 293
9.4.1.2. USGS/USBM Classification Scheme 294
9.4.1.3. UNFC Scheme 295
9.4.1.4. JORC Classification Code 296
9.4.2. Mineral Oil and Gas 296
9.5. Mineral Economics 298
9.5.1. Stages of Investment 298
9.5.2. Investment Analysis 299
9.5.3. Order of Magnitude Study/Scoping Study 300
9.5.4. Prefeasibility Study 302
9.5.5. Feasibility Study 302
9.6. Overview 302
Further Reading 304

10. Hazards of Minerals—Rocks and Sustainable Development

10.1. Definition 306
10.2. Natural Hazards 306
10.2.1. Earthquake 306
10.2.2. Volcano and Volcanism 307
10.2.3. Glacier and Avalanche 309
10.2.4. Lightning 309
10.2.5. Forest Fire 310
10.3. Hazards of Minerals 310
10.3.1. Apatite 311
10.3.2. Arsenic 311
10.3.3. Asbestos 311
10.3.4. Bauxite 311
10.3.5. Cinnabar 311
10.3.6. Clay 311
10.3.7. Coal 311
10.3.8. Corundum 312
10.3.9. Feldspar 312
10.3.10. Fluorite 312
10.3.11. Galena and Cerussite 312
10.3.12. Graphite 312
10.3.13. Gypsum 312
10.3.14. Mica 312
10.3.15. Pyrite 312
10.3.16. Radon Gas 313
10.3.17. Silica 313
10.3.18. Talc 313
10.3.19. Wollastonite 313
10.4. Hazards of Rocks 313
10.4.1. Granite 313
10.4.2. Limestone 313
10.4.3. Sandstone 313
10.4.4. Slate 314
10.4.5. Rock-Fall 314
10.4.6. Balancing Rocks 314
10.4.7. Rock Fault 315

10.5. Hazards of Exploration 315
10.6. Hazards of Mining 316
10.6.1. Baseline Monitoring 316
10.6.2. Surface Land 316
10.6.3. Mine Waste 317
10.6.4. Mine Subsidence 317
10.6.5. Mine Fire 318
10.6.6. Airborne Contaminations 319
10.6.7. Noise 319
10.6.8. Vibration 320
10.6.9. Water Resources 320
10.7. Hazards of Mineral Beneficiation 321
10.8. Hazards of Smelting 321
10.9. Hazards of Refining 322
10.10. Sustainable Mineral Development 322
10.10.1. Mineral/Mining Sustainability 322
Further Reading 323

References 325
Index 327

Preface

"Arise! Awake! And Not Stop Till The Goal Is Reached."—Swami Vivekananda

It was the summer of 2012 and I was exceptionally busy finalizing the publication of *Mineral Exploration – Principles and Applications*. The book was primarily addressed and dedicated to my students from the past, present and future. In the midst of that busy schedule, I glanced at a request from Elsevier through LinkedIn. They wanted someone to take on the role of the lead-author to finalize an incomplete manuscript on *Introductory Mineralogy and Petrology—Oil and Gas Sediment Collectors* by Prof Josip Tišljar. Prof Tišljar, Fellow of the Croatian Academy of Sciences and Arts, had drafted the initial seven chapters, submitted the proposal and passed on in 2009. I commiserated with the loss to the family and to the world. I heard an inner voice, and instantly consented to participate and complete the project in a spirit of homage to the departed author and to help accomplish his wishes and work.

I was greatly impressed by the rich knowledge of Prof Josip and the immense value the book would provide to the students and professionals. I generalized the title of the book to *Introduction to Mineralogy and Petrology*. I expanded five of the first seven chapters and included tables, colored field photographs and photomicrographs of the rocks. I added three new chapters on metallic and nonmetallic miner deposits, nature of occurrences, genetic model, estimation and classification of mineral resources including oil and gas. The new chapters also cover economic aspects of mineral deposits, hazards and sustainable development to make it more meaningful applications and uses of minerals and rocks for the development of human society.

The target readers are aimed at undergraduate and postgraduate students of Geology, Mining, and Civil Engineering, metallic and nonmetallic minerals, petroleum and gas. This book will serve as a manual for professionals in metal–nonmetal mining, petroleum engineering, geotechnical and forestry. Experts in these areas can comfortably understand and solve complex problems of drilling, development and exploitation of oil and gas deposits and geothermal energy sources.

This book gives readers basic information about the general mineralogy, petrology, minerals and rocks that undergo development and exploitation of deposits of minerals, oil and gas and geothermal energy, helps understanding of geological structure, ground and processes of soil formation—*pedogenesis*. It provides an understanding of the primary formation process of igneous, metamorphic, and sedimentary rocks in particular, which have a key role in the forming of the Earth's crust. Special focus is on the sedimentary rocks as the core subject of interest in oil and gas mining, both in terms of reservoir and isolator. The study of sedimentary rocks and rock-strata provides information about the subsurface that is useful for civil engineering. The knowledge can be used in the construction of roads, houses, tunnels, canals or other Earth works. Sedimentary rocks are also important sources of natural resources like coal, fossil fuels, drinking water and ore minerals.

The book is divided into 10 chapters in an orderly manner such as Rocks and Minerals, Basic Mineralogy, Basic Petrology, Igneous rocks,

Sedimentary rocks, Metamorphic rocks, Precipitation Systems of Major Sedimentary Bodies—Collector Rocks for Oil and Gas, Mineral Deposits—Host Rocks and Origin, Resource Assessment, Resource Classification System, and Economic Aspects, and Hazards of Mineral Deposits and Sustainable Development.

We are thankful to many of our colleagues for supporting us during the development of this book and each one has been acknowledged at appropriate pages inside the book. The valuable and timely supports of all the reviewers are appreciated. It was a delight to work with Ms Louisa Hutchins, Editorial Project Manager, Elsevier Limited and I am thankful for her very positive attitude toward any critical issue and help in resolving with alternative solutions. I accomplished truthful happiness while finalizing the proof with Mr Poulouse Joseph, Project Manager-Book Publishing Division, Elsevier and his able team members who accepted all the changes repeatedly with great humility.

On behalf of Prof. Josip Tišljar, I extend our sincere thanks to the families in both Croatia and India for their support during our professional journey. I specially mention the names of Mr Mladen Tišljar, son of Prof Tišljar, for his quick response to my each and every question; Swapna, my wife, for her effortless encouragement; Srishti and Srishta; our two little grand children who are the light of my life.

Traveling is my passion and learning is my wisdom. I love traveling in different countries, seeing diverse landscapes, awe-inspiring nature and meeting people from different cultures. I capture them in my memory and snap their images in my camera. Those images are frequently shared in my books. My wife, Swapna, often takes me out from my routine. My daughter, Soumi, and son-in-law, Surat, took me to see various parts of USA of geological interest. My grand children continue to teach me various aspects of nature. Thanks to all of them. Let my journey continue for eternity.

"The woods are lovely, dark and deep,
But I have promises to keep,
And miles to go before I sleep,
And miles to go before I sleep."
Robert Frost

S.K. Haldar
1st July, 2013
Presidency University, Kolkata

List of Acronyms used in this Book

General

CAPEX Capital Expenditure
EUR Estimated Ultimate Recovery
IMX IMX Resources Limited, Perth, Australia
JORC (Australasian) Joint Ore Reserves Committee
MVT Mississippi Valley Type
OPEX Operating Expenditure
PVT Pressure-Volume-Temperature
RSM Reservoir Simulation Model
SEDEX Sedimentary Exhalative
Sp. Gr Specific Gravity
UNFC United Nations Framework Classification
USGS United State Geological survey/
USBM United State Bureau of Mines
STB Stock Tank Barrel

Minerals

Ch Chert
Cp Chalcopyrite
Cpx Clinopyroxene
Ga Galena
m Microcline
Po Pyrrhotite
Py Pyrite
Q Quartz
S Sericite/seritic/sericitization
Sp Sphalerite

Metals/Semimetals/Nonmetals

Ag Silver
Al Aluminium
As Arsenic
At Astatine
Au Gold
B boron
Bi Bismuth
Br Bromine
C Carbon
Ca Calcium
Cd Cadmium
Ce Cerium
Cl Chlorine
Co Cobalt
Cr Chromium
Cu Copper
F Fluorine
Fe Iron
Ge Germanium
H Hydrogen
He Helium
Hg Mercury
I Iodine
K Potassium
La Lanthanum
Li Lithium
Mg Magnesium
Mn Manganese
Mo Molybdenum
N Nitrogen
Na Sodium
Nd Neodymium
Ni Nickel
O Oxygen
P Phosphorus
Pb Lead
Pd Palladium
Pm Promethium
Pt Platinum
Te Tellurium
Rb Rubidium
Rn Radon
S Sulfur
Sb Antimony
Se Selenium`
Si Silicon
Sm Samarium
Sr Strontium
U Uranium
Zn Zinc

About the Author

S. K. Haldar

S. K. Haldar (Swapan Kumar Haldar) has been a practicing veteran in the field of Mineral Exploration and metal mining for the past 4.5 decades. He received his BSc (Hons) and MSc degree from Calcutta University and Doctorate from Indian Institute of Technology, Kharagpur. The major part of his career from 1966 has been focused on base and noble metals exploration/mining with short stopover at ESSO Petroleum, Hindustan Copper Limited and finally Hindustan Zinc Limited where he undertook a varied set of technical roles and managerial responsibilities. Since 2003 he is associated as Emeritus Scientist with Department of Applied Geology, Presidency University, Kolkata and teaching mineral exploration to postgraduate students of the department and often at Indian School of Mines, Dhanbad. He is consultant with international exploration entities, namely, Goldstream Mining NL/IMX Resources Ltd, Australia and BIL Infratech Ltd, India. His profession has often required visits and interaction with experts of zinc, lead, gold, tin, chromium, nickel and platinum mines and exploration camps of Australia—Tasmania, Canada, USA, Germany, Portugal, France, Italy, The Netherlands, Switzerland, Saudi Arabia, Egypt, Bangladesh and Nepal. He is a life fellow of The Mining Geological and Metallurgical Institutes of India and Indian Geological Congress. Dr Haldar is recipient of "Dr J. Coggin Brown Memorial (Gold Medal) for Geological Sciences" by MGMI. He authored "Exploration Modeling of Base Metal Deposits," 2007, Elsevier and "Mineral Exploration—Principles and Applications," 2013, Elsevier. Dr Haldar has a unique professional blend of mineral exploration, evaluation and mineral economics with an essence of classroom teaching of postgraduate students of two celebrity Universities over the last 1 decade.

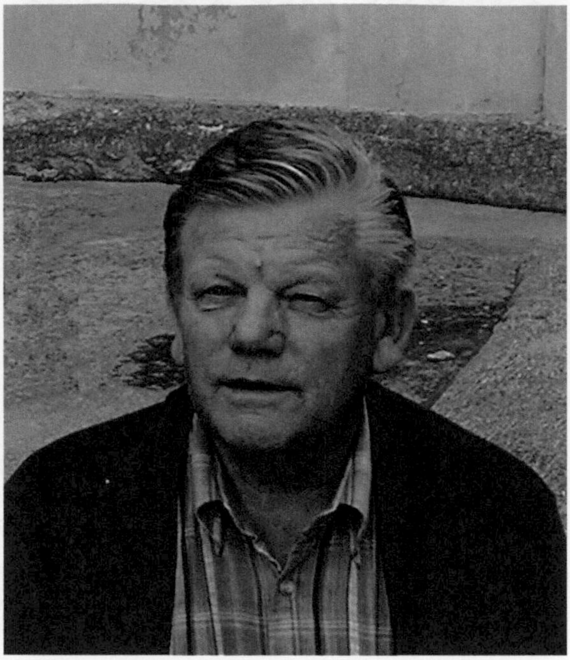

Prof. Josip Tišljar (1941–2009)

Professor Josip Tišljar, obtained MSc and PhD from Faculty of Mining, Geology and Petroleum Engineering, University of Zagreb, Croatia. Since 1965 Prof. Josip held graduate and post-graduate teaching of mineralogy, petrology and petroleum engineering at various capacities as Professors, Head, Director, President and Dean of the parent institute until demise. Academician Professor Josip published +170 research papers focused on Tertiary clastic deposits, carbonates and marls of the Croatian part of the Pannonian Basin, especially glauconitic sandstones and their association with volcano-clastic deposits. Prof. Tišljar was through and through an academician engaged in University teaching and research of mineralogy and petrology including petroleum engineering, He authored five books:

1. J. Tišljar, Petrology of sedimentary rocks, Mining, Geology and Petroleum Engineering Faculty, Zagreb, 1987, p. 242.
2. J. Tišljar, Sedimentary rocks, University of Zagreb, 1994, p. 422.
3. J. Tišljar, Petrology with the basics of mineralogy, University of Zagreb, 1999, p. 211.
4. J. Tišljar, Sedimentology of carbonates and evaporates, Institute of Geology, Zagreb, 2001, p. 375.
5. J. Tišljar, Sedimentology of clastic and siliceous sediments, Institute of Geology, Zagreb, 2004, p. 426.

The mutual knowledge of Academic Institutions and rich experience from mineral industries of the two authors are most appropriate to write this book.

CHAPTER

1

Rocks and Minerals

OUTLINE

1.1. Introduction 1

1.2. Importance of Minerals and Rocks to Society 1

1.3. Minerals 5

1.4. Rocks 12

1.5. Mineral Resources 14

Further Reading 37

1.1. INTRODUCTION

The crust of the Earth and underlying relatively rigid mantle make up the lithosphere. The crust is composed of a great variety of minerals and rocks. More than 80% of all raw materials that are used in various sectors of economy, society and the environment are of mineral origin, and demand for them is greater every day. In most countries, the values of raw materials used for the metal industry and building materials exceed the value of the funds allocated for oil and gas, although, we hear more about oil and gas.

The deposits of raw materials (minerals and rocks) have to be found, investigated, explored and determined their potential of actual reserves/resources and quality/grade. Geological studies of rock formations are extremely significant consequences for major construction projects (roads, railway tracks, airports, tunnels, canals, dam sites, high-rise buildings, industrial and inhabited settlements and many more areas). Not a single such object can be constructed without adequate geological research and documentation on the types of rock and their petrological, engineering, hydrogeological and geotechnical characteristics.

1.2. IMPORTANCE OF MINERALS AND ROCKS TO SOCIETY

All engineering and technical works, roads (Fig. 1.1), tunnels (Fig. 1.2), bridges (Fig. 1.3), dams (Fig. 1.4), buildings, and numerous monuments (Fig. 1.5(A) and (B)) of man's spiritual culture through long-lasting temples (Fig. 1.6), obelisks (Fig. 1.7) and inscriptions on walls (Fig. 1.8) are built of rock, minerals, metals or materials that are either part of the rock or obtained from the rocks. An in-depth knowledge

Introduction to Mineralogy and Petrology
http://dx.doi.org/10.1016/B978-0-12-408133-8.00001-8

FIGURE 1.1 The "Sela Pass", located at Arunachal Pradesh, India, is a high-altitude (13,700 ft or 4,176 m) mountain pass connecting Guwahati (340 km)/Tezpur (155 km)/Bomdila (42 km) in the south and Tawang (78 km) in the north by main access road NH 229. The Pass experiences heavy snow in winter and landslides during rains posing tremendous geological and engineering problem. The road is maintained by the Indian Border Security Force.

of mineralogy, petrology, texture, structure, in situ rock quality, and effect of weathering is essential for planning, execution and optimum uses of natural mineral/rock resources.

The rocks depict the direct evidences and speak the events that happened in the geologic past of Earth (both volcanic and tectonic activities, and interactions between land and sea).

FIGURE 1.2 One of the long-tunnel roads in Europe keeps away from extended high-altitude road travel distance. In situ rock conditions and structures, excessive rains and snow are the main hazards of concern.

FIGURE 1.3 The "Tower/London Bridge" is a combination of cable suspension and moveable type over river "Thames" built between 1886 and 1894 using concrete and steel connects. The bridge is 244 m in length, connects main city and Southwark, and enjoys heritage status. Rock type and structures on either side of the banks, soil condition on river bed, water flow and nature silting are important in designing the Tower Bridge.

Fossils in Latin (fossus = being dug up) are the well-preserved remains of animals, plants, and other organisms from the past. People have always noticed and gathered fossils, pieces of rock and minerals with the remains of biologic organisms. Fossils and their occurrence within the sequence of Earth's rock strata is referred to as the fossil record.

FIGURE 1.4 The "Maithon Dam", 48 km from Dhanbad coal belt town, India, is constructed on "Barakar River". The dam is 4,789 m long, 50 m high and over 65 km^2 water reservoir. It was specially designed, based on in situ rock competency and related structural features, for flood control and generate 60,000 kW hydroelectric power since 1957.

FIGURE 1.5 (A) "The Great Pyramid" of Giza (Cheops) is the oldest (2,560 Before Christ or BC), the tallest (146.5 m) and the largest monument made by the Egyptian Pharaoh (Khufu/King) as tomb. This pyramid consists of 2.3 million limestone blocks from nearby quarry, each varies between 2.5 and 6 t making a total weight of 7.3 million tonnes. It is the oldest of the "Seven Wonders" of the ancient World and the only one to remain largely intact. (B) One of the entries into the Pyramid where the king/queen/high priest was buried along with treasures. There are several false-entry doors to misguide the miscreants. Milk, wine, beer and small piece of bread offered during burial are still preserved in the scientific laboratories in Cairo.

The fossil records are one of the early sources of data relevant to the study to reliably determine the boundaries of sea and land, and the existence of lakes and rivers in different periods of geological history. These are the "rock records" that geologists need to learn to "read" the geological events dated during billions (1,000 million)/ millions of year earlier. According to these indicators, it is clear that the boundaries of land and sea in the past have frequently changed. Many areas that are land now were submerged marine areas in the past, and vice versa.

The fossils in sedimentary rocks have a great significance presenting the development

FIGURE 1.6 The "Abu Simbel" temples are twin massive rock structures on the western bank of "Lake Nasser" in Nubia, southern Egypt. The temples are originally carved out of in situ limestone mountain side during the reign of Pharaoh Ramesses II in the thirteenth century Before Christ, as a lasting monument to himself (picture above) and his queen Nefertari positioned few meters in the right. The complex was relocated in its entirety in 1968, on an artificial hill high above the Aswan High Dam reservoir.

documents for the reconstruction of the Earth and life on it. This primarily assists for the age determination of rocks and the time span in which each fossil communities grow and develop and thus the entire sequence of sedimentation.

The rocks in the Earth's crust are mostly disturbed because of tectonic movements that are not present at the place and in their relations as they were at its origin. The study of their age, location and time of origin and initial relations is obtained knowledge of the tectonic movements that allow the reconstruction of the process of formation mountain chains (orogeny—Section 3.4.2).

Stone, or broken part of a rock, served the man from the Stone Age, ranges between 2,000 BC and 3.4 million years before, as the opportunities for his existence and creation. The prehistoric genus "*Homo* (Great Apes)" and their predecessors widely used stones tools, implements, artifacts with sharp ages, pointed and percussion surfaces for haunting food and learned to control fire. Ample of evidences of their life system had been unearthed along the "Awas River" in Ethiopia following the East African Rift System. "Millennium shift of the culture, material goods and spiritual needs of a variety of people remained recorded in stone as a memorial to the past for the future".[8] The civilization advanced with the advent of metal-working passing through the Copper Age (3500–2300 BC), Bronze Age (~3000 BC) and Iron Age (Vedic Civilization, 2000–500 BC). The modern society uses hundreds of minerals, metals and alloys in day-to-day life and impossible to live without it.

1.3. MINERALS

We draw from minerals and rocks, virtually all the resources for our construction and housing. We grow plants on the surface soil and draw water for drinking and cultivation from them. But among the average man, unfortunately, large number of them know very little or nothing about the minerals and rocks.

FIGURE 1.7 Monolithic granite "Obelisks" of 23 m high stood at the entrance to the Luxor temple complex, Egypt, since 1,300 BC. The obelisk symbolized the Sun God "Ra" and bear inscription that refer the king's seizure of goods.

Mineral is a homogeneous body with a highly ordered arrangement of atoms in atomic structure as a result of crystallization. Mineral is an integral part of the Earth's crust, and has a constant chemical composition that can be expressed by chemical formula. In the specific conditions of temperature and pressure, minerals have stable physical properties.

Constancy of the chemical composition of a mineral is reflected in the fact that any mineral of the same kind anywhere on the Earth has molecules of equal composition. If we break it in smaller pieces, it will still have same characteristics and chemical composition as of the parent mineral. For example, every piece of quartz (SiO_2) always contains 46.73% silicon and 53.27% oxygen.

Constancy of chemical and physical properties of minerals is the result of its internal crystalline structure, i.e. permanent arrangement of atoms and ions in the crystal lattice. This means that each crystallized mineral possesses characteristic permanent arrangement of atoms, ions or ionic groups. For example, mineral halite (Fig. 1.9), also known as common salt or rock salt (sodium chloride), is made of sodium and chlorine ions, which are in proper schedule and unchanged at constant intervals between the two

FIGURE 1.8 Inscriptions on the limestone walls of ancient temples at Luxor, Egypt portray the offerings of flower, food, drinks and wealth to the Crowned God/King sitting at the center.

FIGURE 1.9 Halite, commonly known as rock salt, is the mineral form of sodium chloride (NaCl). Halite forms isometric crystal. The mineral is typically colorless or white, but may also be with shades of light blue, dark blue, purple, pink, red, orange, yellow or gray depending on the amount and type of impurities.

ions arranged along the edges of the cube (Fig. 1.10).

The distance between the two chlorine ions is always 4.12×10^{-10} m or 4.12 Å. The angstrom (symbol Å) is an international unit of length equal to 1×10^{-10} m. It was named after Anders Jonas Ångström.

External crystal form is a reflection of its internal structure. For example, halite crystals have shape of hexahedron; calcite rhombohedra, diamond octahedral and quartz have the form of hexagonal prisms closed by bipyramids (Fig. 1.11).

In all crystals of the same minerals, at the same temperature and pressure, the angles between corresponding sides are equal. This is the result of proper and regular internal structure of crystals. For example, the angles between the surfaces are the same for all the quartz crystals (Fig. 1.11) or all of the plagioclase crystals (Fig. 2.18). Well-developed crystals that have the form of regular polyhedra with properly developed crystal forms, surfaces, edges and peaks (Fig. 1.11, Figs 2.9, 2.11 and 2.13, and Section 2.2.5) in nature are not common. On the contrary, minerals are often found in the form of irregular grains. Minerals with a properly developed external crystal forms occur only where there was enough space for their uninterrupted growth on all sides, such as crystallization in solution or crystallization of lava in the volcanic rocks. In such conditions, the crystals may grow into a regular polyhedron, defined by flat surfaces, which reflects their arrangement of atoms within the crystal lattice (Fig. 2.11).

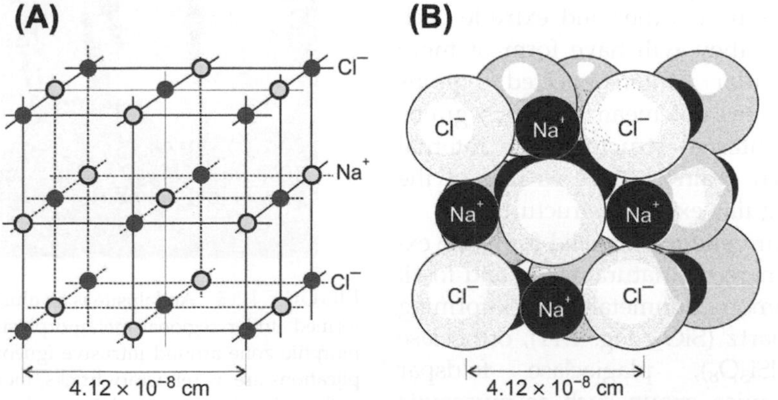

FIGURE 1.10 Halite NaCl crystal structure: (A) schematic representation of the spatial arrangement of sodium and chlorine ions and (B) halite structure with mutual arrangement and size relationships of Na^+ and Cl^- ions.

FIGURE 1.11 Pure natural twined quartz crystals are transparent portraying in general hexagonal prisms and closed by bipyramids. It is the most abundant mineral in the Earth's continental crust. There are different varieties of quartz such as α-quartz (trigonal crystal system) and β-quartz (hexagonal crystal system). Color varies between colorless to black through various shades that are used as semiprecious gemstone.

FIGURE 1.12 Alabaster (calcite) is medium to hard carbonate ($CaCO_3$) mineral of Ancient Egyptians, primarily used as decorative and ornamental objects such as vases and sculptures. The name 'alabaster' is often applied to gypsum ($CaSO_4.2H_2O$) and used for the same purpose.

The minerals may not have enough free space for uninterrupted growth and begin to develop a larger number of crystals in a small space at the same time. This is the case in many rocks and they will have to develop within the available space. They will naturally grow and meet one another and extrude each other. Therefore, they will have form of more or less irregular interconnected grains (Fig. 6.4). This does not mean that such grains have irregular internal structure. The internal structure of each grain is same as that of the crystals with regular external structure.

Minerals occur generally in solid form, the exceptions being mercury, natural water and fossil fuel. The common nonmetallic rock-forming minerals are quartz (SiO_2; Fig. 1.11), orthoclase feldspar ($KAlSi_3O_8$), plagioclase feldspar ($CaNaAlSi_3O_8$), mica group such as muscovite (H_2KAL_3 $(SiO_4)_3$) and biotite ($H_2K(MgFe)_3Al$

FIGURE 1.13 Andalusite is an aluminum silicate mineral formed under regional metamorphism or at contact metamorphic zone around intrusive igneous rocks, primary applications are as refractory bricks/monolithic blocks in iron and steel industry, porcelain spark plugs and transparent variety as gemstone and jewelry.

FIGURE 1.14 Calcite is a carbonate mineral ($CaCO_3$) with trigonal–rhombohedral crystal system. It is colorless and white with occasional gray, yellow and green shades. Calcite is the main constituent of limestone, marble and shells of marine species. *Source: Ref. 24.*

FIGURE 1.15 Fluorite/Fluorspar is a fluoride (CaF_2) mineral with isometric crystal system, colorless, white, purple, blue, green, yellow, orange, red pink, brown and bluish black. Massive variety is suitable for economic mining and industrial uses. *Source: Ref. 24.*

FIGURE 1.16 Garnet is widely distributed in metamorphic terrain such as mica schist. The image presents large crystal of pink almandine fractured garnet (center) poikilitically enclosed in amphibolites host rock at Khetri copper belt, India Brilliant bright red colors and transparency make attractive semiprecious gemstone (necklace) and inferior quality as abrasive, steel cutting, leather and wood polishing and water filtration media.

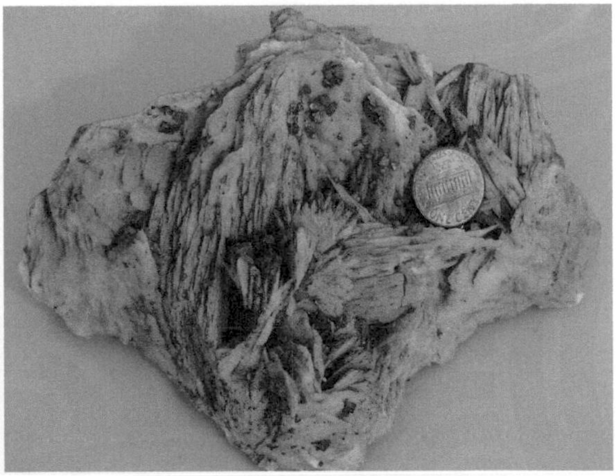

FIGURE 1.17 Gypsum is a soft sulfate mineral ($CaSO_4$) with monoclinic-prismatic crystal system, colorless to white with various shades of gray, yellow, blue, and brown due to impurities, collected during visit to zinc–lead polymetallic ore deposit, Megan Mine, Germany. *Source: Ref. 24.*

FIGURE 1.18 Tourmaline is complex borosilicate of aluminum, magnesium, iron, manganese, chromium and lithium formed during the pneumatolytic process of igneous rock formation. Primary applications are as semiprecious gems, piezoelectricity, therapeutic for relaxation of body and mind, and hair care products.

FIGURE 1.19 Bauxite is a primary aluminum ore often with pisolitic structure. The sample is collected from Bagru Hill/Group of open-pit mines, operating since 1933, Jharkhand, India. *Source: Ref. 24.*

$(SiO_4)_3$), alabaster ($CaCO_3$; Fig. 1.12), albite ($NaAlSi_3O_8$), andalusite (Al_2SiO_5; Fig. 1.13), calcite ($CaCO_3$; Fig. 1.14), fluorite (CaF_2; Fig. 1.15), garnet ($Ca_3Al_2(SiO_4)_3$; Fig. 1.16), gypsum ($CaSO_4 \cdot 2H_2O$; Fig. 1.17), tourmaline (complex borosilicate of Al, Mg, Fe, Mn, Cr, Li; Fig. 1.18), and so on.

The common metallic/nonmetallic ore forming minerals are apatite ($Ca_5(PO_4)_3$ (F, Cl, OH)), baryte ($BaSO_4 \cdot 2H_2O$), bauxite ($Al_2O_3 \cdot 2H_2O$; Fig. 1.19), chalcopyrite ($CuFeS_2$; Fig. 1.20), galena

FIGURE 1.20 High-grade chalcopyrite (brassy-golden) and pyrrhotite-rich ore in chlorite–amphibole ± garnet schist host rock at Kolihan Section, Khetri Copper Mine, India.

FIGURE 1.21 Massive galena (steel-gray) primary lead ore (PbS) with isometric crystalline texture in calc-silicate host rock at Rajpura-Dariba Mine, India. *Source: Prof. Martin Hale.*

FIGURE 1.23 Chromite ($FeOCr_2O_3$) is the primary ore of chromium with isometric-hexoctahedral crystal system, Sukinda layered Igneous Complex, Orissa, India. This complex hosts + 90% of chromite ore in the country. *Source: Ref. 24.*

(PbS; Fig. 1.21), cassiterite (SnO_2), cerussite ($PbCO_3$; Fig. 1.22), chromite ($FeCr_2O_4$; Fig. 1.23), cinnabar (HgS; Fig. 1.24), hematite (Fe_2O_3; Fig. 1.25), pyrite (FeS_2; Fig. 1.26), pyrrhotite (Fe_nS_{n+1}; Fig. 1.27), rhodochrosite ($MnCO_3$), skutterudite ($CoAs_3$; Fig. 1.28), sphalerite (ZnS; Fig. 1.29), stibnite (Sb_2S_3; Fig. 1.30), wolframite ($(Fe,Mn)WO_4$; Fig. 1.31), etc.

Natural brilliant colored crystals of pyrite-fluorite (Fig. 1.32), amythist (Fig. 1.33) often

FIGURE 1.22 Cerussite (lead carbonate), also known as "white lead ore", is the important secondary source of lead metal. Sample collected during a visit to zinc–lead poly-metallic deposit, Megan Mine, Germany.

FIGURE 1.24 Cinnabar with bright cochineal-red color and extremely high specific gravity occurs as vein-filling by recent volcanic activity and acid-alkaline hot spring is the primary source of mercury.

FIGURE 1.25 Hematite (steel-gray, Fe_2O_3) and Jasper (red) the primary ore of iron, Trigonal/Hexagonal crystal is the primary source of iron ore, Jharkhand, India. *Source: Ref. 24.*

FIGURE 1.26 Pyrite (FeS_2), pale brass-yellow fast tarnishing color with Cubic Isometric crystal is a metallic mineral usually associated with sulfide ore, Rajpura-Dariba Mine, India. *Source: Ref. 24.*

originate by hydrothermal veins, volcanic and sub-volcanic and sedimentary re crystallization.

A list of common nonmetallic and metallic minerals with diagnostic characteristics are given in Table 1.1.

FIGURE 1.27 Massive pyrrhotite, iron sulfide mineral, brownish-bronze color with inclusions of quartz and rock fragments hosted by quartz chlorite ± garnet schist from Kolihan Section, Khetri Copper Mine, India. *Source: Prof. Martin Hale.*

FIGURE 1.28 Skutterudite (pin-head dots), a cobalt arsenide mineral with variable proportion of nickel and iron, occurs as hydrothermal ore found in moderate- to high-temperature veins. The sample is collected from the hanging wall shear zone of copper lode at Kolihan Section Mine, Khetri Copper Mine, India..

1.4. ROCKS

Rock or stone is a geological body of specific mineral composition, structure and texture, i.e. mineral aggregate of the same or different with wide variation.

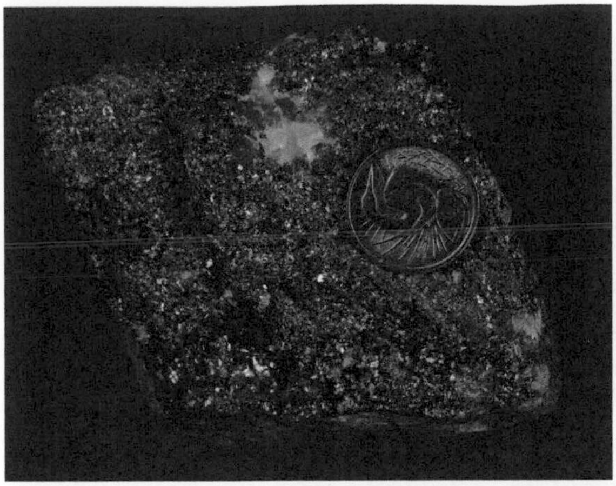

FIGURE 1.29 Sphalerite (ZnS), dark-brown color with isometric-hextetrahedral crystal, is the primary ore of zinc, Zawar Mine, India. *Source: Ref. 24.*

FIGURE 1.31 Wolframite, $(Fe,Mn)WO_4$, steel-gray to brownish color and monoclinic system, is the primary source of tungsten, Degana Mine, Rajasthan, India. *Source: Ref. 24.*

Granite is an igneous rock which contains granular minerals of quartz, microcline and/or orthoclase, Na-plagioclase feldspar and muscovite (Figs 3.5, 4.6(A), 4.10 and 4.12) that is crystallized from molten rock masses (magma) and cools deep in the rocky crust of the Earth.

Sandstone is a clastic sedimentary rock formed through transportation, deposition, compaction and cementation of different mineral composition of sand grains (Figs 5.14 and 5.27).

Limestone is a chemical and biogenic sedimentary rock that is composed mostly of calcite

FIGURE 1.30 Stibnite, also known as "antimonite", occurs as soft-gray color, needle-like radiating acicular crystals. Small hydrothermal veins of stibnite are common and big deposits are rare. The mineral is potentially toxic primary sulfide ore of antimony.

FIGURE 1.32 Cubic overgrowth and twined crystals of fluorite (vitreous and light rose color) resting on perfect crystalline twined pyrite (shining black) often form in hydrothermal veins.

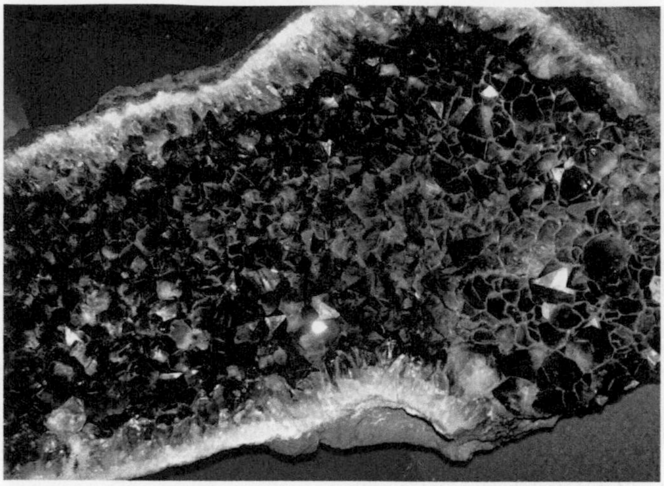

FIGURE 1.33 Twined crystalline overgrowth of amethyst forms as geode or vug from the cavities in fissures and veins out of gas bubbles in basaltic lava and rounded cavities in sedimentary formations. The most distinct features of amethyst are extreme hardness and colorless at the inside surface to brilliantly sparkling purple or violet towards the hollow chamber that make it suitable for jewellery.

(Figs 5.46–5.48). Calcite was formed from the secretion of the sea or fresh water, with the participation of plants and/or animals (biochemical limestone) or by inorganic chemical processes (chemogenic limestone), as explained in detail in Section 5.7.1.

Marble, for example, is a metamorphic rock composed of mineral calcite (Figs 6.4(A) and 6.17) formed at high temperature and pressure deep beneath the Earth's surface by metamorphic transformation of sedimentary limestone rock (Section 6.5.2).

It is important to distinguish between the terms "rock" and "stone". The term "stone" means a smaller or larger part of the rock, which is broken off of some rocks by a natural or technical process. A list of common rocks and characteristic features is given in Table 1.2.

1.5. MINERAL RESOURCES

Mineral resource is the mineral deposit consisting of useful concentration that may or may not exceed economic cost for obtaining the valuable minerals. The technological process, the needs of the economy and prices in the market, depends on whether and when the rock/mineral becomes raw material. For example, for road construction mineral raw material is stone recovered from the quarry, and the stone blocks broken away from the rock mass for the construction of stone structures or processing into polished slab. Rock bauxite is mineral raw material for obtaining aluminum, hematite for iron, and kaolin for porcelain. Less-pure clay is mineral raw material for manufacture of ceramics, and impure clay for production of tiles or bricks. Ore deposits are formed in the Earth's crust by different geological processes and accumulates minerals or ores in such quantities that it is technologically possible to mine and economically profitable to gain. In such condition, the mineral resource becomes mineral or ore reserve. The mineral reserves/resources have been discussed in detail in Chapter 9.

TABLE 1.1 List of Common Nonmetallic and Metallic Minerals, Diagnostic Features and Uses

Name (Formula)	Crystal System	Color/Streak	Luster	Hardness */Sp. Gr	Cleavage/Fracture	% Content of Valuable Component	Origin or Occurrence	Major Uses
Alabaster ($CaCO_3$ or $CaSO_4.2H_2O$) (Fig. 1.12)	Alabaster is applied to two distinct minerals with prefix to "calcite" and "hydrous gypsum" representing the individual properties of each.							Decorative material and ornamental objects, e.g. vases.
Andalusite (Al_2SiO_5) (Fig. 1.13)	Orthorhombic	White, pink, red, brown/whitish	Vitreous	6.5–7.5 3.16–3.20	Perfect on (110) Uneven to conchoidal	36.8 SiO_2 63.2 Al_2O_3	Regional metamorphism and contact metamorphic zone around intrusive igneous rocks	Major application goes into refractory bricks/monolithic blocks used by iron and steel industry, heated to form mullite for porcelain spark plugs and transparent variety as gemstone and jewelry.
Amphibole ($(Mg,Fe,Ca,Na)_{2-3}(Mg,Fe,Al)_5(Si,Al)_8O_{22}OH_2$)	Monoclinic, blocky crystals, fibrous	Black, green, white, gray/pale gray	Vitreous to dull	5.0–6.0 2.9–3.4	Two at acute angle/uneven	–	Occurs as hornblende, tremolite of igneous or metamorphic origin	Hornblende as semiprecious gems and tremolite as asbestos.
Apatite ($Ca_5(PO_4)_3(FClOH)$)	Hexagonal prism, tabular	White, green, brown/white	Vitreous	5.0 3.17–3.23	Imperfect/uneven to conchoidal	41–42 P_2O_5	Pegmatite and metamorphosed limestone	Primarily as fertilizer, occasionally gemstone and as index mineral of Mohs hardness scale of "Five".
Argentite (Ag_2S)	Cubic, octahedral	Lead-gray/shining	Metallic	2.0–2.5 7.2–7.4	Traces, subconchoidal	87.0 Ag 13.0 S	Galena and other sulfide association	Primary source of silver, jewelry, photoprocessing, currency and investment bars.
Arsenopyrite (FeAsS)	Orthorhombic	Steel-gray to silver white/grayish black	Metallic	5.5–6.0 5.9–6.2	Faint traces/ uneven, brittle	46.0 As 34.3 Fe 19.7 S	Hydrothermal veins, pegmatite, contact metamorphism, metasomatism	Major source of arsenic, minor ore of gold as well as herbicide, alloys, wood preservative, medicine, insecticide, rat poison.

(Continued)

TABLE 1.1 List of Common Nonmetallic and Metallic Minerals, Diagnostic Features and Uses (*cont'd*)

Name (Formula)	Crystal System	Color/Streak	Luster	Hardness */Sp. Gr	Cleavage/Fracture	% Content of Valuable Component	Origin or Occurrence	Major Uses
Barite ($BaSO_4$)	Orthorhombic, tabular	Colorless, white, light blue, yellow/white	Vitreous, pearly	3.3–5.0 4.3–5.0	Perfect on (001) face uneven, irregular	65.7 $Ba(OH)_2$ 34.3 SO_3	Hydrothermal, biogenic and evaporation in lead–zinc veins, in limestone	Common applications are drilling fluids in oil, gas, mineral exploration, filler, paper, rubber industry, automobiles, sugar refining and radiology X-ray.
Bauxite ($Al_2O_3 \cdot 2H_2O$) (Fig. 1.19)	Amorphous, massive, oölitic, pisolitic	White, grayish, yellow, red, brown/white	Dull, earthy	1.0–3.0 2.0–2.5	None	73.9 Al_2O_3 26.1 H_2O	Leaching of silica and other soluble in wet tropical or subtropical climate	Primary source of aluminum, construction, transport, consumer durables, packaging, electrical, machinery equipment, refractory bricks, and abrasives.
Bentonite ((Mg,Ca,Na,K) $O \cdot Al_2O_3 \cdot 5SiO_2 \cdot nH_2O$ with $n = 5–7$)	Massive clay like	White, grayish, rose-red, bluish	Feeble	Very soft and tender 2	Unctuous	66.7 SiO_2 28.3 Al_2O_3	Interbeds in marine shale formed from alteration of glassy tuffs	Extraordinary power of swelling by 3–10 times when immersed in water. Drilling mud, geotechnical, pellets, bricks, tiles, pottery, metal casting and medical.
Beryl ($Be_3Al_2Si_6O_{18}$)	Hexagonal, prismatic	Emerald green, blue, yellow/white	Vitreous to resinous	7.5–8.0 2.63–2.80	Imperfect and indistinct, conchoidal to brittle	67 SiO_2 19 Al_2O_3 14 BeO	Granitic pegmatite, mica schist, limestone, tungsten ore	Primary source of beryllium, verities of color and transparency make it attractive gemstones, astrology, alloys, electronics, and ceramics.
Bismuthinite (Bi_2S_3)	Orthorhombic	Lead-gray, tin-white/silver to white	Metallic	2–2.5 6.8–7.2	Perfect on (010) brittle	81.2 Bi 18.8 S	Tourmaline bearing hydrothermal veins in granite ± Cu, Au	Cosmetics, medicine, pharmaceutical, glazes, soldering. alloy (Mn–Bi) create strong permanent magnet, thermocouple, acrylic fiber and carrier for U-235 or U-233 fuel in nuclear reactors.

Bornite (Cu_5FeS_4)	Orthorhombic	Copper-red, bronze-brown, purple/black	Metallic	3–3.25 4.9–5.3	Imperfect on (111) conchoidal	63.3 Cu 11.1 Fe 25.6 S	In zone of secondary supergene enrichment, source of rich copper metal	Source of rich-grade copper, major applications in electrical wires, cables, plumbing, currency, utensils, machinery, alloy, architecture, nutritional supplements and fungicides in agriculture.
Braggite ($(Pt_5Pd_2Ni)S_8$)	Tetragonal	Steel-gray	Metallic	1.5 9.38	None	64 Pt 27 Pd 10 Ni	Layered mafic and ultramafic intrusion at high magmatic temperature	Source for platinum, palladium and nickel used in vehicle emission control, in jewelry, electrodes, spark plug, anticancer drugs, investment, nickel steel and electroplating.
Calcite ($CaCO_3$) (Fig. 1.14)	Rhombohedral	White, colorless, yellow tint/white	Vitreous to pearly	3 2.71	Perfect on (101) conchoidal, brittle	56 CaO 44 CO_2	Main constituents of limestone, marble and shells of marine species.	Dimension stones, mortar, blocks of pyramids, monuments, statuary, alabaster for sculpture, flooring, titles, architecture, acid neutralizer, medicine, antiaircraft weaponry and as index mineral of Mohs hardness scale of "Three".
Cassiterite (SnO_2)	Tetragonal	Brown or black/white, gray, brown	Adamantine, metallic, greasy	6–7 6.8–7.1	Imperfect on (100) subconchoidal, brittle	78.6 Sn 21.4 O	Hydrothermal veins, alluvial and placer as resistant weathered grains.	The chief tin ore throughout ancient history and remains primary source of tin metal as plate, cans, container, solder and polishing compounds and alloys.

(Continued)

TABLE 1.1 List of Common Nonmetallic and Metallic Minerals, Diagnostic Features and Uses (*cont'd*)

Name (Formula)	Crystal System	Color/Streak	Luster	Hardness */Sp. Gr	Cleavage/Fracture	% Content of Valuable Component	Origin or Occurrence	Major Uses
Cerussite ($PbCO_3$) (Fig. 1.22)	Orthorhombic, pseudohexagonal, fibrous form	White, gray, black, green/white	Adamantine, resinous, pearly	3–3.5 6.46	Good on (110) and (021) conchoidal, brittle	83.5 PbO 16.5 CO_2	Delicate acicular ore at Broken Hill mine, Australia, Friedrichssegen, Germany.	Secondary source of lead metal that constitutes as key ingredient in paints, plumbing, bullets, automobile battery, alloys, sheet/brick as radiation shield and cosmetics. Environmentally sensitive, health hazards, slow poisoning.
Chalcocite (Cu_2S)	Orthorhombic	Black lead-gray/ shiny black	Metallic	2.5–3 5.5–5.8	Indistinct on (110) conchoidal	79.8 Cu 20.2 S	Zone of secondary supergene enrichment, source of rich copper metal.	Source of rich copper with major applications in electrical wires, cables, plumbing, currency, utensils, machinery, alloy, architecture and nutritional supplements and fungicides in agriculture.
Chalcopyrite ($CuFeS_2$) (Fig. 1.20)	Tetragonal	Brass-yellow, often tarnished/ greenish black	Metallic	3.5–4 4.1–4.3	Indistinct on (011) uneven, brittle	34.5 Cu 30.5 Fe 35.0 S	Large massive, irregular veins, disseminated and porphyry deposit at granitic/dioritic intrusive and SEDEX type	Primary source of copper metal with major applications in electrical wires, cables, plumbing, currency, utensils, machinery, alloy, architecture, nutritional supplements and fungicides in agriculture.
Chromite ($FeCr_2O_4$) (Fig. 1.23)	Isometric, octahedron, massive	Black, brown/ brown	Submetallic	5.5 4.1–4.9	None Uneven, brittle	68.0 Cr_2O_3 32.0 FeO	Layered mafic–ultramafic intrusion at high magmatic temperature, differential segregation, and crystallization	Primary source of chromium and applications in hard rustles steel, chrome plating, anodizing of aluminum, superalloys, refractory bricks, pigments and dyes, synthetic ruby, wood preservative, leather tanning and catalysts for hydrocarbon processing.

Cinnabar (HgS) (Fig. 1.24)	Rhombohedral, trapezohedral	Cochineal-red, brownish-red/ scarlet	Adamantine, metallic and dull	2.0–2.5 8.0–8.2	Perfect, prismatic Subconchoidal	86.2 Hg 13.8 S	Vein-filling by recent volcanic activity and acid alkaline hot spring	Primary source of mercury, industrial chemicals, electrical, electronic, thermometers, medicine, cosmetics, pigment, fluorescent lamp. Environmentally sensitive due to health and safety regulations.
Coal C, O and H, N, S (Anthracite and bituminous coal)	Compact, massive	Black	Dull to brilliant	0.5–2.5 1.0–1.8	Nil Conchoidal, uneven	+91.5 C −3.75 H −2.5 O −1 S 7–12 volatile	Gradual change of vegetation (wood) buried under sediments	Significant uses are in the form of fuel and energy, electricity generation, gasification, metallurgical purposes like production of steel, cement and liquid fuel.
Cobaltite (CoAsS)	Orthorhombic, pseudocubic	Reddish silver white, violet steel-gray/ grayish black	Metallic	5.5 6.0–6.3	Perfect on (001) Uneven	35.5 Co 45.2 As 19.3 S	High-temperature hydrothermal and contact metamorphic deposit with magnetite, sphalerite	Industrially useful metal, high-temperature superalloy, steel tools, lithium cobalt oxide battery, pigments and coloring, radioisotope and electroplating owing to its attractive appearance, hardness and resistance to oxidation.
Corundum (Al_2O_3)	Hexagonal, six-sided prism	Colorless, gray, brown, pink to pigeon-blood-red/ white	Adamantine to vitreous	9 3.95–4.10	Absent Conchoidal to uneven	52.9 Al 47.1 O	In mica schist, gneiss, low silica nepheline syenite, lamprophyre dyke	Colorful gemstones, tiny jewelry, abrasive, grinding media and as index mineral of Mohs hardness scale of "Nine".
Covellite (CuS)	Hexagonal	Indigo-blue, brass-yellow, deep-red/ Lead gray	Submetallic, resinous, dull	1.5–2.0 4.6–4.8	Perfect on (0001) Flexible in thin lamina	66.4 Cu 33.6 S	Zone of secondary supergene enrichment, source of rich copper metal	Natural superconductor chips, electrical wires, cables, plumbing, currency, utensils, machinery, alloy, architecture, nutritional supplements and fungicides in agriculture, insecticide.

(Continued)

TABLE 1.1 List of Common Nonmetallic and Metallic Minerals, Diagnostic Features and Uses (*cont'd*)

Name (Formula)	Crystal System	Color/Streak	Luster	Hardness */Sp. Gr	Cleavage/Fracture	% Content of Valuable Component	Origin or Occurrence	Major Uses
Cuprite (Cu_2O)	Isometric plagiohedral	Cochineal-red, crimson-red, black/shining brown, red	Adamantine, submetallic, earthy black	3.5–4.0 5.85–6.15	Interrupted on (111) Conchoidal, uneven	88.8 Cu 11.2 O	Zone of oxidation, secondary enrichment, source of rich copper metal	Applications in electrical wires, cables, plumbing, currency, utensils, machinery, alloy, architecture, nutritional supplements and fungicides in agriculture.
Diamond (C)	Isometric, octahedral, hexoctahedral	Colorless, yellow, orange, blue, green/colorless	Adamantine to greasy	10 3.516–3.525	Perfect on (111) Conchoidal, brittle	Pure carbon	Kimberlite, potassic volcanic pipes, conglomerate and alluvial deposits	Unique properties of diamond make it suitable for super quality gemstone, jewelry (Koh-i-Noor, Millennium Star), abrasive, cutting tool, drill bit and as index mineral of Mohs hardness scale of "Ten".
Epidote ($Mg_6Fe_6Al_{10}Si_4(OH)_8O_{10}$)	Monoclonic	Pistachio-green, yellow, brown/ grayish white	Vitreous to resinous	6–7 3.3–3.6	Perfect (001). Imperfect (100) Flat to uneven	–	Product of hydrothermal alteration and metamorphic origin in schist and marble	Perfect transparency and deep colors suitable for semiprecious gemstone.
Feldspar ($NaAlSi_3O_8$-$KAlSi_3O_8$-$CaAl_2Si_2O_8$)	Monclinic or triclinic	Pink, white, gray, brown/white	Vitreous	6.0–6.5 2.55–2.76	Two or three/ along cleavage plane	18.4 Al_2O_3 16.9 K 64.7 SiO_2	In most igneous rocks and felsic lavas	Primary use in ceramics, glass manufacture, fillers, paints, plastics, rubber and as index mineral of Mohs hardness scale of "Six".
Fluorite (CaF_2) (Fig. 1.15 and 1.32)	Isometric with cubic habit	White, purple, blue, green, yellow/white	Vitreous	4 3.01–3.25	Indistinct Glassy	51.1 Ca 48.9 F	Vein with metallic minerals, occasionally of hydrothermal origin	Flux in steel manufacture, opalescent glass, enamels for cooking utensils, hydrofluoric acid, high-performance telescopes, camera lens and as index mineral of Mohs hardness scale of "Four".

Galena (PbS) (Fig. 1.21)	Isometric, granular cubes	Lead gray and silvery/lead gray	Metallic	2.5–2.75 7.2–7.6	Perfect cubic on (001) Subconchoidal	86.6 Pb 13.4 S	Individually or associated with zinc and copper sulfide deposit	Primary source of lead metal and constitutes as key ingredient in paint, plumbing, bullets, automobile battery, alloys, sheet, radiation shield, electrodes, ceramic-glazes, stained glass and cosmetics. Environmentally sensitive and health hazards.
Garnet ($Ca_3Al_2(SiO_4)_3$) (Fig. 1.16)	Isometric	Pink, red, brown, green/white	Vitreous	6.5–7.5 3.1–4.3	Indistinct/ conchoidal to uneven	–	Widely distributed in metamorphic rocks, e.g. mica schist	Brilliant bright red colors and transparency make attractive semiprecious gemstone and inferior quality as abrasive, steel cutting, leather and wood polishing and water filtration media.
Graphite (C)	Hexagonal	Iron black, dark steel-gray/black	Metallic, dull, earthy	1.0–2.0 2.09–2.23	Basal perfect, on (001) Flaky	70–85 C	Reduction of sedimentary carbon compounds during metamorphism	Steel making, crucibles, refractory, batteries, break lining, foundry facings, lubricants, pencil and electrodes.
Gypsum ($CaSO_4 \cdot 2H_2O$) (Fig. 1.17)	Monoclinic, prismatic and flattened	Colorless to white, yellowish/white	Vitreous, pearly, silky, waxy	1.5–2 2.31–2.33	Perfect (010) Conchoidal, splintery	32.5 CaO 46.6 SO_3 20.9 H_2O	As evaporite beds deposited in lake, sea, hot spring and by-product of sulfide oxidation	Plaster-board for walls and ceilings, surgical splints, casting molds, modeling, fertilizer and soil conditioner, cement, insulation, alabaster for sculpture, wood substitute and as index mineral of Mohs hardness scale of "Two".

(Continued)

TABLE 1.1 List of Common Nonmetallic and Metallic Minerals, Diagnostic Features and Uses (*cont'd*)

Name (Formula)	Crystal System	Color/Streak	Luster	Hardness */Sp. Gr	Cleavage/Fracture	% Content of Valuable Component	Origin or Occurrence	Major Uses
Halite (NaCl) (Fig. 1.9)	Isometric, usually in cubes	Colorless, white, purple, red, pink, yellow/white	Vitreous	2.5 2.1–2.6	Cubic perfect on (001) Conchoidal, brittle	39.4 Na 60.6 Cl	Vast beds of sedimentary evaporate, salt domes and pipes essentially "squeezed up" from underlying salt beds	Extensively used in cooking as a flavor enhancer and preservative to cure a wide variety of foods such as fish, meat and pickles, soda ash for glass, soap and bleaching industry. Salt aids in managing ice and spreading salt on walkways and driveways after snow storm.
Hematite (Fe_2O_3) (Fig. 1.25)	Rhombohedral	Black or steel-gray/cherry-red	Metallic, splendent, dull	5.5–6.5 4.9–5.3	Indistinct Subconchoidal to uneven, brittle	70.0 Fe 30.0 O	Large volume of banded hematite–quartzite	Primary source of iron and steel industry, ancient "Crypto-Minoan script", red/black carvings, casting, ornamental jewelry, paints and cosmetics, alloys.
Ilmenite ($FeTiO_3$)	Trirhombohedral	Iron-black/black to brown red	Submetallic to metallic	5.0–6.0 4.5–5.0	Absent Conchoidal	31.6 Ti 36.8 Fe 31.6 O	Accessory to igneous rock especially gabbros and diorites	Alloy for high tech in aerospace and medical application, deoxidizer in stainless steel, often alloyed with copper, iron, aluminum, vanadium, molybdenum for sheet, plate, bar, wire, forgings, castings and pigments.
Kaolin ($Al_4Si_4O_{10}(OH)_8$) (Fig. 2.16)	Monoclinic, rarely as crystal	White, red, blue, brown/white	Pearly to dull earthy	2.0–2.5 2.6–2.63	Basal, perfect Flexible, inelastic	46.5 SiO_2 39.7 Al_2O_3	Decomposition of aluminous minerals, e.g. feldspar of granites and gneisses.	Paper, rubber manufacture, coating clay, linoleum, paints, inks, leather, refractory, pottery, stoneware, bricks, insecticide, plastics and fertilizers.

Kayanite ($3Al_2O_3$, $2SiO_2$)	Triclinic, bladed form	Blue, white, green, pink/white	Vitreous	5.0–7.0 3.53–3.65	Two perfect in (100) and (010) plane/ splintery	36.8 SiO_2 63.2 Al_2O_3	Occurs in gneiss, schist, pegmatite, quartz veins resulting from high-pressure metamorphism	Heating element, electrical insulation, electronics, ceramic and refractory industry, porcelain plumbing fixtures, gemstone.
Lepidolite (Li-mica $(OH,F)_2KLiAl_2Si_3O_{10}$)	Monoclinic, tabular to prismatic	Red-rose, violet, lilac/white	Translucent	2.5–4.0 2.8–3.3	Basal highly perfect on (001) Uneven	–	High-temperature quartz veins in greisens (altered) granites and pegmatite	Source of lithium, battery, flux and coloring of ceramics and glass, electrical and electronics, lubricating greases, alloys, air purification, medicine for bipolar disorder, rocket propellant and nuclear fusion.
Magnesite ($MgCO_3$)	Rhombohedral	White, yellow, gray, transparent/ white	Vitreous	3.5–4.5 3.0–3.12	Perfect Conchoidal, flat	47.6 MgO 52.4 CO_2	Alteration product of magnesium-rich ultramafic rocks	Refractory bricks in metallurgical furnaces, cement industry, slag former in steel making, catalyst and filler in synthetic rubber, chemicals and fertilizers, and dyed beads.
Magnetite ($FeO \cdot Fe_2O_3$)	Isometric and common in octahedrons	Black, gray with brownish tint/black	Metallic	5.5–6.5 5.17–5.18	Indistinct Subconchoidal to uneven, brittle	72.4 Fe 27.6 O	Common in igneous rocks, placer-type aggregate and beach sand	Source of sulfur, low-grade iron ore and occasionally ornamental stone. Commercially less attractive.
Marcasite (FeS_2)	Orthorhombic	Pale-bronze yellow/black	Metallic	6.0–6.5 4.85–4.99	Distinct on (110) Uneven, brittle	46.6 Fe 53.4 S	Shale, coal, limestone, and hydrothermal veins	Iron and steel industry.
Marmatite ((ZnFe)S)	Isometric, tetrahedral	Dark brown to black/brown	Metallic, resinous	3.5–4.0 3.9–4.05	Perfect Conchoidal, brittle	46–56 Zn <20 Fe Rest S	An opaque black iron-rich variety of sphalerite.	Source of zinc metal.
Mica (Aluminosilicate of K/Na, Fe/Mg and rarely Li or Cr)	Monoclinic sheets, books	Colorless, brown, black, green/white	Pearly	2.0–2.5 2.76–3.2	Perfect and parallel to base	–	Widely occurs in igneous, metamorphic, sedimentary rocks	Primary use in insulator, electrical, electronic, heat proof windows, optical filters, thermal regulators, microwave ovens, well drilling fluids, and asphalt roof-shingles.

(Continued)

TABLE 1.1 List of Common Nonmetallic and Metallic Minerals, Diagnostic Features and Uses (cont'd)

Name (Formula)	Crystal System	Color/Streak	Luster	Hardness */Sp. Gr	Cleavage/Fracture	% Content of Valuable Component	Origin or Occurrence	Major Uses
Millerite (NiS)	Rhombohedral	Pale brass or bronze yellow/greenish black	Metallic	3.0–3.5 5.3–5.65	Perfect, uneven, brittle	64.7 Ni 35.3 S	Radiating cluster of acicular needles in ultramafic serpentinite bodies	High-grade source for Ni, used for stainless steel, superalloys, electroplating, alnico magnets, coinage, rechargeable batteries, electric guitar strings, microphone capsules, and green tint in glass.
Molybdenite (MoS_2)	Hexagonal	Black, lead-silver-gray/bluish gray	Metallic	1.0–1.5 4.7–4.8	Perfect on (001) Flexible lamellae	60.0 Mo 40.0 S	High-temperature hydrothermal ore of chalcopyrite, pyrite, molybdenite	Primary source of molybdenum, corrosion resistance ferroalloy, Mo metal and superalloys, stainless steels, lubricant, tools and high-speed steels, cast iron, electrodes, fertilizers, and pollution control in power plants.
Monazite ((CaLaTh)PO_4)	Monoclinic	Hyacinth-red, clove brown, reddish brown/white	Resinous to adamantine	5.0–5.5 4.9–5.3	Distinct on (100) Poor on (010) Conchoidal, uneven, brittle	48 Ce 24 La 17 Nd	Mainly as placer deposit and beach sand	Important source rare earth metals like thorium, lanthanum, cerium, neodymium etc. used for radioactive dating and gaslight mantle.
Olivine ($(Mg,Fe)_2SiO_4$)	Orthorhombic, Tabular	Yellow, green, blue, brown/white	Vitreous	6.5–7.0 3.27–3.37	Poor in one direction	–	Constituents of basic and ultrabasic intrusive magma	Spectacular green colored verities as gems. Aluminum foundry industries utilize olivine sand mold to cast objects in aluminum.
Niccolite or nickeline (NiAs)	Hexagonal	Pale copper-red/Pale brownish black	Metallic	5.0–5.5 7.33–7.67	Massive, reniform-columnar, Uneven, brittle	43.9 Ni 56.1 As	Layered mafic–ultramafic intrusion at high magmatic temperature, differential segregation	Rarely used due to presence of arsenic, deleterious to smelting and milling, except blending with "clean" ore, which the mill and smelter can handle with acceptable recovery.

Pentlandite ((Fe, Ni)$_9$S$_8$)	Isometric	Pale-bronze yellow/ bronze-brown	Metallic	3.5–4.0 4.6–5.0	Absent, octahedral parting Uneven	22 Ni 42 Fe 36 S	Layered mafic–ultramafic intrusion at high magmatic temperature, differential segregation	Primary source of nickel associated with PGE, tarnish-resistant stainless steel, superalloys, electroplating, alnico magnets, coinage, rechargeable batteries, electric guitar strings, microphone capsules, and green tint in glass.
Psilomelane (MnO_2)	Massive and botryoidally	Iron black, steel-gray/brownish black	Submetallic, dull	5.0–7.0 3.3–4.7	None Conchoidal, uneven	50.0 Mn	Primarily is sedimentary and less frequently of hydrothermal origin	Source of manganese, essential to iron and steel making, aluminum alloy, additive in unleaded gasoline to boost octane rating, dry cell battery, coinage and drier in paints.
Pyrite (FeS_2) (Fig. 1.26 and 1.32)	Isometric, cubic	Pale brass-yellow, often tarnished/ greenish-brownish black	Metallic, shiny, glossy	6–6.5 4.95–5.10	Indistinct on (001) Very uneven sometime conchoidal	46.6 Fe 53.4 S	Common in all rocks and massive sulfide deposits associated with gold	Main uses are production of sulfur-dioxide for paper and sulfuric acid for chemical industry, rarely mined for iron content due to complex metallurgy and commercially uneconomic. Acid drainage and dust explosion are common hazards with pyrite deposits.
Pyrolusite (MnO_2)	Orthorhombic, usually columnar	Iron black, dark steel-gray/black	Metallic	2.0–2.5 4.73–4.80	Perfect on (110)	63.0 Mn	Primarily is sedimentary and less frequently of hydrothermal origin	Source of manganese, essential to iron and steel making, aluminum alloy, additive in unleaded gasoline to boost octane rating, dry cell batteries, coloring in bricks, and decoloring in glass and pottery.
Pyroxene (silicates of Fe/Mg, Ca/Al rarely with Na/Li)	Monoclinic, orthorhombic	Green, brown, blue/white	Vitreous, pearly, resinous	5.0–6.0 3.2–3.6	Perfect/irregular, uneven, conchoidal	–	Occurs as augite, diopside, hypersthene	Gems and ornamental stones, ceramics and glass-ceramics.

(*Continued*)

TABLE 1.1 List of Common Nonmetallic and Metallic Minerals, Diagnostic Features and Uses (*cont'd*)

Name (Formula)	Crystal System	Color/Streak	Luster	Hardness */Sp. Gr	Cleavage/Fracture	% Content of Valuable Component	Origin or Occurrence	Major Uses
Pyrrhotite (Fe_nS_{n+1}) (Fig. 1.27)	Hexagonal	Bronze-yellow to copper-red/black	Metallic	3.5–4.5 4.58–4.64	Absent Uneven	60.4 Fe 39.6 S	Common in mafic and layered intrusive and sulfide deposits	No specific application other than source of sulfur for making sulfuric acid, rarely for recovering iron due to complex metallurgy and often nickel bearing.
Quartz (SiO_2) (Fig. 1.11)	Hexagonal	Colorless to black through various shades/white	Vitreous, waxy to dull when massive	7.0 2.65	Indistinct Conchoidal	46.7 Si 53.3 O	Occurs universally in all rocks except pure limestone, marble, gabbro, basalt, and peridotite	Source of silicon compounds, polymers. Due to thermal-chemical stability and abundance widely used as building material, mortar, ceramics, cement, foundry, abrasives, clock, oscillators, gemstone, porcelain, glass, paint, acid flux in smelting furnaces and as index mineral of Mohs hardness scale of "Seven".
Rhodochrosite ($MnCO_3$)	Trigonal, hexagonal	Brilliant pink, cherry red, yellow/white	Vitreous and pearly	3.5–4.0 3.7	Perfect Uneven, conchoidal and brittle	61.7 MnO 38.3 CO_2	Hydrothermal veins with other low-temperature manganese minerals	An ore of manganese, aluminum alloys, brilliant transparent verities as decorative stone and jewelry.
Rhodonite ((Mn^{2+} Fe^{2+} Mg,Ca) SiO_3)	Triclinic	Rose-pink, red, yellow/white	Vitreous to pearly	5.5–6.5 3.57–3.76	Perfect/conchoidal to uneven	54.1 MnO 45.9 SiO_2	Associated in manganese and iron ore deposits	Used mainly as ornamental and decorative stones.
Rutile (TiO_2)	Tetragonal, acicular to prismatic	Wine-red, reddish brown/pale brown	Metallic, adamantine	6.0–6.5 4.18–4.25	Perfect on (110) Twining common, subconchoidal, uneven	60.0 Ti 40.0 O	Heavy mineral in beach sand	Source of titanium. Rutile used as refractory, ceramic, welding electrode cover, sunscreen to protect UV-induced skin damage and brilliant white pigment in paint, plastics and paper.

Scheelite ($CaWO_4$)	Tetragonal–pyramidal	White, yellow, brown, green, red/white	Vitreous, adamantine	4.5–5.0 5.9–6.1	Perfect on (111) Uneven, brittle	80.6 WO_3 19.4 CaO	Contact metamorphic skarn, in high-temperature hydrothermal veins and greisens, less commonly in granite and pegmatite	Source of strategically important tungsten metal, filaments for light bulbs-electronic tubes and furnace, abrasives, super heavy alloyed with nickel, cobalt and iron as kinetic energy penetrators (small arms bullets designed to penetrate), armor, cannon shells, grenades and missiles to create supersonic shrapnel.
Sillimanite ($Al_2O(SiO_4)$)	Orthorhombic	Off-white, gray, brown/white	Vitreous	6–7 3.23–3.24	Perfect in (010) Splintery	36.8 SiO_2 63.2 Al_2O_3	Gneiss, schist, pegmatite, quartz veins resulting from high-pressure metamorphism	Glass industry, high-alumina refractory, quality porcelain.
Skutterudite ($CoAs_3$) (Fig. 1.28)	Isometric-octahedral-pyritohedral	Tin-white, lead-gray/black	Metallic	5.5–6.0 6.5–6.9	Distinct on (100) Conchoidal, uneven	–	Hydrothermal ore found in moderate- to high-temperature veins with other Ni–Co minerals	Strategically and industrially useful, high-temperature superalloy, steel tools, lithium cobalt oxide battery, pigments and coloring, radioisotope and electroplating owing to its attractive appearance, hardness and resistance to oxidation.
Smithsonite ($ZnCO_3$)	Rhombohedral	White, green, brown, pink, yellow, white	Vitreous, pearly	4.5–5.5 4.3–4.4	Perfect Uneven, subconchoidal	64.8 ZnO 35.2 CO_2	A secondary mineral in weathering/oxidation zone of zinc-bearing ore deposits	Secondary source of zinc, main applications in galvanizing, alloys, cosmetics, pharmaceutical, micronutrient for human, animals and plants.
Sperrylite ($PtAs_2$)	Cubic-pyritohedral	Tin-white/black	Metallic	6.0–7.0 10.58	Indistinct conchoidal	57.0 Pt	Layered igneous complex	Primary source of platinum, automobile emission controls devices, jewelry, catalyst, electrode, anticancer drug, oxygen sensors, spark plug and turbine engine.

(*Continued*)

TABLE 1.1 List of Common Nonmetallic and Metallic Minerals, Diagnostic Features and Uses (*cont'd*)

Name (Formula)	Crystal System	Color/Streak	Luster	Hardness */Sp. Gr	Cleavage/Fracture	% Content of Valuable Component	Origin or Occurrence	Major Uses
Sphalerite (ZnS) (Fig. 1.29)	Isometric-tetrahedral	Brown, black, honey yellow/ brownish, pale yellow	Adamantine, resinous, greasy	3.5–4 3.9–4.1	Perfect on (110) Uneven to conchoidal	67.0 Zn 33.0 S	Majority as large SEDEX-type deposits associated with galena, chalcopyrite and silver	Primary source of zinc, main applications in galvanizing, alloys, cosmetics, pharmaceutical, micronutrient for human, animals and plants.
Stannite ($Cu_2S \cdot FeS \cdot SnS_2$)	Tetragonal	Steel-gray to iron-black/black	Metallic	3.5–4.0 4.3–4.52	Cubic, indistinct Uneven	27.5 Sn 29.5 Cu 13.1 Fe 29.9 S	Hydrothermal vein deposit containing Sn, Cu, Zn, W, Fe, Ag, and As	Primary source of tin and copper.
Staurolite ($Fe_2Al_9O_7(OH)(SiO_4)_4$)	Orthorhombic, prismatic	Brown with tinge of red or orange/ white to gray	Vitreous	7.0–7.5 3.74–3.83	Poor cleavage on (010)/ subconchoidal	–	High-grade regional metamorphic rocks, e.g. garnet mica schist	Index mineral to estimate the temperature, depth, and pressure at which a rock undergoes metamorphism.
Stibnite (Sb_2S_3) (Fig. 1.30)	Orthorhombic	Lead-gray, tarnishing black/ lead-gray	Metallic, splendent	2 4.52–4.62	Highly perfect on (010) Subconchoidal	71.7 Sb 28.3 S	Hydrothermal deposits associated with other sulfide minerals	Primary source of antimony, flame retardant, textiles and coatings, fiber, alloy with lead for batteries, plain bearings and solders.
Sulfur native (S)	Orthorhombic	Yellow, straw, greenish, reddish/ white	Resinous	1.5–2.5 2.05–2.09	Imperfect Conchoidal to uneven	100.0 S	Natural elemental form, sulfide and sulfate minerals	Sulfuric acid, fertilizer chemicals, fungicide and pesticide, bactericide in wine making and food preservation.
Sylvite (KCl)	Isometric	Colorless, white, blue, yellow/white	Vitreous	2.0 1.97–1.99	Perfect on (100) (010) (001) Uneven	52.4 K 47.6 Cl	Evaporite mineral precipitates out of solution in very dry saline areas	Source of potash and principal use as fertilizers.
Sylvanite ($(AuAg)Te_2$)	Monoclinic	Steel-graysilver-white, yellow/ steel-gray	Metallic brilliant	1.5–2.0 7.9–8.3	Perfect on (010) Uneven	24.5 Au 13.4 Ag 62.1 Te	Most commonly in low-temperature hydrothermal veins	Sylvanite represents a minor source of silver, gold and tellurium.

Talc (3MgO, $4SiO_2H_2O$)	Orthorhombic, monoclinic Granular and fibrous	White with gray, green and brown tinge/white, pearl green	Wax like or pearly	1 2.7–2.8	Perfect on (001) basal cleavage Uneven pattern	31.7 MgO 63.5 SiO_2 4.8 H_2O	Metamorphism of magnesium minerals, e.g. serpentine, pyroxene, olivine from ultramafic rocks	Cosmetics, paint and coating, plastic, paper making, rubber, ceramics, pharmaceutical, electric cable, detergents, food additive and as index mineral of Mohs hardness scale of "One".
Topaz ($(AlF)_2SiO_4$)	Orthorhombic	Straw-yellow, wine-yellow, colorless, green, blue, white	Vitreous	8 3.4–3.6	Perfect on (001) Subconchoidal, uneven	–	Commonly associated with silicic igneous rocks of granite, pegmatite and rhyolite type	Brilliancy and transparency rank it attractive as gemstone, birthstone, jewelry, astrology and as index mineral of Mohs hardness scale of "Eight".
Tourmaline (complex borosilicate of Al, Mg, Fe, Mn, Cr, Li) (Fig. 1.18)	Rhombohedral, hexagonal	Black, brown, green, violet/ white	Vitreous to resinous	7–7.5 3–3.2	Indistinct Glassy, conchoidal, brittle	–	Product of pneumatolytic process of igneous rock formation	Transparent verities as semiprecious gems, piezoelectricity, therapeutic application as relaxation of body and mind, hair care products.
Uraninite (UO_3)	Isometric	Steel-velvet-brown black/black, green	Submetallic, greasy	5.0–6.0 10.63–10.95	Indistinct Conchoidal to uneven	88.0 U	Hydrothermal colloform veins in granitic and syenitic pegmatite and quartz-pebble conglomerates	Fuel for nuclear reactor to generate sustainable electricity for civilian purposes, for propulsion of naval warships for military and nuclear powered icebreaking.
Wolframite ($(Fe,Mn)WO_4$) (Fig. 1.31)	Monoclinic, tabular, prismatic	Dark grayish or brownish black/ reddish brown	Submetallic, resinous	5.0–5.5 7.0–7.5	Perfect on (010) Uneven, rough	76.0 W	Granite and pegmatite veins formed under pneumatolytic condition	Main source of tungsten, strong and dense metal with high melting temperature used for electric filaments, bulb, alloy, cutting material, armor piercing ammunitions like shot and shell in defense.

(*Continued*)

TABLE 1.1 List of Common Nonmetallic and Metallic Minerals, Diagnostic Features and Uses (*cont'd*)

Name (Formula)	Crystal System	Color/Streak	Luster	Hardness */Sp. Gr	Cleavage/Fracture	% Content of Valuable Component	Origin or Occurrence	Major Uses
Wollastonite ($CaSiO_3$)	Triclinic, monoclinic,	White, gray, colorless/ white	Vitreous, dull, pearly	4.5–5.0 2.86–3.09	Perfect in two directions at 90°	48.3 CaO 51.7 SiO_2	Thermally metamorphosed impure limestone	Principal ingredient in ceramics industry, paint, paper, polymers and metallurgical applications.
Zincite (ZnO)	Hexagonal-hemimorphic	Deep-red, orange-yellow/orange-yellow	Submetallic	4.0–4.5 5.43–7.7	Perfect, prismatic Conchoidal	80.3 Zn 19.7 O	Both natural and synthetic	Rich ore of zinc. Zincite crystals are significant as semiconductor in early development of crystal radios before the advent of vacuum tubes.
Zircon ($ZrSiO_4$)	Tetragonal	Red, brown, yellow, green/white	Vitreous, adamantine, greasy	7.5 4.6–4.7	Indistinct on (110) and (111) Conchoidal, uneven	67.2 ZrO_2 32.8 SiO_2	Common trace mineral in granite and felsic igneous rocks	Alloy in nuclear reactors, as a pacifier in the decorative ceramics, refractory and foundry industries, gemstone and radiometric age dating.

* *Mohs hardness scale.*
Source: Refs 33, 9 and internet.

TABLE 1.2 Lists of Common Rocks and Diagnostic Features under Overall Classification

Name	Color	Composition	Texture	Major Uses
IGNEOUS ROCKS				
Granite (Figs 4.8, 4.9, 4.10, 4.12, 4.14 and 4.15)	White, gray, black, pink to red	Quartz, feldspar biotite mica ± amphibole	Massive plutonic intrusive, granular and crystalline formed as batholiths	Building and decorative stones, tiles, kitchen counter, ancient and modern sculptures, engineering, curling and rock climbing.
Pegmatite	Mix of red, white, gray, cream, silvery and dark	Same as granite ± tourmaline, topaz, beryl	Exceptionally large crystals intrusive dyke, veins in and near granite	Source of rare earth and gemstone viz. aquamarine, tourmaline, topaz, beryl, fluorite, apatite, corundum, mica-books, lithium, tin and tungsten.
Syenite (Fig. 4.16)	Typically light color of white, gray, and pink	Same as granite with quartz <5% + nepheline	Coarse-grained intrusive igneous rock	Better fire-resistant qualities suitable for dimension stone for building facings, foyers and aggregate in road industries.
Monzonite (adamellite)	Typically light color of white, gray, pink, brown, and bronze	Equal amount of orthoclase and plagioclase with <5% quartz	Medium- to coarse-grained intrusive igneous rock	Seldom as host rock for gold and silver deposits, primarily used as building stone for monuments (The Mormon temple, Salt Lake City, Utah), mountaineering.
Granodiorite	Light gray	Plagioclase exceeds orthoclase, and +20% quartz	Large phaneritic crystal due to slow cooling	Most often used as crushed stone for road building and occasionally as ornamental stone.
Diorite	Typically speckled black and white with bluish, greenish and brownish tinge	Principally of plagioclase feldspar and ferromagnesian minerals (biotite, hornblende and pyroxene)	Intrusive igneous rock intermediate between granite and gabbro with medium to coarse phaneritic texture	Aggregate, fill in construction and road industries, cut and polished for dimension stone for building facings and foyers, statue and vase made during ancient Inca, Mayan and Egyptian civilization.

(Continued)

TABLE 1.2 Lists of Common Rocks and Diagnostic Features under Overall Classification (cont'd)

Name	Color	Composition	Texture	Major Uses
Gabbro (Fig. 4.17)	Dark gray, black, greenish and rarely reddish	Chiefly ferromagnesian Fe-rich clinopyroxene (augite) at greater than equal to plagioclase	Coarse-grained intrusive mafic igneous rock	Often contains Cr, Ni, Co, Cu Au, Ag, Pt and Pd. Common usages are ornamental facing, paving, graveyard headstone at funerary rites and kitchen countertops.
Norite (Figs 4.22 and 4.23)	Light to dark gray, brownish	Ca-rich plagioclase (labradorite), Mg-rich orthopyroxene (enstatite) and olivine	Mafic intrusive igneous rock, indistinguishable from gabbro, other than type of pyroxene under microscope	Occurs in association with mafic (gabbro)/ultramafic layered intrusion e.g. Bushveld (South Africa) and Stillwater (Montana, USA) with large platinum group of deposits. Usages are ornamental facing, paving, graveyard headstone at funerary rites and kitchen countertops.
Anorthosite (Fig. 4.24)	White, yellowish to brown, gray, blush, smoky pigment	Predominance of plagioclase feldspar (90–100%) and mafic components of pyroxene, magnetite, ilmenite (0–10%) ± olivine	Phaneritic intrusive igneous rock	Source of titanium, aluminum, gemstones, building material and scientific research of similar composition of Moon, Mars, Venus and meteorites.
Peridotite (Fig. 4.27)	Dark green and greenish gray	Magnesium-rich olivine and pyroxene, <45% silica	Dense coarse-grained layered ultramafic igneous intrusive (plutonic) rock	Layered intrusive variety is most suitable host rock of chromium, nickel, copper and platinum–palladium ore bodies, and glassy green type as gem and ornamental stones.
Pyroxenite	Dark green, gray and brown	Essentially pyroxene (augite and diopside), hypersthene (bronzite and enstatite)	Dense coarse-grained layered ultramafic igneous intrusive (plutonic) rock	Source of MgO as flux in metallurgical blast furnace, refractory and foundry applications, filtering media and filler, building materials and sculptures and often host deposits of Cr–Ni–Cu–Platinum group of minerals.

Dunite (Figs 4.29 and 4.31)	Usually light to dark green with pearly or greasy look	+90% olivine, typically Mg/Fe ratio at 9:1	Igneous plutonic ultramafic layered coarse-grained or phaneritic texture	Finely grounded dunite used as sequesters of CO_2 and mitigate global climate change, source of MgO as flux in metallurgical blast furnace, refractory and foundry applications, filtering media and filler and often host deposits of Cr–Ni–Cu–Platinum group of minerals.
Rhyolite (volcanic equivalent of granite) (Fig. 4.33)	White, gray, pink	Predominantly quartz, alkali feldspar (orthoclase/microcline)	Igneous is felsic extrusive (volcanic) rock with glassy, aphanitic or porphyritic texture	Suitable as aggregate, fill in construction, building material and road industries, decorative rock in landscaping, cutting tool, abrasive and jewelry.
Dacite (volcanic equivalent of granodiorite)	Black, dark gray, pale brown, yellow and pink	Mostly of plagioclase feldspar with quartz, biotite, hornblende, augite ± enstatite	Felsic extrusive rock with aphanitic and porphyritic texture, composition between rhyolite and andesite	Suitable as aggregate, fill in construction, building material and road industries, decorative rock in landscaping, cutting tool, abrasive and jewelry.
Andesite (volcanic equivalent of diorite)	White, gray, black, pale brown, green	Dominated by plagioclase with pyroxene, hornblende, biotite and garnet	Extrusive igneous rock with aphanitic and porphyritic texture, composition between dacite and basalt	Suitable mainly for naturally slip-resistant tiles, bricks, water or landscape gardens, aggregates, and fill in construction.
Basalt (volcanic equivalent of gabbro/norite) (Figs 4.3, 4.4, 4.36 and 4.37)	Dark gray to black and green, rapidly weathered to brown and rust-red	Plagioclase feldspar (labradorite), pyroxene, olivine, biotite and hornblende	Common extrusive igneous rock with aphanetic texture due to rapid cooling on surface, very fine-grained and firmly detectable under microscope	Used most commonly as construction materials (building blocks, flooring titles and aggregates, road surface and railway track), cobblestone in pavement (columnar variety), architecture, statues, stone–wool fiber as excellent thermal insulator.

(*Continued*)

TABLE 1.2 Lists of Common Rocks and Diagnostic Features under Overall Classification (*cont'd*)

Name	Color	Composition	Texture	Major Uses
Dolerite or diabase (equivalent to plutonic gabbro or volcanic basalt) (Figs 4.19 and 4.20)	Dark gray, black and greenish	Elongated lath-shaped euhedral plagioclase (~60%) in fine matrix of pyroxene (~30% augite), olivine (~10%), magnetite and ilmenite	Fine- to medium-grained subvolcanic rock occurs as dyke and sill easily recognized by style of occurrence in the field	Used as crushed stone in road making, concrete mixture in rough masonry, block paving and ornamental stone in monumental purposes.
Komatiite	Light shades of gray, brown, green, yellow	Extremely high magnesium-rich forsteritic olivine, calcic and chromian pyroxene and chromite	Rare and area restricted mafic subvolcanic intrusive rock	Massive nickel, copper sulfide and gold deposits are hosted by Komatiite in S. Africa, Australia and Canada.
SEDIMENTARY ROCKS				
Mudstones/Claystone	Grey, black, chocolate red	Extremely minute clay particles	Fine-grained sedimentary rock, finely bedded	Brick and ceramics, fillers, bleaching agents, pigments in paint and suspending media in drilling.
Siltstone	White, gray, crimson, red	Quartz and clay	Clastic sediments, grain size coarser than mudstone and finer than sandstone	Road and building material.
Argillites	Black, gray, violet, blue	Lithified mud and oozes	Hard, compact, indurated clay	Carvings, helmets, masks, totems, fetishes, stylized, souvenirs, amulets, brooches, candle holders, containers (bowls and boxes), flutes, medallions, pendants, plates, platters and poles.
Shale	White, gray with shades of red, brown, yellow-ocher, blue and black	Argillaceous sediments of aluminosilicates and clay minerals (kaoline, montmorillonite, illete and chlorite)	Minutely fine-grained, soft, homogeneous, thinly laminated	Used as filler in paint, plastic, roofing cement, bricks; dimensional stone for landscaping, paving, driveway material, and reservoir for oil and gas.
Graywacke (Fig. 5.31)	Dark gray, black, yellow, brown	Angular grains of quartz, feldspar and rock fragments set in compact clay-matrix	Poorly sorted immature sedimentary rock with fine clay to assorted fragments	Widely used as aggregate, fill in construction, road industries, armor rock for sea walls and sculpture by power tools.

Sandstone (Figs 5.27 and 5.30)	Yellow, brown, white, red, gray, pink, tan and black	Quartz and/or feldspar and other durable minerals interspace cemented	Fine to coarse sand size grains cemented by very fine matrix	Building material for domestic houses, palaces, temples, cathedrals, ancient forts, ornamental fountains, statues, roof tops, grindstone, blades and other equipments. Significant collector rocks for water, oil and gas.
Limestone (Figs 5.46, 5.47, 5.48 and 5.63)	White, gray, black, buff, yellow and shades of brown, purple, orange, cream and scarlet	Mainly calcite and aragonite, skeletal fragments of marine organism (coral and foraminifera), and silica (chert and flint)	Chemogenic sedimentary rock with soft, fine to coarse crystalline shelly open and chalky texture	Architecture and sculpture (pyramid, monuments, historical buildings, artifacts, statues), aggregates, manufacture of quick-lime, cement, mortar, soil and water conditioner, petroleum reservoir, flux in blast furnace, medicine, cosmetics, toothpaste, paper, plastics, paint, and tiles. Significant collector rocks for water, oil and gas.
Dolostone (dolomite rock) (Fig. 5.64)	Gray, white, buff and brown color	Predominantly dolomite (calcium–magnesium carbonate) $\pm$ silica	Chemogenic sediments, soft, fine to coarse grained with sugary and greasy texture	Source of magnesium metal and magnesia (MgO), refractory bricks, aggregate for cement and bitumen mixes, flux in blast furnaces of iron and steel industry, important host rock for zinc–lead–silver deposits.
Conglomerate (Fig. 5.22)	Various colors depending on preexisting source material	Clasts of preexisting rocks and minerals within fine-grained matrix	Predominantly coarse (pebbles, cobbles and boulders) in fine cementing material	Dimension stone for decoration of walls, aggregate, fill in the construction and road industries and significant source for placer diamond, gold, and uranium.
Laterite (Fig. 5.41)	Brick-red	Rich in iron and aluminum	Featureless massive residual product of weathering	Regular-sized blocks as building and road construction, aquifer for water supply in rural areas, waste water treatment plant and source of low-grade aluminum, iron and nickel.

(*Continued*)

TABLE 1.2 Lists of Common Rocks and Diagnostic Features under Overall Classification (*cont'd*)

Name	Color	Composition	Texture	Major Uses
METAMORPHIC ROCKS				
Slate	Blue-black shade	Clay or volcanic ash, product of low-grade regional metamorphism	Fine-grained, homogeneous and foliated	Building materials, roof-shingles, tiles, gravestone, electric insulator, fireproof, switchboard, laboratory bench and billiard table top and blackboards.
Phyllite	Gray, shades of brown, red, blue, green	Quartz, sericite, mica and chlorite	Phyllitic texture with silky glossy appearance	Decorative objects such as pendants and beads.
Schist (Figs 6.2, 6.3 and 6.10)	Silvery gray, brown, green	Micas, chlorite, talc, quartz, feldspar, garnet, kyanite, staurolite	Medium-grade metamorphic rock with well-developed schistocity	Dimension and decorative stone as building material, walls, garden, road industries and paving.
Gneiss (Figs 6.1, 6.8, 6.11, 6.13 and 6.14)	Variegated of black and white, light brown	High-grade regional metamorphic process of existing igneous/ sedimentary rocks	High-grade metamorphic rock showing gneissose texture, medium to coarse foliated	Building material, roads and curbs.
Amphibolite (Figs 6.5 and 6.6)	Dark colored, green, gray, brown	Hornblende, actinolite, plagioclase ± quartz	Weakly foliated or schistose structure	Attractive textures, dark color, hardness and polishing ability suits as dimensional stone in construction, paving and facing of buildings.
Serpentinite	Dark to light green	Serpentine group, e.g. antigorite, chrysolite and lizardite ± chromite	Hydrated and regional metamorphic transformation of ultramafic rocks	Decorative and curving stone in architecture and sculptures. Rich in elements toxic to plants such as chromium and nickel.
Quartzite (Figs 6.15 and 6.16)	Gray, off-white, yellow, light brown, red	Monomineralic and dominantly of quartz	Massive, extremely hard, nonfoliated metamorphic rock	Extreme hardness and angular shape is suitable for railway ballast, roads, walls, roofing/ flooring, stair steps, high-purity ferrosilicon, industrial silica sand and silicon carbide.
Marble (Fig. 6.17)	White, pink, and green	Monomineralic and dominantly of calcite	Nonfoliated metamorphic equivalent of limestone	Sculpture, ancient and present-day monuments, statues, dimensional and decorative stone, construction material, tiles, and flooring.

FURTHER READING

Reading of text books authored Dana,[9] Vrkljan,[60] Vrkljan et al.[61] will be a smart approach to step into the theme of Mineralogy and Petrology. Haldar[23] and Haldar[24] can be the foundation study material to understand the definition of minerals, rocks and ore, fundamentals to start any branch of Geology. Tišljar[49] will be helpful in reading for Croatian readers. The fundamental concepts of mineralogy and petrology are explained by Cornelis Klein.[27]

CHAPTER

2

Basic Mineralogy

OUTLINE

2.1. Introduction 39

2.2. Internal Structure of Crystals and Their Properties 40

2.2.1. Crystallized and Amorphous Solid Minerals 40

2.2.2. Formation of Minerals 41

2.2.3. Crystal Lattice 42

2.2.4. Crystallographic Axes, the Crystal Planes and Elements of Crystal Symmetry 43

2.2.5. Crystal Systems, Crystal Forms, Single Crystals and Crystal Twinning 45

2.3. Chemical and Physical Properties of Minerals 47

2.3.1. Chemical Properties of Minerals 47

2.3.2. Physical Properties of Minerals 47

2.4. Polymorphism and Isomorphism 50

2.5. Overview of the Main Rock Forming Minerals 51

2.5.1. Autochthonous Elements 52

2.5.2. Sulfides 52

2.5.3. Oxides and Hydroxides 53

2.5.4. Carbonates 55

2.5.5. Halides 57

2.5.6. Sulfates 57

2.5.7. Phosphates 58

2.5.8. Silicates 58

2.5.8.1. Nesosilicates $[SiO_4]^{4-}$ 59

2.5.8.2. Sorosilicates—$[Si_2O_7]^{6-}$ 61

2.5.8.3. Cyclosilicates—$[Si_nO_{3n}]^{2n-}$ 62

2.5.8.4. Inosilicates 63

2.5.8.5. Phyllosilicates—$[Si_{2n}O_{5n}]^{2n-}$ 68

2.5.8.6. Tectosilicates 73

Further Reading 79

2.1. INTRODUCTION

Mineralogy is the systematic study that deals with the characteristics of minerals. The mineralogy has more scientific branches such as the following:

1. Crystallography studies crystal forms, i.e. forms in which the minerals crystallize, as well as their internal structure, relations and distribution of atoms, ions or ionic groups in the crystal lattice.

Introduction to Mineralogy and Petrology
http://dx.doi.org/10.1016/B978-0-12-408133-8.00002-X

2. Physical mineralogy is the study of physical properties of minerals, such as cohesion (hardness, cleavage, elasticity, and density; refer Table 1.1), optical, thermal and magnetic properties, electrical conductivity, and radioactivity, and so on.
3. Chemical mineralogy is the study of chemical formula (Table 1.1) and chemical properties of the minerals.
4. Environmental mineralogy studies complex and very different conditions of the origin of minerals, understand element behavior in echo-systems, natural and industrial effects of minerals, and mitigates potential contamination problems.
5. Descriptive mineralogy deals with the classification of minerals into groups based on their common properties, mostly chemical and structural properties.

2.2. INTERNAL STRUCTURE OF CRYSTALS AND THEIR PROPERTIES

2.2.1. Crystallized and Amorphous Solid Minerals

The constancy of chemical composition and physical properties of minerals are the outcome of their internal crystal structure. The exact and unique arrangement of atoms, ions or ionic groups is the characteristic feature of each crystallized mineral. The minerals are mostly found as crystallized substance and less frequently as amorphous solid in Earth's crust (Fig. 3.2).

A crystal or crystalline solid is a solid material, whose constituent atoms, molecules, or ions are arranged in an orderly repeating pattern extending in all three spatial dimensions. During crystallization, where there was enough space for their uninterrupted growth in all directions, the crystals can have a regular polyhedral shape. That is, for example, often the case with minerals halite (Fig. 1.9), quartz (Fig. 1.11), calcite (Fig. 1.14 and Fig. 2.5) or garnet (Fig. 1.16 and Fig. 2.9).

Crystallized minerals have specific and constant physical properties. Same minerals always have a constant melting point or crystallization point. If we increase the temperature of crystallized mineral, when it reaches melting point, it will stop to heat as long as the mineral does not convert to mineral melt. It is because that all the heat energy is spent on the decomposition of the crystal lattice, or melting of minerals. It is called *melting point* or *crystallization point* in reverse process.

Melting point or crystallization point is always constant for a mineral in same pressure.

Physical properties of crystallized minerals are always exactly equal in a particular direction: for example, all minerals which are not part of the cubic system are double refracting, where ordinary light passing through them is broken up into two plane polarized rays that travel at different velocities and refracted at different angles. They are anisotropic. Minerals which are part of the cubic system and amorphous solid are isotropic which means that light behaves the same way no matter which direction it is traveling in the crystal. The isotropy and anisotropy of crystallized minerals will be discussed in more detail in Sections 2.2.3 and 2.3.2.

Amorphous minerals from Greek amorphous "shapeless" (from a "without" + morphē "form") in which there is no long-range order of the positions of the atoms and in fact are not minerals, but mineraloids. Therefore, never assume a regular polyhedral shape. Mineraloids possess chemical compositions that vary beyond the generally accepted ranges for specific minerals. Mineraloids unlike the crystallized minerals do not have a specific melting point and crystallization point. With the gradual increase in temperature of mineraloids, for example, glass, gradually becoming softer and softer until it finally softens enough to become liquid. All mineraloids are optically isotropic. Agate is

FIGURE 2.1 Agate-nodule—concentric secretion of light and dark zones of amorphous silicon hydroxide from edges toward center. Agate is a variety of chalcedony formed from layers of quartz showing multicolor bands. Most agates occur as rounded nodules or veins in volcanic rocks or ancient lavas.

considered a "Mineraloids" because of its lack of crystallization (Fig. 2.1).

Mineraloids are not so common in the Earth's crust as crystallized minerals and usually occur during spending of minerals on Earth surface or secretion from lava. A good example of a mineraloid is opal, semiprecious stone. It usually occurs by excretion of minerals substances from the edge to the center of fissures of almost any kind of rock, being most commonly found with basalt, rhyolite, limonite, sandstone and marl. It is basically hydrated silica with variable amounts (1–21%) of water (Section 2.5.3).

2.2.2. Formation of Minerals

Minerals, as integral part of the rocks, are forming in different ways by complex processes such as the following:

1. Crystallization of magma (silicate composition), "pyrogenesis" processes, such as olivine, pyroxenes and plagioclase (Section 4.1.2).
2. Crystallization from gases and vapors, "pneumatolysis" processes, such as tourmaline.
3. Crystallization from the hot solution, "hydrothermal" processes, such as fluorite and galena (Section 4.1).
4. Crystallization and deposition of minerals from aqueous solutions, "hydatogenesis" processes, such as secretion of aragonite and calcite from seawater (Section 5.7.1.1).
5. "Vaporization" of highly concentrated aqueous solutions due to the strong evaporation, evaporation processes, such as gypsum, anhydrite and halite (Section 5.8.1).
6. "Dynamic metamorphism" is associated with zones of high to moderate strain such as fault zones. Cataclasis, crushing and grinding of rocks into angular fragments, occurs in dynamic metamorphic zones, giving cataclastic texture (Section 6.5).
7. "Contact metamorphism" occurs typically around intrusive igneous rocks as a result of the temperature increase caused by the intrusion of magma into cooler country rock. The area surrounding the intrusion (called aureoles) where the contact metamorphism effects are present is called the *metamorphic aureole*. Contact metamorphic rocks are usually known as hornfels. Rocks formed by contact metamorphism may not present signs of strong deformation and are often fine grained (Section 6.4).
8. The action of aqueous solutions and the atmospheric conditions on solid minerals can create "authigenic" minerals. Chemical wear can create new minerals under the influence of water and CO_2 occurs mild carbon acid $[H]^+ + [HCO_3]^-$ under whose effects of spending feldspars can create kaolinite (Section 5.2.1.2).
9. Life processes of organisms, or "biochemical" processes, are biogenic minerals such as secretion of calcite or aragonite for building shells or coral skeletons (Section 5.7.1.1).

Pyrogenesis is the process of mineral/rock formation derived by solidification of intrusive or extrusive magma/molten lava/and its derivatives emplaced by volcanism.

Pneumatolysis is the alteration of rock or mineral crystallization affected by gaseous emanations from solidifying magma.

Hydrothermal process/synthesis is crystallizing single crystal/substances from high-temperature aqueous solution at equilibrium vapor pressure.

2.2.3. Crystal Lattice

The mineral as a homogeneous body is defined with a regular arrangement of atoms, ions or ionic groups in the crystal lattice. This means that all the same crystallized minerals have same formations of ions. Specifically, in all the three dimensions, every mineral has the same pattern of a set of atoms arranged in particular way according to their type. This can be thought of as forming identical tiny boxes, called unit cells that fill the space of the lattice. Such unique arrangement of atoms or molecules is called *crystal lattice* (Fig. 1.10 and Fig. 2.2). This is the homogeneity of the crystals.

Lattice constants can be determined using X-ray diffraction or with an atomic microscope. The structure of each mineral species, i.e. their proper internal structure is determined by using X-rays. Three-dimensional grid or lattice of each mineral was determined by regular—periodic—sorting of ions (or ionic groups) in one direction and at equal distances, forming a so-called long-range order (Fig. 2.2). Two dimensional planar network results by repeating the pattern periodically in two different directions and similarly three dimensional forms of lattice is obtained by repeating in three different directions (Fig. 2.2). The basic unit of the crystal lattice of the unit cell in the crystal periodically repeated in three directions, two of which lie in the same plane. The unit cell has defined edges a_0, b_0 and c_0 and angles α, β and γ between them, respectively. For example, the unit cell halite, which crystallizes in the cubic system (Fig. 1.10), has the same intervals in all the three directions ($a_0 = b_0 = c_0$ and amounts to 4.12×10^{-10} m), and right angles ($\alpha = \beta = \gamma = 90°$).

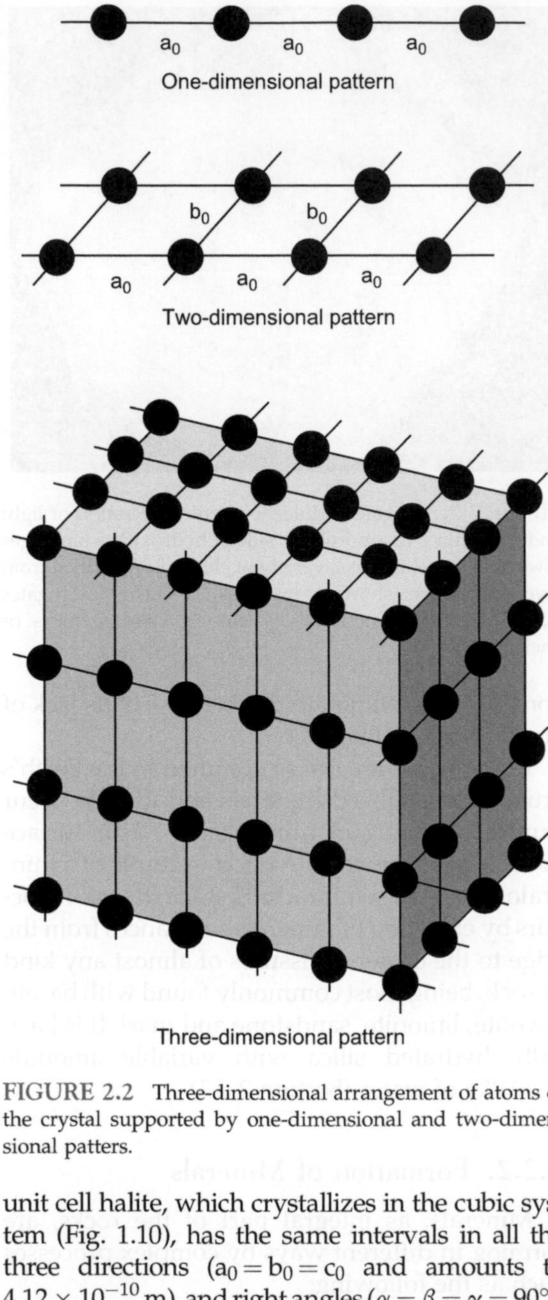

FIGURE 2.2 Three-dimensional arrangement of atoms of the crystal supported by one-dimensional and two-dimensional patters.

By the lattice constant of crystallized minerals is defined a constant chemical composition and chemical formula of mineral. Lattice constant also determines the other very important properties of crystallized minerals homogeneity isotropy, anisotropy and symmetry.

Homogeneity of minerals is reflected by the fact that in parallel direction of crystal lattice, atoms have same interval, and in the different directions, they do not.

Isotropy is uniformity in all directions. The term is made up from the Greek words iso (equal) and tropos (direction). Isotropy means that physical properties of mineral are identical in all crystallographic directions. These are isotropic minerals which belong only to minerals of cubic system and mineraloids. Only minerals which crystallize in the cubic system in the crystal lattice have in all three directions equivalent arrangement of atoms, (Fig. 1.10) such as Na and Cl in a halite lattice (distance of 4.12×10^{-10} m).

Anisotropy is the property of being directionally dependent. The term is made up from the Greek words aniso (without) and tropos (direction). Anisotropy means that physical properties of mineral are not identical in all crystallographic directions. For example, a_0, b_0, and c_0 have different values. Anisotropy has all the minerals that crystallize in the tetragonal, orthorhombic, hexagonal, rhombohedral, monoclinic and triclinic system (Section 2.2.5).

Each arrangement of atoms has a certain number of elements of symmetry, i.e. changes in the orientation of the arrangement of atoms seem to leave the atoms unmoved. One such element of symmetry is rotation; other elements are translation, reflection, and inversion. The elements of symmetry present in a particular crystalline solid determine its shape and affect its physical properties (Section 5.2.4).

2.2.4. Crystallographic Axes, the Crystal Planes and Elements of Crystal Symmetry

All planes in the crystal can be placed in the imaginary coordinate system of crystallographic axes, which is known as a *common point*. These are imaginary lines that intersect at the center of the crystal (Fig. 2.4). Crystal planes in relation to the crystallographic axes are placed so that at some distance from the center of the crystal intersect one or more axes. Distance to the crystallographic axes are called parameters and they exactly determine the position of each plane.

Crystal planes are a reflection of proper internal structure of crystals and are part of the network plane, therefore, are not random phenomena in crystals. Each plane in the crystal occupies a specific position in relation to a particular crystallographic axis: it cuts it or is parallel with it (Fig. 2.3). Each crystal plane can be described parametric relationship. For a plane that intersects all three crystallographic axes in their unit distances parametric relationship is (1a:1b:1c) to a surface that is parallel to the crystallographic axes and c (∞ a:1b:∞ C).

These parameters are knows as Weiss parameters. Today in the mineralogy, however, to indicate the position of the crystal surface is in use Miller indices. Miller indices were introduced in 1839 by the British mineralogist William Hallowes Miller. Specifically, Miller indices are much easier for writing and marking on the surface of a crystal that has a large number of surfaces.

1. Planes with Weiss parameter (1a:1b:1c) has a Miller indices (111).
2. Planes with Weiss parameter (1a:1b:∞ C) has a Miller indices (110).
3. Planes with Weiss parameter (∞ a:1b:∞ C) has a Miller indices (010).

By convention, negative integers are written with a bar, as in 3 for $\bar{3}$, for example, if the axis is cut on the negative arm of the Miller Index is ($\bar{1}$11).

In crystals are available seven planes of different positions with respect to crystallographic axes, which can show the seven different types of Miller index. By its position according to the crystallographic axes to be surface bipyramid, prism and pinacoid (Fig. 2.3):

1. Bipyramid, (111) intersects all three axes.
2. First-order prism, (011) plane parallel to the axis of "a".

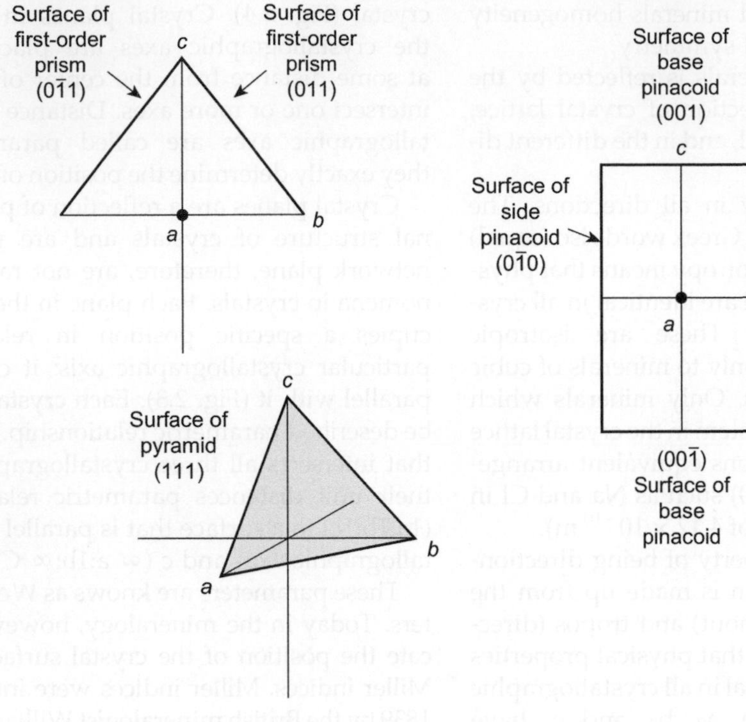

FIGURE 2.3 Location of the crystal planes and their Miller indices in relation to the crystallographic axes *a*, *b*, and *c*.

3. Second-order prism, (101) plane parallel to the axis "b".
4. Third-order prism, (110) plane parallel to the axis "c".
5. Front pinacoid, (100) plane intersects a front axis "a".
6. Side pinacoid, (010), plane intersect side axis "b".
7. Base or basal pinacoid, (001) plane intersects the vertical axis or base axis "c".

Crystals can occur in simple forms or in combinations of different forms. If crystal has developed, for example, in six same planes, pinacoid of the same side shape and symmetry. They will form in cubic system a geometric body shape of cube which is called in crystallography, regular hexahedron (Fig. 1.10). Such crystal is a simple form. If, however, the crystal is composed of various planes, two or more different simple shapes will be obtained. Such crystal is a crystal combination (Fig. 1.11, Figs 2.5, 2.11, 2.13, 2.18(A) and (B)). Figure 1.11 shows the crystal combination in which the quartz crystals occur. A combination of six-sided prisms, and four different types of bipyramid crystal combinations are among the minerals much more widespread than the simple forms. And they have lot more planes than simple forms (Fig. 1.11; Figs 2.11, 2.13 and 2.18(B)).

Crystals are symmetry bodies that have one or more planes of symmetry, one or more axes of symmetry and a center of symmetry. There are crystals that have not got a single plane of symmetry or any of the axes of symmetry, or are without center of symmetry. The plane of symmetry divides crystal on two mirrors same parts (mirror plane). Axis of symmetry is the direction in a crystal around which crystals can turn and repeat two, three, four or six times within 360°.

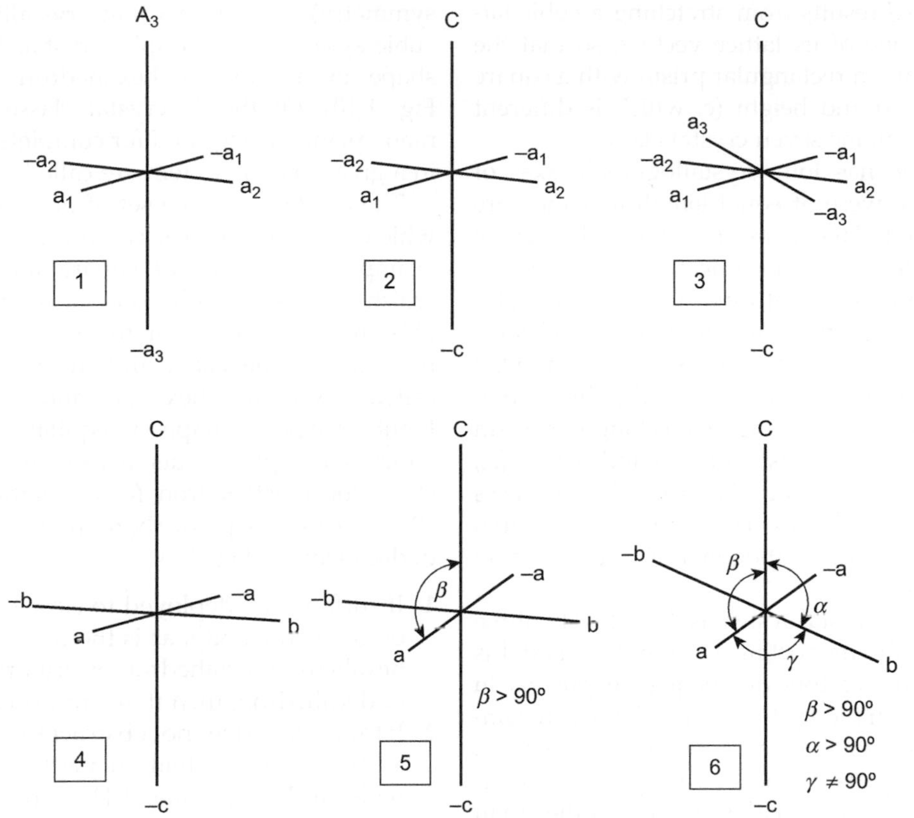

FIGURE 2.4 The relationship between the position of the crystallographic axes: 1, cubic; 2, tetragonal; 3, hexagonal and trigonal; 4, orthorhombic; 5, monoclinic; 6, triclinic.

In a complete 360° rotation, the grain will repeat itself in appearance two times, once every 180°, it is called twofold rotational symmetry; three times, once every 120°, it is called threefold rotational symmetry; four times, once every 90°, it is called fourfold rotational symmetry; and six times, once every 60°, it is called sixfold rotational symmetry.

2.2.5. Crystal Systems, Crystal Forms, Single Crystals and Crystal Twinning

Based on the size of the crystallographic axes, there are different six or seven lattice point groups. In fact, there are two different grids—hexagonal and trigonal, but the same common point—hexagonal (Fig. 2.4). All known crystals have only 32 possible combinations of planes of symmetry, which due to the mutual relations of crystallographic axis, their lengths and angles between them can be grouped into six or seven lattice point groups (Fig. 2.4).

Cubic crystal lattices have three mutually perpendicular crystallographic axes of equal length (axis a_1:a_2:a_3) from which the first horizontal axis (front axis) is directed toward the observer, the second horizontal axis extending from left to right (lateral axis) and the third axis is vertical (vertical axis). Contains five crystal classes.

Tetragonal results from stretching a cubic lattice along one of its lattice vectors, so that the cube becomes a rectangular prism with a square base (a by a) and height (c, which is different from a). Contains seven crystal classes.

Hexagonal has four crystallographic axes of which three were of equal length and they are in horizontal plane (axes a1:a2:a3), each at angle 120°. Fourth (c axis or vertical axis) is longer or shorter than the rest of three and perpendicular to them. This system contains 12 crystal classes.

Orthorhombic results from stretching a cubic lattice along two of its orthogonal pairs by two different factors, resulting in a rectangular prism with a rectangular base (a by b) and height (c), such that a, b, and c are distinct. All three bases intersect at 90° angles. The three lattice vectors remain mutually orthogonal. Contains three crystal classes.

Monoclinic crystal system is described by three vectors. In the monoclinic system, the crystal is described by vectors of unequal length, as in the orthorhombic system. They form a rectangular prism with a parallelogram as its base. Hence two pairs of vectors are perpendicular, while the third pair makes an angle other than 90°. Contains three crystal classes.

Triclinic crystal system is described by the three basis vectors. In the triclinic system, the crystal is described by vectors of unequal length, as in the orthorhombic system. In addition, all three vectors are not mutually orthogonal ($a \neq b \neq c$ and $\alpha \neq \beta \neq \gamma$). Contains three crystal classes.

The triclinic lattice is the least symmetric. It has (itself) the minimum symmetry all lattices have: points of inversion at each lattice point and at seven more points for each lattice point: at the midpoints of the edges and the faces, and at the center points. It is the only lattice type that itself has no mirror planes although they are characterized by proper internal structure.

Combination of the largest number of symmetry elements (9 planes of symmetry, 13 axis of symmetry) has crystals that crystallized in the cubic system, for example, crystals having the shape of a cube or hexahedron (see halite Fig. 1.10). Of the 32 crystal classes, those as many planes as required for complete symmetry in a given crystal system are called holohedral.

The crystal form is a set of uniform surfaces, which as a whole has certain symmetry. For example, six planes of square form, each of which intersects only one axis (100), (010), (001), make a crystal form that geometrically corresponds to the cube, and in crystallography called hexahedra (hex = six and hédra = flat). Eight planes in shape of equilateral triangle, which each plane cuts across all three axes (111), form octahedron (octa = eight), and the 12 surface in shape of rhombus forms rhombic dodecahedron (Fig. 2.9).

1. If surfaces are deployed to close some space or geometric body, as is the case in a hexahedron, octahedron or orthorhombic dodecahedron, then they are closed forms.
2. If the surface does not obstruct the space, such as prism surfaces that intersect two axes and with third are parallel (011), (110) or (101), then they are open forms.
3. Prism is an open form, and may have a different number of planes.
4. Pyramid is an open form. Pyramid has a different number of planes and is usually combined with another symmetric side of the pyramid, creating a so-closed form—bipyramid, or is combined with a prism, as for example the case of quartz (Fig. 1.11).

Separate crystals can be completely free (Fig. 2.9) or they grow on some base (Fig. 2.5). When two separate crystals share some of the same crystal lattice points in a symmetrical manner, the result is an intergrowth of two separate crystals in a variety of specific configurations. A twin boundary or composition surface separates the two crystals. It is called *crystal twinning* (Fig. 2.18(B), (C), (D) and (E)).

FIGURE 2.5 Crystal of calcite composed of planes that form the rhombohedral shape.

2.3. CHEMICAL AND PHYSICAL PROPERTIES OF MINERALS

2.3.1. Chemical Properties of Minerals

Minerals have a defined chemical composition which can be determined by various methods of analytical chemistry and determine their chemical formula. The results of quantitative chemical analysis of minerals, usually expressed in two ways:

1. Ratio of chemical elements (for example, quartz, 46.73% silicon and 53.27% oxygen).
2. Ratio of oxides (for example, forsterite 57.11% MgO and 42.89% SiO_2 or fayalite 70.57% FeO and 29.43% and SiO_2).

Minerals may include water as the following:

1. Constitutional water in form of hydroxide ions (OH) which are an integral part of the crystal lattice and have a steady position in the grid. By the loss of water, crystal will fall apart and the water cannot get back into the grid.
2. Crystalline water or adsorption water in the form of H_2O molecules.

Crystal water have certain points in the crystal lattice (for example, gypsum $CaSO_4 \cdot 2H_2O$). By heating, a portion of such water is lost (= process of dehydration), but the crystal lattice is not destroyed and may again receive such water (= hydration).

Adsorption water has not got strict location in the crystal lattice. In some clay minerals can cause plasticity or swelling which can make landslides.

In minerals with a layered grid, for example, smectite, it is the interlayer water located between the layers. In zeolite group of minerals, the water is in holes and channel grids, and it is called *zeolite water*. By heating those kind of minerals, they lose water, but their lattice will not break apart, but the unit cell is reduced. By receiving that water again their unit cell is increased, for example, montmorillonite will increase in size by 2.5 times compared to the dry mineral.

Content of constitutional, crystal and adsorption water is expressed in chemical formula of mineral, for example, kaolinite $Al_2Si_2O_5\ (OH)_4$ or gypsum $CaSO_4 \cdot 2H_2O$.

Minerals can also contain hygroscopic water and mechanically incorporated into the water.

Hygroscopic water is actually the humidity that is located on the surface of minerals or which fills cracks and gaps in it, so that it is not related to the crystal lattice. It can be removed by heating to 110 °C.

Mechanically blended water is located in the minerals in the form of inclusion, drops of water embedded in mineral during its growth in fluid environments (fluid inclusions).

2.3.2. Physical Properties of Minerals

Classifying minerals can range from simple to very difficult. A mineral can be identified by several physical properties such as:

1. category (oxide, sulfide, silicate, carbonate, etc.);

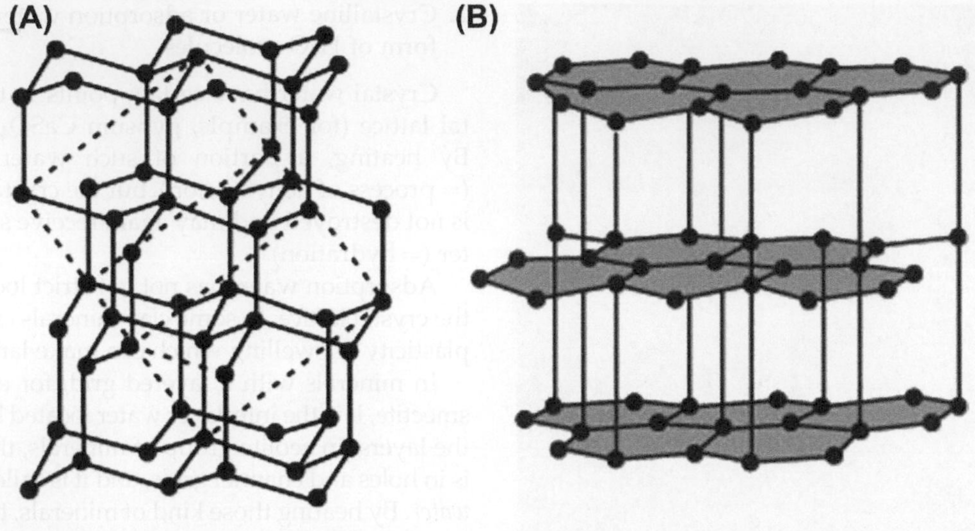

FIGURE 2.6 Crystal lattice of polymorphic modifications in carbon minerals: (A) diamonds; (B) graphite.

2. crystal system (cubic, tetragonal, hexagonal, etc.);
3. cohesion properties (hardness in Mohs scale 1–10, cleavage, tenacity, and fracture);
4. density/specific gravity;
5. macroscopic optical properties (color, brightness, luster, and streak); and
6. microscopic optical properties (refractive index, birefringencea, and pleochroism).

The minerals can be analyzed for chemical composition and broadly grouped under oxide, sulfide, silicate, etc.

Crystal system/family represents an identical mathematical and geometrical three-dimensional space for a group of minerals.

Hardness of minerals can be seen in the resistance of minerals to the encroachment of a solid object in its surface and interior. Hardness of minerals is dependent on the structure, i.e. the distribution and density of packing of atoms, ions and ionic groups in the crystal lattice, as it can be seen in structure of graphite and diamond (Fig. 2.6). The mineralogy defines and determines the hardness of minerals and applied as relative hardness. Higher relative hardness still has the mineral that can rip another, i.e. a softer mineral. The relative hardness of minerals is determined by using and comparing Mohs' scale of hardness, which includes 10 lined-up minerals from softest to the hardest mineral (Table 2.1).

TABLE 2.1 Mohs' Scale of Mineral Hardness

Mineral	Hardness
Talc	1
Gypsum	2
Calcite	3
Fluorite	4
Apatite	5
Orthoclase feldspar	6
Quartz	7
Topaz	8
Corundum	9
Diamond	10

For example, corundum (9) is twice as hard as topaz (8), but diamond (10) is almost four times as hard as corundum.

Cleavage is the tendency of minerals to split along crystallographic planes as a result of structural locations of atoms and ions in the crystal, creating planes of relative weakness. Mineral graphite, for example, has an excellent cleavage with basal pinacoid plane (0001) because its structure consists of planes of carbon atoms arranged in parallel with the basic pinacoid. Diamond which crystallizes cubic and has significantly denser packing of carbon atoms has no cleavage (Fig. 2.6). Cleavage is an important property for determination of minerals even when the minerals are found as irregular grains (Fig. 2.12).

Minerals with a large difference in cohesion in different directions have great cleavage (e.g. micas, calcite, gypsum, feldspar, pyroxenes, and amphiboles), and minerals with small difference in cohesion have poor or no cleavage (e.g. quartz, apatite, and olivine).

Mineral density is defined as mass of unit volume, i.e. the mass of 1 cm^3 expressed in grams (g/cm^3) and at a certain temperature and pressure is constant for all the minerals of the same mineral species. As a unit of measure is usually taken the density of water at +4 °C which is 1 g/cm^3 so for mineral is used relative density which indicate how much is mineral denser than water at +4 °C, for example, the relative density of quartz is 2.65. The most common petrogenic minerals (minerals that are the main constituents of rocks) have a density between 2.0 and 4.5 g/cm^3. The densities of minerals are easily determined by pycnometer, small glass vials and a thermometer following the steps given below:

1. The pycnometer is filled with water of +4 °C, seals and measures the weight of pycnometer with water (p).
2. Measure the weight of mineral crushed to dust (m).
3. Dust of mineral is put inside the pycnometer and check the temperature which must be +4 °C.
4. Measure the weight of pycnometer with mineral dust and water (v).
5. Calculate density (γ) of mineral with formula:

Density = (Weight in air)/(Weight in air − Weight in water)

or

$$\gamma = \frac{m(gr)}{p + m - v(gr)}$$

Color indicates the appearance of the mineral in reflected light or transmitted light for translucent minerals.

Color of petrogenic silicate minerals are usually derived from isomorphic mixed Fe ions. The minerals with a small proportion of isomorphic admixtures of iron are bright green, while those with more isomorphic admixtures are dark green and black color (Section 2.5.8.4.2). The color of minerals is decisive for colors of rocks. The proportion of mineral aggregate is accountable for different interrelated rock color. The colorless or white petrogenic or leucocratic minerals give rise to white to light gray color rocks (such as most granite). The dark green and black or melanocratic minerals give rise to rocks of dark gray, dark green or black color (e.g. gabbro and peridotite).

Idiochromatic minerals have constant characteristic color and that characteristic color does not change even with small amount of various additives or impurities, as for instance the case of golden-yellow pyrite, green malachite or blue azurite.

Allochromatic minerals, that occurs most frequently, show colors of their impurities, such as quartz, which otherwise should be clear, colorless and often translucent like glass. It can be milky white, purple, yellow or black when it contains impurities.

Shine of minerals depends on the ability of refraction, i.e. the size of refraction. Minerals that are characterized by excellent cleavage and smooth cleavage planes (e.g. micas) and minerals that are characterized by high index of refraction (e.g. sphalerite, galena and diamond) have high glossy shine.

Optical properties of minerals, perceived with polarization microscope, are the most significant characteristics for identification of minerals, both in mineralogy and petrology. This is specifically significant if the rocks are composed mostly of fine grains or microcrystalline mineral aggregates and the mineral ingredients are mixed together. The identification of the mineral composition is not easy with the naked eye or a magnifying glass, but can only be possible through a microscope.

Polarizing microscope is used to examine the optical properties of minerals, determines the type and quantity of certain mineral components in the rock as well as structural and genetic features of rocks in the linearly polarized light that passes through the mineral or rock. The optical properties of minerals are the result of propagation and behavior of light as an electromagnetic wave in the mineral. The propagation of light in a mineral depends on the structure of minerals, i.e. about its internal structure. Therefore, it is unique for every mineral.

Isotropic minerals have the same properties in all directions. This means that light passes through the minerals in the same way, no matter in what direction the light is traveling.

Anisotropic minerals have different properties, i.e. light travels through them in different ways and with different velocities, depending on the direction of travel through a grain. Anisotropic minerals cause polarized light to be split into two rays as it travels through a grain. The rays may not travel at the same velocity or follow exactly the same path. Birefringence is a value that describes the difference in velocity of the two rays. When the rays emerge from the grain, they combine to produce interference colors. The colors repeat as the birefringence increases. Minerals with low birefringence show only white, gray and black interference colors.

2.4. POLYMORPHISM AND ISOMORPHISM

Although the crystal structure of each mineral is constant and characteristic, it is a common phenomenon that chemically the same substance is found in two or more crystalline forms. It crystallizes in two or more crystal systems with small or very large differences in crystallographic and physical properties. This phenomenon is called *polymorphism* (from the Greek. "Poly" = "more" and "morph" = "form").

The good examples of polymorphs are graphite and diamond (Fig. 2.6). The chemical composition of both the minerals consists of the same elemental carbon with the formula of "C". While diamond is the hardest mineral (Mohs hardness of 10), translucent, high glossy shine, density 3.52 g/cm^3, an excellent electrical insulator, and known as the precious gem, graphite is very soft (Mohs hardness of 1), black and completely opaque, density 2.1 g/cm^3 and a good conductor of electricity. Diamond crystallizes in cubic form, usually in the form of an octahedron and graphite crystallizes in hexagonal structure. These differences are caused by a very different way of stacking carbon atoms in the crystal lattice. The carbon atoms of diamond have very dense and compact arrangement, while graphite atoms are not. The most common examples of polymorphs in nature are given in Table 2.2.

Isomorphism (from the Greek. "Izos" = "same" and "morph" = "form") is a phenomenon that represents the minerals of different, but analogous chemical composition in the same crystal system and shapes. The minerals can blend in crystalline state, and form isomorphic compounds or mixed crystals. One of the basic conditions for this is that the cations are replaced in the

TABLE 2.2 Examples of Polymorphic Minerals that are Most Often Seen as Essential Ingredients of Rock

Mineral	Crystal System	Relative Density	Properties
α and β quartz SiO_2	Hexagonal (two different classes)	2.65	Stable to 870 °C
α and β tridymite SiO_2	Rhombic and hexagonal (two different classes)	2.32	Stable from 870 to 1470 °C
α and β cristobalite SiO_2	Tetragonal and cubic	2.32	Stable above 1470 °C
Calcite $Ca[CO_3]$	Trigonal	2.72	Stable modification
Aragonite $Ca[CO_3]$	Rhombic	2.94	Unstable modification
Silimanite $AlSiO_5$	Rhombic (different unit cells)	3.25	Two modifications stable at lower, and silimanite at higher temperatures
Andalusite $AlSiO_5$		3.10	
Kianite $AlSiO_5$	Triclinic	3.56	

crystal lattice, for example, Mg^{2+} can be substituted with Fe^{2+}, Ca^{2+} with Mg^{2+}. Similarly, Si^{4+} can be replaced with Al^{3+} at simultaneous installation of one cation (K^+ or Na^+). Two Si^{4+} ions can be replaced with two Al^{3+} ions with installation of Ca^{2+} to fulfill difference in valence (Section 2.5.8.6.1.1).

There are many examples of mixed crystals, especially among petrogenic silicate minerals. A simple example of isomorphism and formation of mixed crystals is olivine $(Mg,Fe)_2SiO_4$. The olivine is essentially a mixture of isomorphous minerals, forsterite (Mg_2SiO_4) and fayalite (Fe_2SiO_4). It is evident from the formula of olivine, forsterite and fayalite that Fe^{2+} and Mg^{2+} ions mix and form mixed crystal in which exact ratio of Fe^{2+} and Mg^{2+} cannot be determined. Therefore, the chemical composition of such compounds cannot be accurately expressed by the formula or the stoichiometric ratio, but ratio must always be 2 (Mg + Fe):1-Si:4O. The crystal formula of mixed crystals are written as cations and isomorphic replaced elements are separated by commas and placed in parentheses, for example, olivine $(Mg,Fe)_2SiO_4$. However, the contribution of different end-members that make up the crystal half-breed, we indicate their proportion, for example, olivine mixture of 80% forsterite and 20% fayalite (Table 2.12). The numerous examples of isomorphic mixtures, with olivine, are certainly most important isomorphic mixture of plagioclase.

2.5. OVERVIEW OF THE MAIN ROCK FORMING MINERALS

Mineralogy today accounts for more than 4000 different minerals, but only small numbers are essential ingredients in the composition of rocks. The scientific researches indicate that the crust materials are representing by primarily feldspars, quartz, pyroxenes, amphibole and olivine, as shown in Table 2.3.

The classification of minerals is based on their chemical composition and structure. Certain number of chemical elements in nature is found in elemental form. Those elements which are in crystallized state constitute a special group of minerals or elements. The majority of minerals are in the form of chemical compounds: oxides, hydroxides, sulfides, sulfates, chlorides, fluorides, carbonates, phosphates and silicates in particular (Table 2.4).

TABLE 2.3 Percents of Main Rock Forming Minerals

Minerals	%
Minerals from feldspars group	57.9
Pyroxenes, amphiboles and olivine	16.4
Quartz	12.6
Fe oxides (magnetite and hematite)	3.7
Mica	3.3
Calcite	5.0
Clay minerals	1.0
All other minerals	3.6

TABLE 2.4 The Most Important Group of Minerals by Their Chemistries

Elements	Graphite, diamond, gold and sulfur
Sulfides	Pyrite, marcasite, and pyrrhotite
Oxides and hydroxides	Quartz, opal, chalcedony, corundum, gibbsite, boehmite, diaspore, magnetite, hematite, goethite, chromite, limenite, rutile, pyrolusite, psilomelane, and spinel
Carbonates	Aragonite, calcite, magnesite, siderite, and dolomite
Halogenides	Halite, silvite, and carnallite
Sulfates	Gypsum, anhydrite, and barite
Phosphates	Apatite and phosphorite
Silicates	Nesosilicates Sorosilicates Cyclosilicates Inosilicates Phyllosilicates Tectosilicates

2.5.1. Autochthonous Elements

The minerals that can be found in crystallized state as autochthonous elements are graphite, diamond, gold and sulfur. They are the most economic and important members.

Graphite (C) is a stable hexagonal polymorphic modification of carbon with a layered lattice (Fig. 2.6(B)). Graphite is a layered compound and in each layer, the carbon atoms are arranged in a hexagonal lattice. It is soft, black in color and leaves a black mark on fingers. It can be found in pegmatites and granites, and particularly in the crystalline schists, and it is an essential ingredient of graphite schist (Table 6.1). It is an important raw material due to high electrical conductivity.

Diamond (C) is a stable cubic polymorphic modification of carbon in the form of octahedra (Fig. 2.6(A)). The properties include hardness of 10, relative density of 3.52, transparent, colorless, with impurities can be white, gray, yellowish, bluish and rarely black. The impure varieties of diamond are used as abrasive due to the extreme hardness and cutting material, making the crown for rock drilling. The clean and pure varieties are treated as a high-value gemstone. The diamonds crystallizes as the primary ingredient in the olivine-rich ultramafic rock kimberlite, and it can be usually found in sand–gravel river deposits due to the exceptional resistance to physical and chemical weathering.

Gold (Au) is found as autochthonous cubic mineral in hydrothermal ore veins or strings, and as resistant mineral in the debris. Color is golden yellow, metallic shine, the relative density of 15.5–19.3.

Sulfur (S) is often found in nature as autochthonous mineral that crystallizes in the orthorhombic system. The most common form of occurrences is aggregates of granular, fibrous or kidney-shaped structure. Sulfur often forms around volcanic craters and on the outbreaks of sulfur and water vapor around the volcanoes and hot springs. It can also occur through organic processes of bacteria that reduce sulfate. It is an important raw material in chemical industry.

2.5.2. Sulfides

Sulfides are compounds of transition metals with sulfur. Sulfides are very frequent and widely distributed as ore and petrogenic minerals with most significant are pyrite, marcasite and pyrrhotite.

Pyrite (FeS_2) is a widespread mineral in many rocks (Figs 1.26 and 1.32) and belongs to the most widespread sulfide minerals in the lithosphere. Pyrite crystallizes in the cubic system, and has a brass-yellow color. It is found as rock-forming mineral in regular cubic grains, clusters of fine-grained aggregates. Pyrite can turn into limonite and hematite by processes of oxidation (Section 5.2.1.2). It occurs from crystallization of magma, from hydrothermal solutions, sediments in reducing conditions and metamorphic processes.

Marcasite (FeS_2) is an orthorhombic modification of the substance FeS_2 and is generally associated with sedimentary rocks in the form of spherical aggregates. Marcasite crystallizes at low temperatures from solutions containing ferrous sulfate, and is never found as a primary mineral in igneous rocks.

Pyrrhotite (FeS) (Fig. 1.27) is usually found in basic and ultrabasic igneous rocks in the form of dense aggregates. Pyrrhotite is often magnetic, has metal shine, opaque, and brownish bronze in color. The mineral frequently contains a small amount of nickel in basic and ultrabasic rocks. Pyrrhotite is an important component of some nickel bearing meteorites.

2.5.3. Oxides and Hydroxides

An oxide is a chemical compound containing at least one oxygen atom as well as at least one other element. The oxides result when elements are oxidized by oxygen in air. Hydroxides are compounds in which the anion OH− groups contain OH− ion as the OOH− group. Some minerals from the group of oxide and hydroxide minerals, particularly, silica, iron and aluminum, are very important and widespread mineral components of rocks. The most abundant among the group are listed in Table 2.5.

TABLE 2.5 Overview of Main Rock Forming Minerals from the Group of Oxides and Hydroxides

	Oxides	Hydroxides
Silicon	Quartz SiO_2 Chalcedony = fibrous quartz	
Aluminum	Corundum Al_2O_3	Gibbsite γ-$Al(OH)_3$ Boehmite γ-AlOOH Diaspore α-AlOOH
Iron	Hematite Fe_2O_3 Magnetite $FeO \cdot Fe_2O_3$	Goethite α-FeOOH Limonite
Iron and chrome	Chromite $FeO \cdot Cr_2O_3$	
Iron and titanium	Limenite $FeO \cdot TiO_2$	
Titanium	Rutile TiO_2	
Manganate	Pyrolusite MnO_2	Psilomelane $MnO \cdot MnO_2 \cdot nH_2O$
Magnesium and aluminum	Spinel $MgO \cdot Al_2O_3$	

Ice (H_2O) is the solid phase of water at a temperature of 0 °C and crystallizes as a hexagonal mineral with density of $0.9175\ g/cm^3$, and floats on water. The snowflakes are formed by sublimation (from water vapor and not from water) and crystallizes in the hexagonal crystal in the form of six-sided stars.

Quartz (SiO_2) (Figs 1.11 and 8.1) is the most common polymorphic modifications of silicon dioxide in rocks. The crystalline silicon dioxide occurs in several polymorphic alpha modifications in rocks: tridymite and cristobalite, each with another unstable beta-modification. The different crystallized SiO_2 occurs in as many as 12 polymorphic modifications depending on temperature. Transformation of one modification into another is an extremely slow process. Sometimes higher temperature and lower temperature polymorphic modification can be both stable during a certain time. Quartz crystallizes in the hexagonal system and is often found in nature in crystal form as shown in Fig. 1.11. Quartz is very common and abundantly present in the sediments of silicon rocks, acidic intrusive and vein rocks, as well as most metamorphic rocks (mylonite, quartzite, phyllite, mica, and green schists, gneiss and granulite).

Opal ($SiO_2 \cdot nH_2O$) is a mineraloid (amorphous mineral) of irregular shapes and white in color. The color changes to yellow, gray, brown or red in the presence of impurities. A special type of opal becomes precious, which is characterized by awesome flows of bluish gray and white colors. Opal is excreted from hot springs and geysers. In sedimentary rocks, particularly limestone, opal is formed by carbonate and excreted from solutions containing silicon. Opal may occur as secondary mineral during the weathering process of primary silicate minerals, basic and ultrabasic igneous rocks. The skeletons of algae, diatoms, silicon sponges and radiolarians are consisting of opal, and with their deposition and diagenesis occurring siliceous sedimentary rocks.

Chalcedony is a fibrous type of cryptocrystalline quartz, which forms very dense kidney-shape clusters. It is composed of thin parallelly and linearly arranged aggregates of fibrous crystals that usually show X-ray structural features of β-quartz. It has waxy and glossy appearance, and can be white, gray, pink, yellowish, dark brown to black color. The yellowish and red color chalcedony is considered as semiprecious stones. Chalcedony is a frequent ingredient in silicon sediment, i.e. chert (Table 5.7) and radiolarite (Section 5.9.1). Chalcedony usually occurs by extraction from aqueous solutions containing silicon acid and from opal in processes of recrystallization. It occurs in the form of irregular masses, concretions, lenses and nodules in limestone as filling voids in the rock appears in the basic volcanic or extrusive rocks.

Agate is a variety of chalcedony with modified thin lamina or layers of different colors (Fig. 2.1). Agate with the changes of black and white laminae is known as *onyx*. Dense and opaque types of chalcedony with admixtures of iron brownish red color are called Jasper and used for making jewelry.

Corundum (Al_2O_3) is a crystalline form of aluminum oxide and have traces of iron, titanium and chromium. It is a rock-forming mineral, and clear transparent natural material. It can have different colors in the presence of impurities. The transparent red color varieties are known as *ruby* and are used as high-value gems. Corundum with all other colors is called *sapphire*. It can scratch almost every other mineral due to extreme hardness (pure corundum has harness of 9 in Mohs hardness scale). It is commonly used as an abrasive, on everything from sandpaper to large machines, machining metals, plastics and wood. Corundum occurs as a mineral in mica schist, gneiss, and some marbles in metamorphic terranes. It also occurs in low-silica igneous syenite and nepheline syenite intrusives. Other occurrences are as masses adjacent to ultramafic intrusives, associated with lamprophyre dikes and as large crystals in pegmatites.

Gibbsite ($Al(OH)_3$) is an aluminum hydroxide, sometimes known as hydrargillite and crystallizes in monoclinic system. It is rarely found in a pure state and often blended with boehmite, kaolinite, hematite and limonite in bauxites and laterites, i.e. sedimentary rocks that serve as the ore to obtain aluminum (Section 5.5.6).

Boehmite γ-AlO(OH) or Böehmite is an aluminum oxide hydroxide mineral and alumogel is an amorphous gel $Al(OH)_3$. It occurs with gibbsite as main mineral constituents of bauxite and laterite (Section 5.5.6).

Hematite (Fe_2O_3) (Fig. 1.25) and *magnetite* ($FeO \cdot Fe_2O_3$) are very common, but usually minor constituents of many rocks. However, magmatic and hydrothermal processes may form large deposits of these minerals suitable for iron ore mining.

Goethite (FeO(OH)) is the iron hydroxide, formed as amorphous clusters and known as limonite (brown iron ores formed due to wearing of iron minerals). Limonite includes amorphous Fe-hydroxides with variable amounts of water.

Limonite is often found in many rocks that give the brown, yellow or tan color. It usually develops as a product of chemical weathering of many minerals containing Fe^{2+}, or ferrous iron. In the oxidation process oxidizes in the trivalent iron Fe^{3+}, which is known as a process of *limonitization*. Limonitization processes are well visible in yellow-brown color in freshly broken-off greenish-gray rocks along and around cracks. The brown, yellow or reddish-brown color of clays and many other rocks is derived from limonite. Finely dispersed limonite in the rock is a natural pigment that causes a yellow-brown color of rocks. Dense clusters of limonite are known as limonite ocher and pigment. Limonite above the hematite and siderite deposits often forms a crust of weathering and is known as the "iron hat". Limonite and goethite are important as a source of low-grade iron and nickel ore.

Chromite ($FeO \cdot Cr_2O_4$) (Fig. 1.23), is an iron chromium oxide, and *ilmenite* ($FeO \cdot TiO_2$) is a titanium–iron oxide minerals. These minerals are important ores of chromium and titanium, and are often the ingredients of rock in minor quantities. Chromite and ilmenite crystallize at high temperatures of intrusive magma in the initial stage of crystallization, and present as the regular ingredients of ultrabasic and ultramafic rocks (Section 4.3.1.4).

Rutile (TiO_2) crystallizes in tetragonal system and is commonly found in the form of rod or needle crystals. The color varies between dark red and black. It occurs usually in regional metamorphism, and associates with crystalline schists (gneiss, mica schists, phyllites and amphibolites, Table 6.1). It is a common mineral in clastic sediments and sedimentary rocks (sand/sandstones), with high resistance to weathering. Rutile is often occurring in the river and offshore deposits along with the gold concentrate in significant quantities. Rutile is common in acid igneous rocks as secondary mineral (Section 4.3.1.1).

Pyrolusite (MnO_2) is manganese oxide and crystallizes in tetragonal system. It is usually located in strip and needle aggregates with dark gray to black in color. It is formed by hydrothermal and sedimentary origin. The pyrolusite is very widespread and the main ore of manganese.

Psilomelane ($MnO \cdot MnO_2 \cdot nH_2O$) is a colloidal modification of manganese oxide with water. It occurs as kidney-like clusters. It is often found in the form of thin crusts, coatings, and dendrites in layer surfaces or crevices of different rocks, especially limestone. The mineral colors are iron-black or blue-black and the rocks which contain psilomelane are dark gray or black. Psilomelane is originated from aqueous solutions enriched with manganese as a product of the surface weathering of various minerals containing manganese.

Spinel ($MgO \cdot Al_2O_3$) is a member of isomorphic mixtures of different Al^{3+}, Fe^{3+} and Cr^{3+} spinel. The mineral is a typical product of contact metamorphism. It is formed from clay sediments in contact with the magma, and is located in skarns and hornfels (Table 6.1).

2.5.4. Carbonates

Carbonates are salts of carbonic acid, characterized by the presence of the carbonate ion, CO_3^{2-}. The minerals from the group of carbonates form isomorphic series of Ca–Mg–Fe–Mn–Zn trigonal carbonate and Ca–Sr–Ba–Pb rhombic carbonate (calcite and aragonite group). The carbonates also include the dolomite group (Table 2.6).

The most important petrogenic minerals from carbonate group are calcium, magnesium and iron carbonates or aragonite, calcite and dolomite, less frequent siderite and very rare magnesite.

Aragonite ($CaCO_3$) is a carbonate mineral, one of the two common, naturally occurring crystal forms of calcium carbonate (the other form is

TABLE 2.6 Minerals of Calcite and Aragonite Groups

Calcite Group	Aragonite Group
Calcite $CaCO_3$	Aragonite $CaCO_3$
Magnesite $MgCO_3$	Strontianite $SrCO_3$
Siderite $FeCO_3$	Witherite $BaCO_3$
Rhodochrosite $MnCO_3$	Cerussite $PbCO_3$
Smithsonite $ZnCO_3$	
DOLOMITE GROUP	
Dolomite $CaMg(CO_3)_2$	

the mineral calcite.) It is formed by biological and physical processes, including precipitation from marine and freshwater environments. It is essential, and with high-magnesium calcite, practically the only carbonate mineral component of shallow marine limestone deposits of warm and tropical seas.

Calcite ($CaCO_3$) (Fig. 1.14) forms isomorphic series with magnesite (Table 2.6) and can contain up to 28 mol% $MgCO_3$ and there are also low-magnesium calcite (<4 mol% $MgCO_3$) and high-magnesium calcite (4–28 mol% $MgCO_3$). In the lattice of calcite, Ca^{2+} can be replaced with Fe^{2+}, Mn^{2+} and Zn^{2+}, thus resulting magnesium carbonate ($MnCO_3$) and zinc carbonate ($ZnCO_3$) (Table 2.6). Petrogenic significance have only calcite and magnesium calcite and in some limestone, especially in the form of cement or iron calcite (calcite, in which part of the Ca^{2+} isomorphic is replaced by Fe^{2+}).

Calcite crystallizes in the trigonal crystal system, excretes in the deeper and/or colder sea and freshwater. It often occurs with contact and regional metamorphism of limestone and shale. Large crystal grains are distinctly visible from, the color and transparency like quartz. It is very significant and extensive petrogenic mineral or the main mineral component of limestone, marl and marble. It is also an important ingredient, especially as cement, to virtually all clastic sedimentary rocks (Sections 5.7.1.1, 5.5.4.2 and 5.5.5.1 and Section 6.5.2). The various types of occurrences are the following:

Low-magnesium calcite is very common in biogenic origin. It builds skeletons of planktonic organisms and regularly as a very stable mineral in limestone preserves during all diagenetic changes.

High-magnesium calcite, together with the metastable aragonite, is an essential ingredient of the carbonate mineral deposits of shallow marine limestone. The transformation into calcite is usually not found in limestone due to low stability. High-magnesium calcite is most commonly found in many recent carbonate skeletons, especially coralline algae, calcareous sponges, bryozoans and serpulite, as in many cement and ooids (Section 5.7.1.1). It is often found in some freshwater sediment and their cements, such as travertine limestone.

Dolomite ($CaMg(CO_3)_2$) (Fig. 5.64) with calcite is the most common mineral in carbonate rocks. The mineral dolomite crystallizes in the trigonal–rhombohedral system. It develops white, gray to pink, commonly curved crystals, usually in the massive forms. It has physical properties similar to those of the mineral calcite. Small amounts of iron in the structure give the crystals a yellow to brown tint. Unlike calcite, Mg calcite, aragonite and dolomite are not formed by biochemical processes or by direct precipitation from seawater, but they are generally secondary mineral formed by process of dolomitization. Replacement of Ca^{2+} ions with Mg^{2+} ions with suppression of calcite and aragonite in the limestone sludges or in the already-tough limestone with dolomite (Section 5.7.2.1).

Siderite ($FeCO_3$) is relatively poorly represented petrogenic minerals, but occasionally can be accumulated in large quantities to constitute a reservoir of iron ore. The crystals belong to the hexagonal system, and are rhombohedral in shape, typically with curved and striated faces. The color ranges between yellow and dark brown or black.

Magnesite ($MgCO_3$) usually occurs in the form of dense or granular masses, such as porcelain white. The mineral possesses hardness of four and relative density of three. Smaller amounts of magnesia can occur by deposition from seawater, and in larger quantities that have economic importance (magnesite ore deposits), mainly caused by the effect of

1. hot Mg bicarbonate solution to limestone to form calcite and dolomite in the first stage and followed by magnesite;
2. hot solutions containing CO_2 to ultramafic igneous rocks rich in olivine from which the serpentine group of minerals occur (Section 2.5.8.5.6) and magnesite.

Magnesite is the raw material for production of refractory matter and special types of cement (sorel cement).

2.5.5. Halides

Halides are the salts of sodium fluoride and hydrochloric acid. The minerals halite, sylvite and carnallite from this group only contain chloride and have petrogenic significance.

Halite (NaCl) is the mineral form of sodium chloride, and commonly known as rock salt. Halite forms isometric crystals. The mineral is typically colorless or white, but may also be light blue, dark blue, purple, pink, red, orange, yellow or gray depending on the amount and type of impurities (Fig. 1.9). It is an essential mineral component of evaporite sediments (deposits of rock salt, see Section 5.8.1).

Sylvite (KCl) unlike halite, has a bitter taste that makes these two minerals distinct. Sylvanite occurs with halite as essential mineral ingredient of evaporite sediments (salt deposits).

Carnallite ($KMgCl_3 \cdot 6(H_2O)$) crystallizes orthorhombic, has no cleavage, and is colorless or white to pale reddish. Carnallite occurs together with halite and sylvite and is the essential mineral ingredient evaporite sediments and salt deposits.

2.5.6. Sulfates

Sulfates are salts of sulfuric acid (H_2SO_4). The minerals gypsum and anhydrite, and rarely baryte, kieserite and polyhalite from the group of sulfate have petrogenic importance and occur as the main ingredients of evaporites sediments.

Gypsum ($CaSO_4 \cdot 2H_2O$) (Fig. 1.17) crystallizes in monoclinic system and is usually located in dense aggregates of fine or coarse granulated transparent aggregates called alabaster. It is very soft (hardness 2), white in color or monochrome, and from the admixture of organic matter and clay may be gray or brown gray. The mineral loses 75% water at a temperature of 120–130 °C and becomes so-called burnt gypsum. This feature is widely applied in construction. Gypsum is a much abundant mineral that can independently form a rock, because it is secreted in large masses of seawater in evaporation conditions. It can also be secreted from pore water in the desert. Gypsum is an essential ingredient of evaporite sediments (Section 5.8.1). It occurs as secondary mineral in many rocks by the oxidation of iron sulfide.

Anhydrite ($CaSO_4$) is an anhydrous calcium sulfate. It is in the orthorhombic crystal system, with three directions of perfect cleavages parallel to the three planes of symmetry. Anhydrite crystallizes from seawater with its strong evaporation, and in large quantities is the evaporite sediment, usually in association with early diagenetical dolomites (Section 5.8.1).

Baryte, or barite ($BaSO_4$), crystallizes orthorhombic system, commonly found in tabular crystals or granulated, fibrous and radial clusters of white or bluish-white color. It has a high relative density (from 4.3 to 4.7) and excellent cleavage of plane basal pinacoid (001). It can be found, almost always, together with sulfide ores formed in hydrothermal processes, and also forms in crystallization from seawater in reductive terms. Baryte is widely used in paper-making, paint, insulation material for protection against radiation and in addition to drilling mud to increase their density.

2.5.7. Phosphates

Phosphates are salts of phosphoric acid (H_3PO_4). The minerals from the apatite group only have petrogenic significance.

Apatite (Ca_5 (F, Cl, OH) $(PO_4)_3$) is a group of phosphate minerals, usually referring to hydroxyapatite, fluorapatite, chlorapatite and bromapatite, named for high concentrations of OH^-, F^-, Cl^- or Br^- ions, respectively, in the crystal. Apatite is the main representative of isomorphic minerals of the apatite group that crystallizes in hexagonal system. The fluorapatite, a part of F isomorphic replaced with Cl and OH− group, is mostly located in the rocks. These are mostly minor minerals, elongated prismatic or needle-shaped, glassy shine, brown, red, green and emerald green color. Apatite is a common mineral, and often a secondary mineral ingredient in almost all igneous rocks, especially in pegmatite and mafic igneous rocks forming as veins. The primary use of apatite is in the manufacture of fertilizer and a source of phosphorus. It is occasionally used as a gemstone.

Phosphorite, the amorphous colloid and cryptocrystalline variety of Ca phosphate, is mostly of sedimentary origin. Most of it is from the organogenic origin because phosphorus is derived from the bones, feces and other organic remains of animals. Phosphorite occurs in effect of ammonium phosphate from bones or excrement on calcium carbonate (limestone). About 90% of phosphate rock production is used for fertilizer and animal feed supplements and the balance for industrial chemicals.

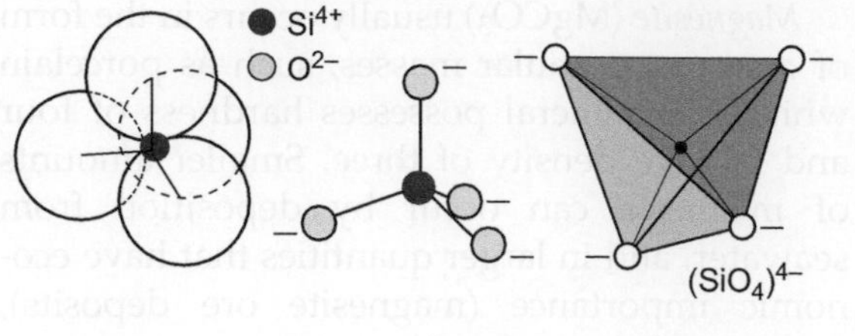

FIGURE 2.7 The basic structural unit of silicate minerals—SiO_4 tetrahedron.

2.5.8. Silicates

Silicates are the most common and most important petrogenic minerals, particularly feldspars, amphiboles, pyroxenes, olivine, micas and clay minerals, as indicated in Table 2.3. Their main characteristics are difficult to melt and often have very complex chemical composition because of isomorphic replacement. Most silicate minerals are formed by crystallization of magma at high temperatures, and in metamorphic processes at high temperature and high pressure.

Silicate minerals are classified according to the structure with main feature of strong relationship between major oxygen ions, and minor silicon ions. Four oxygen ions are arranged in close form of the tetrahedron with a small silicon ion in the center (Fig. 2.7). Therefore, the basic structural unit of silicate minerals is SiO_4 tetrahedra. The distance between the centers of two ions of oxygen is always 2.6×10^{-8} cm (2.6 Å). The distance between the center of silicon ions and each of the relatively large oxygen ions is only 1.6×10^{-8} cm (1.6 Å), as shown in Fig. 2.7.

Silicate minerals are put together by binding silicon−oxygen tetrahedra to each other and to other ions in a fairly small number of ways. Even this number represents only variations on the theme of combining ionic and covalent bonds.

The ionic bonding of tetrahedra involves another atom, a cation which usually carries a +2 charge. This ion is situated between the corners of two tetrahedra where it can receive one electron from the nearest oxygen in each.

The covalent bonding of tetrahedra involves actually sharing one oxygen atom between two adjacent tetrahedra. One of the extra electrons of the shared oxygen is used by one silicon, and the other electron is used by the other.

Between these two extreme cases, there are a number of different cases of bonding two, three, four, six or more of the SiO_4 tetrahedra, so that there are seven different major structural types of silicate minerals. These are the following:

1. Nesosilicates (lone tetrahedron)—$[SiO_4]^{4-}$
2. Sorosilicates (double tetrahedra)—$[Si_2O_7]^{6-}$
3. Cyclosilicates (rings)—$[Si_nO_{3n}]^{2n-}$
4. Inosilicates (single chain)—$[Si_nO_{3n}]^{2n-}$
5. Phyllosilicates (sheets)—$[Si_{2n}O_{5n}]^{2n-}$
6. Tectosilicates (3D framework)—$[Al_xSi_yO_{2(x+y)}]^{x-}$

2.5.8.1. *Nesosilicates* $[SiO_4]^{4-}$

In the structure of nesosilicates, SiO_4 tetrahedra are not directly connected with mutual oxygen ion, only by interstitial cations. The simplest structure in nesosilicates have mineral forsterite $Mg_2[SiO_4]$. The most important minerals from the nesosilicates are shown in Table 2.7 and Fig. 2.8.

Olivine with little iron is closer to forsterite with greenish color. The same with more iron is closer to fayalite with dark green color. Olivine crystallizes in orthorhombic system and hardness of 7–6.5 (depending on the isomorphous replacement of Mg with Fe). It forms by crystallization of magma at high temperatures (pyrogen minerals). In normal atmospheric conditions, it has low resistance to weathering and easily subjected to metamorphism in the mineral serpentine (olivine serpentinization), talc or actinolite. Olivine is the important mineral constituent of basic igneous rocks (gabbro, norite, basalt) and ultramafic rocks, as well as some crystalline schists formed in the deep rock layers. Olivine is extensively present in meteorites.

TABLE 2.7 The Most Important Petrogenic Minerals from Nesosilicates Group

Olivines Group	Al_2SiO_5 Group
Forsterite—Mg_2SiO_4	Andalusite—Al_2SiO_5
Fayalite—Fe_2SiO_4	Kyanite—Al_2SiO_5
	Sillimanite—Al_2SiO_5
Garnet Group	**Zircon Group**
Pyrope—$Mg_3Al_2(SiO_4)_3$	Zircon—$ZrSiO_4$
Almandine—$Fe_3Al_2(SiO_4)_3$	Titanite—$CaTiSiO_{52}$
Spessartine—$Mn_3Al_2(SiO_4)_3$	
Grossular—$Ca_3Al_2(SiO_4)_3$	
Andradite—$Ca_3Fe_2(SiO_4)_3$	
Uvarovite—$Ca_3Cr_2(SiO_4)_3$	

Garnets (Fig. 1.16) consist of the free SiO_4 tetrahedra interconnected by ions of various divalent and trivalent metals. Garnets have a complex composition due to the high possibility of isomorphic substitution of these ions. The end members are known as series: pyrope–almandine–spessarite and uvarovite–grossular–andradite. The general chemical formula is as follows:

$$M^{2+}{}_3\,M^{3+}{}_2\,Si_3O_{12}$$

where, $M^{2+} = $ Ca, Mg, Fe or Mn

$M^{3+} = $ Al, Fe or Cr.

Members of each of the two series of crystals form a cross-breed with each other, and the isomorphic replacement between the two series are limited. In divalent cations, there is unlimited possibility of isomorphic substitution of Mg^{2+} with Fe^{2+} and Mn^{2+} with Fe^{2+}. Isomorphic replacement of Mg^{2+} with Mn^{2+} is limited, and Mg^{2+}, Mn^{2+} and Fe^{2+} may still be up to about 20% replaced with Ca^{2+}. In all garnets part of the Si can be isomorphic replaced with Ti. There are almost no pure members in nature, and garnet gets its name by the main components dominate (Table 2.7). Garnet crystallizes in cubic system (Fig. 1.16), and is commonly found in isometric form, i.e. the regular crystallographic forms, mostly orthorhombic dodecahedron, and their combination (Fig. 2.9 and Fig. 6.6). Garnets do not have cleavage, and have hardness of 6.5–7, and the color depends on their chemical composition.

Garnets may originate in different ways. It may form by crystallization of magma in the pegmatite and acidic intrusive rocks (granites and granodiorites). However, the common

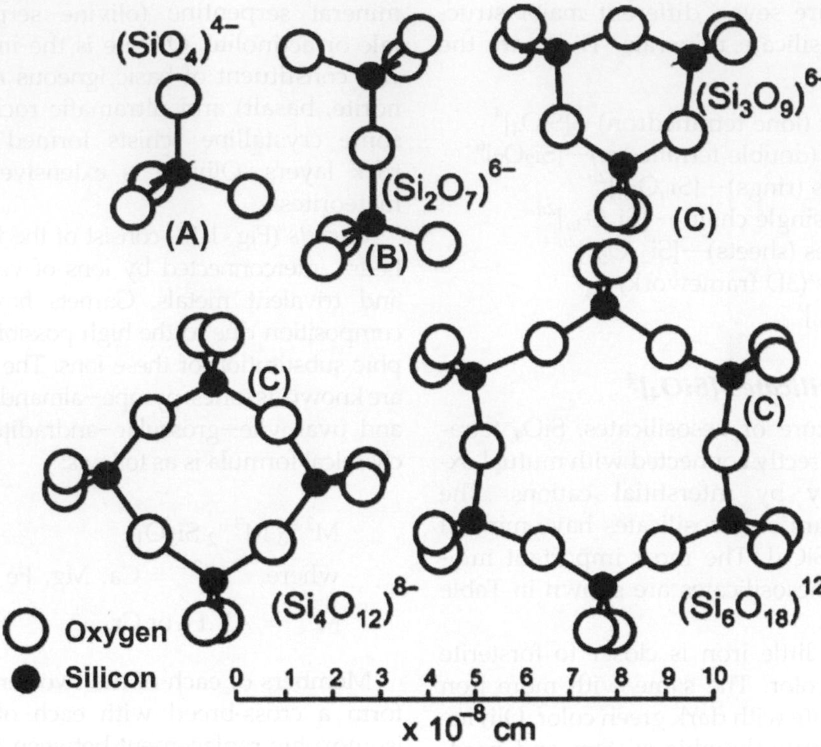

FIGURE 2.8 Representation of free and in separate groups linked SiO_4 tetrahedra in the structures of (A) nesosilicates, (B) sorosilicates, and (C) cyclosilicates.

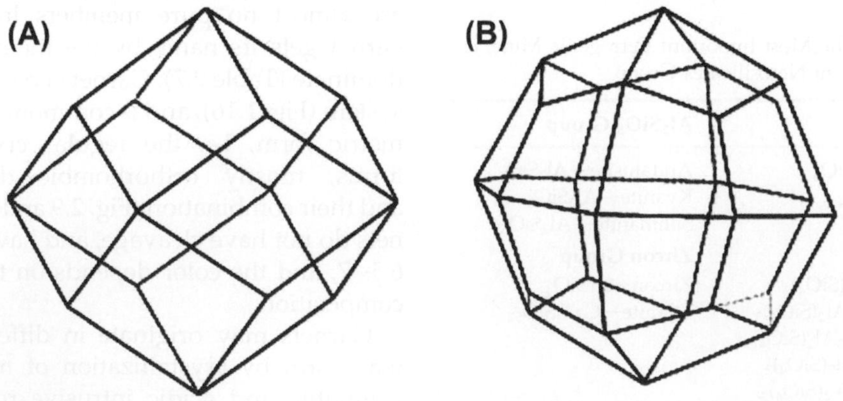

FIGURE 2.9 Crystal forms of garnet: (A) rhombic dodecahedron; (B) deltoid icosahedron.

occurrences are by regional, contact and plutonic metamorphism, and are essential ingredients of a high-degree metamorphic schist (gneiss and mica schist), skarn, hornfels and eclogite (Table 6.1). Garnets are resistant to weathering and regularly found as minor components in the clastic sediments and sedimentary rocks (sandstones). Garnets are used as abrasive material.

Kyanite, *Andalusite* and *Sillimanite* (Al_2SiO_5) are polymorphic modifications of aluminum silicates with the same formula (Table 2.7). Kyanite crystallizes in triclinic system. Andalusite and sillimanite crystallize in orthorhombic system. Sillimanite has the structural features of inosilicates. The crystals are usually elongated. Kyanite hardness is 4.5–5.0 parallel to one axis 6.5–7.0 perpendicular to that axis. The hardness of sillimanite and andalusite is 6–7. Kyanite is a mineral typical of regional metamorphosis under high pressure. Andalusite is a typical mineral for contact metamorphism and sillimanite represents regional and contact metamorphism at high temperatures. All three minerals occur in the metamorphic environment from clay sediments. These are common and essential mineral components of hornfels (Table 6.1).

Staurolite crystallizes in orthorhombic system and occurs mainly as twinned and cruciform patterns in metamorphic rocks. It has complex chemical formula of $(Fe,Mg,Zn)_{3-4}Al_{18}Si_8O_{48}H_{2-4}$. The hardness is 7.0–7.5 in Mohs scale. Staurolite is formed by regional metamorphism of rocks such as mica schists, slates, and gneisses, and generally associated with other minerals like kyanite, garnet, and tourmaline.

Zircon ($ZrSiO_4$) regularly contains a small amount of hafnium, thorium, yttrium, iron or uranium, and generally weakly radioactive. Zircon crystallizes in tetragonal crystal system. It has hardness of 7–8, incomplete cleavage, and relative density of 4.7. It is highly resistant to weathering. The natural color varies between colorless, yellow-golden, red, brown, blue, and green. It is a widespread mineral, but usually in very small quantities. It is a component of igneous and metamorphic rocks, and is a regular ingredient in clastic sediments.

Titanite ($CaTiSiO_5$) crystallizes in monoclinic system. It is found as individual crystal. Cleavage is clear at (110), hardness is 5.0–5.5, and relative density of 3.5. It often occurs as secondary or accessory mineral in neutral and acidic magmatites (alkali syenite and pegmatites, Table 4.1), some crystalline schists, and gneisses in particular. It is usually located associated with the bauxite.

2.5.8.2. Sorosilicates—$[Si_2O_7]^{6-}$

Sorosilicates have isolated double tetrahedra groups with $(Si_2O_7)^{6-}$ or a ratio of 2:7. There are no significant petrogenic minerals among sorosilicates, except epidote, zoisite and vesuvianite (Fig. 2.8).

Epidote ($Ca_2Al_2(Fe^{3+};Al)(SiO_4)(Si_2O_7)O(OH)$) is a calcium aluminum-silicate mineral, in which part of the aluminum is substituted with trivalent iron. If the iron is replaced by only 10%, then the aluminum is white and is known as clinozoisite. If it is replaced by 10–40% aluminum, it is called pistacite which has a typical green color with yellow and dark-gray shades. Epidote crystallizes in the monoclinic system. The crystals are prismatic, the direction of elongation being perpendicular to the single plane of symmetry. It occurs in the presence of water vapor at low and high temperatures (100–450 °C) in hydrothermal conditions, and can occur, in dynamic and contact metamorphism (Section 6.2). Epidote is a constituent of igneous rocks that have undergone hydrothermal changes and metamorphic rocks from the schist group.

Zoisite (Ca_2Al_3 (OH) Si_3O_{12}) is a calcium aluminum-silicate without iron. It crystallizes in orthorhombic system. The most common occurrence is in the form of fine-grained aggregates of light to blue-gray color, usually associated with epidote and albite, in hydrothermal metamorphism of basic plagioclase, in pegmatitic phase crystallization of magma, and regional metamorphism of calcium-rich rocks.

Zoisite is an essential ingredient of green schist and amphibolite eclogites (Table 6.1), a common constituent of secondary pegmatite.

Vesuvianite ($Ca_{10}Al_4(Mg,Fe)_2Si_9O_{34}(OH)_4$) is a mineral of complex composition crystallizing in the tetragonal system because of isomorphic impurities. It was first discovered in blocks or adjacent to lavas on Mount Vesuvius, hence its name. The color is usually green or brown but may be yellow, blue, or red. The hardness is 6.5, and specific gravity is 3.35–3.45. The mineral forms by contact metamorphism of clay limestone, dolomite and marl (Section 6.2). It usually occurs in conjunction with garnets in marble, and during crystallization of magma.

2.5.8.3. Cyclosilicates—$[Si_nO_{3n}]^{2n-}$

Cyclosilicates, or ring silicates, have linked tetrahedra with $(Si_xO_{3x})^{2x-}$ or a ratio of 1:3. These groups of minerals exist as three-member $(Si_3O_9)^{6-}$, four-member $(Si_4O_{12})^{8-}$ and six-member $(Si_6O_{18})^{12-}$ rings (Table 2.8) (Fig. 2.8):

1. Three-member ring
   Benitoite $\{BaTi(Si_3O_9)\}$
2. Four-member ring
   Axinite $\{(Ca,Fe,Mn)_3Al_2(BO_3)(Si_4O_{12})(OH)\}$
3. Six-member ring
   Beryl/Emerald $\{Be_3Al_2(Si_6O_{18})\}$
   Cordierite $\{(Mg,Fe)_2Al_3(Si_5AlO_{18})\}$
   Tourmaline $\{(Na,Ca)(Al,Li,Mg)_3(Al,Fe,Mn)_6(Si_6O_{18})(BO_3)_3(OH)_4\}$.

TABLE 2.8 The Most Important Petrogenic Cyclosilicates

6-Member Ring	
Tourmaline group	Beryl
Mg–Al:	$Be_3Al_2Si_6O_{18}$
$Al_5Mg_3CaMg\ (OH,F)_4\ Si_6O_{27}B_3$	
Na–Al:	
$Al_7Na_2Mg\ (OH,F)_4\ Si_6O_{27}B_3$	Cordierite
Fe–Al:	$(Mg,Fe)_2Al_3(Al,Si)_5O_{18}$
$(Al,Fe)_5FeCaFe\ (OH,F)_4\ Si_6O_{27}B_3$	

In cyclosilicates, only six-member ring has petrogenic important minerals.

The tourmaline group includes isomorphic series of silicate minerals of highly variable and complex composition that often contains B and Al. However, the tourmaline bearing rocks frequently contain Na and Li, a part of Mg and Fe is isomorphic substitute of Mn and Ca. Tourmaline crystallizes in trigonal system. The crystals typically occur as long, slender to thick prismatic and columnar structure (Fig. 2.18). The mineral has great hardness between 7.0 and 7.5, and changing colors depending on the chemical composition.

Tourmaline (Fig. 1.18) is very abundant secondary mineral, especially in acidic intrusives (granites), pegmatites and rocks affected by pneumatolytic processes. Tourmalines are typical pneumatolytic minerals that crystallize in pneumatolytic phase of magma-rich gases and vapors (Section 2.2.2). The minerals are very resistant to the processes of physical and chemical weathering and regularly found in sand dunes, or nearly all clastic sedimentary rocks.

Beryl ($Be_3Al_2Si_6O_{18}$) hexagonal crystals may be very small or range to several meters in size. Pure beryl is colorless, but often contains impurities. The various common colors are green, blue, yellow, red, and white. Beryl has great hardness between 7.5 and 8.0, and clear cleavage. The precious beryls are known as aquamarine (blue), emerald (deep green), and morganite (pink to red). Beryl forms by pneumatolytic stage of crystallization of magma, and is located in veins in granites, also in the crystalline schists, particularly gneisses and mica schists (Table 6.1).

Cordierite ($(Mg,Fe)_2Al_4Si_5O_{18}$) crystallizes in orthorhombic system, and occurs as crystals and granular aggregates of short form. If it contains more iron than magnesium, then it is known as Fe-rich cordierite, and if it contains more magnesium than iron, it is known as Mg-rich cordierite. The hardness is 7, and specific gravity 2.6. The various colors are greenish-blue, lilac blue, or dark blue. Cordierite occurs in contact or regional metamorphism of argillaceous rocks. It is especially common in hornfels produced by contact metamorphism of pelitic rocks. Sometimes, the share in paragneiss may be so large that it forms a special type known as *cordierite gneisses* (Table 6.1).

2.5.8.4. Inosilicates

Inosilicates, or chain silicates, have interlocking chains of silicate tetrahedra with either SiO_3, 1:3 ratio, for single chains or Si_4O_{11}, 4:11 ratio, for double chains.

2.5.8.4.1. SINGLE-CHAIN INOSILICATES—PYROXENE GROUP

The pyroxenes are important rock-forming inosilicate minerals and often exist in many igneous and metamorphic rocks. They share a common structure of single chains of silica tetrahedra (Fig. 2.10). The group of minerals crystallizes in the monoclinic and orthorhombic systems.

Inosilicates with a single-chain SiO_4 tetrahedron of the pyroxene group are very important and widespread petrogenic minerals (Table 2.3). Pyroxenes constitute a related group of silicate minerals with similar crystallographic, physical and chemical properties. The most important of them are given in Table 2.9.

2.5.8.4.1.1. ORTHOPYROXENES Orthopyroxenes have general formula (Mg,Fe,Ca) (Mg,Fe,Al) $(Si,Al)_2O_6$. The natural compositions are dominated by two major end-member components: enstatite, $Mg_2Si_2O_6$, and ferrosilite, $Fe_2Si_2O_6$. The most common rock is bronzite that contains 10–30% ferrosilite components, and hypersthenes containing 30–50% ferrosilite

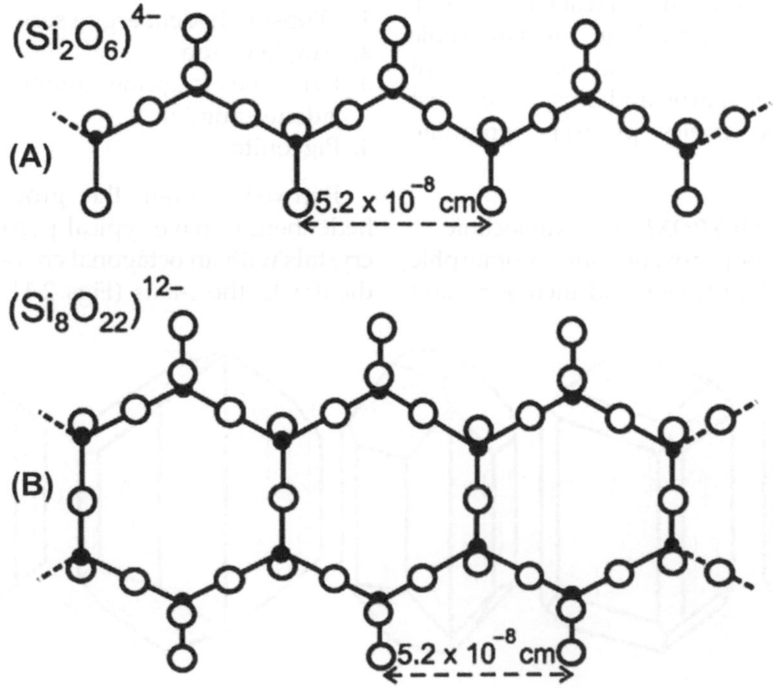

FIGURE 2.10 Inosilicates—(A) single chain pyroxenes; (B) double chain amphibole.

TABLE 2.9 The Most Important Petrogenic Minerals from Pyroxene Group

Petrogenic Important Pyroxenes		
Orthopyroxenes	**Clinopyroxenes**	**Alkaline Pyroxenes**
Enstatite $Mg_2Si_2O_6$	Pigeonite $(Mg,Fe^{2+},Ca)(Mg,Fe^{3+})Si_2O_6$	Jadeite $NaAlSi_2O_6$
Bronzite $(Mg,Fe)_2SiO_6$	Diopside $CaMgSi_2O_6$	Aegirine $NaFeSi_2O_6$
Hypersthene $(Mg,Fe)_2Si_2O_6$	Hedenbergite $CaFeSi_2O_6$	Aegirine–Augite = isomorphic member of aegirine and augite
	Dialage = rich in iron diopside turned into Al–augite	Spodumene $LiAlSi_2O_6$
	Augite $Ca(Mg,Fe^{2+},Al)\ (SiAl)_2O_6$	Omphacite $(Ca,Na)(Mg,Fe^{2+},Al)Si_2O_6$
	Fassaite = augite with $Al_2O_3 > Fe_2O_3$	
Wollastonite $Ca_3Si_3O_9$		

components (Table 2.9). These minerals are pyrogenic origin and regular mineral ingredients of basic and ultrabasic rocks (gabbro, norite, peridotite, and lercololite), and hypersthenes is often found in trachyte and andesite (Table 4.1). Hypersthene may occur in metamorphic process of contact and regional metamorphism as an ingredient of skarns and crystalline schist (especially biotite gneiss pyroxene granulite (Table 6.1)).

2.5.8.4.1.2. CLINOPYROXENES Monoclinic pyroxene or clinopyroxene are isomorphic blends of several different end-members, and regularly have a very complex chemical composition (Table 2.9). The most petrogenic important minerals among the group are the following:

1. Diopside-hedenbergite series
2. Augite group
3. Pyroxenes (aegirine–augite and jadeite–augite)
4. Pigeonite

Pyroxenes from the group of diopside–hedenbergite have typical pyroxene short pillar crystals with an octagonal cross-sections perpendicular to the *c*-axis (Figs 2.11 and 2.12). Their

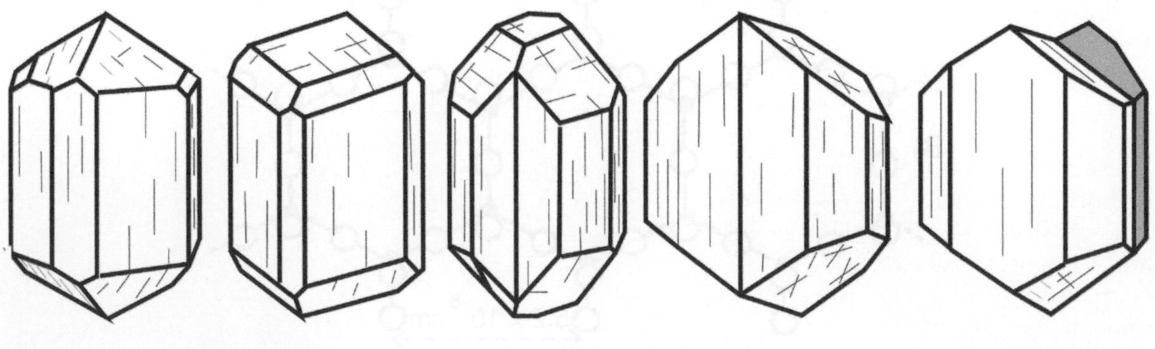

FIGURE 2.11 Typical crystal forms of pyroxene. The diagnostic feature of any pyroxene is two sets of cleavages that intersect at 89° or 91°. Pyroxene show mediocre cleavages that are hard to separate, unlike the feldspars or hornblende.

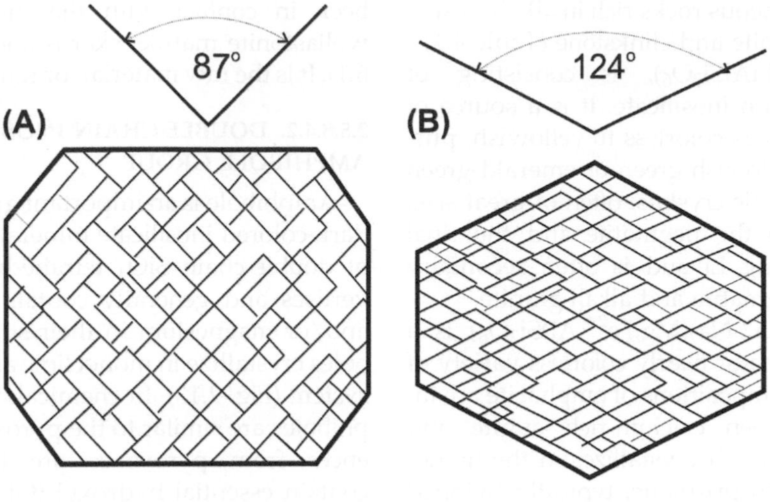

FIGURE 2.12 Typical pyroxene cleavage. The pyroxene (hornblende) always forms oblique cleavages and intersect (A) at 87° in case of octagonal crystals, and (B) between 120° and 124° for hexagonal crystals.

color depends on the amount of iron, magnesium and aluminum. The color of diopside is pale green and hedenbergite of dark green. Diopside may arise from the crystallization of magma and are common ingredients of neutral, basic and ultrabasic igneous rocks (Table 4.1). Diopside occurs in contact and regional metamorphism and is often found in marble, hornfels, green schists and schists of high degree of metamorphism, especially in the mica schists.

Diallage is yellowish brown, greenish mineral in the diopside–hedenbergite series containing pieces of magnetite or ilmenite. Diallage is a regular component of basic and ultrabasic igneous rocks from the group of gabbro and lercolite (Table 4.1).

Augite groups have lattice analogous to diopside, but with the difference that are part of the SiO_4 tetrahedra Si^{4+} is replaced by Al^{3+} ions. These minerals have a complex chemical composition, and involve a very wide range of isomorphic replacement of Ca, Na, Mg, Fe^{2+}, Fe^{3+}, Ti and Al. Most of the augite is dark green to black color. Augite is widely abundant rock-forming minerals in igneous and metamorphic rocks (Tables 4.1 and 6.1). They are the typical ingredients of gabbro, basalt and dolerite.

Augite rich in iron is called *black augite*. It is particularly common in neutral and basic volcanic rocks (diabase, basalts and andesites). Titanium augite (augite rich in titanium, because it is part of the Si replaced by Ti) exists in dolerite, olivine basalt and alkali gabbro.

Jadeite ($NaAlSi_2O_6$) has hardness between 6.5 and 7.0 depending on the composition. The mineral is dense, with a specific gravity of about 3.4. The common colors of Jadeite ranges from white through pale apple green to deep jade green-blue-green, pink, lavender, and a multitude of other rare colors. Jadeite is formed in metamorphic rocks under high pressure and relatively low temperature.

Aegirine ($NaFeSi_2O_6$) occurs as dark green monoclinic prismatic crystals. It has a glassy luster and perfect cleavage. The Mohs hardness varies from 5 to 6 and the specific gravity is between 3.2 and 3.4. It is located only in igneous rocks rich in alkalis, mainly syenite and trachyte, and some alkali granites (Table 4.1). Aegirine–augite (isomorphous mixtures of aegirine and augite components) is characterized by a clear green color. It is a typical magmatic mineral and, in association of leucite forms an important component of igneous rocks rich in alkalis, especially leucite syenite and clinkstone (Table 4.1).

Spodumene ($LiAlSi_2O_6$), is consisting of lithium–aluminum inosilicate. It is a source of lithium. It occurs as colorless to yellowish, purplish or lilac, yellowish-green or emerald-green hiddenite, prismatic crystals, often of great size. It crystallizes in the pegmatite and the final magma containing Li and is often secondary ingredient of pegmatite and alkali granite.

Omphacite ($(Ca, Na)(Mg,Fe^{2+},Al)Si_2O_6$) is a deep to pale green or nearly colorless variety of pyroxene. The compositions of omphacite are intermediate between calcium-rich augite and sodium-rich jadeite. It crystallizes in the monoclinic system with prismatic, typically twinned forms. The hardness is between 5 and 6. It is a major mineral component of eclogite, i.e. metamorphic rocks formed under conditions of high pressure and temperature (Table 6.1). It is a common ingredient in peridotites, in particular kimberlite (ultramafic igneous rocks formed at high temperatures where there are diamonds—Section 2.5.1 and Section 4.3.1.4).

Wollastonite ($Ca_3Si_3O_9$) is in chemical composition calcium silicate which in structure belongs to inosilicates, and the chemistry is similar to cyclosilicates. It is a typical contact-metamorphic mineral and therefore a regular ingredient of changed clayey limestone rocks which have been in contact with the magma, especially wollastonite marble, skarns and kornites (Table 6.1). It is the raw material for refractory materials.

2.5.8.4.2. DOUBLE-CHAIN INOSILICATE—AMPHIBOLE GROUP

Amphibole is an important group of generally dark-colored inosilicate minerals. It is composed of double-chain SiO_4 tetrahedra, linked at the vertices and generally containing ions of iron and/or magnesium in their structures. Amphiboles crystallize in monoclinic and orthorhombic system (Fig. 2.13). In chemical composition, amphiboles are similar to the pyroxenes. The differences from pyroxenes are that amphiboles contain essential hydroxyl (OH) or halogen (F, Cl) and the basic structure is a double chain of tetrahedra. Amphiboles are the primary constituent of amphibolites.

Amphiboles along with pyroxenes and feldspars are the most abundant rock-forming minerals (Table 2.3).

2.5.8.4.2.1. ORTHORHOMBIC AMPHIBOLES

Anthophyllites are isomorphic mixture of magnesium anthophyllite ($Mg_7(OH)_2Si_8O_{22}$) and ferroanthophyllite ($Fe_7(OH)_2Si_8O_{22}$). Anthophyllite is the product of metamorphism of magnesium-rich rocks especially ultrabasic igneous rocks and impure dolomitic shales.

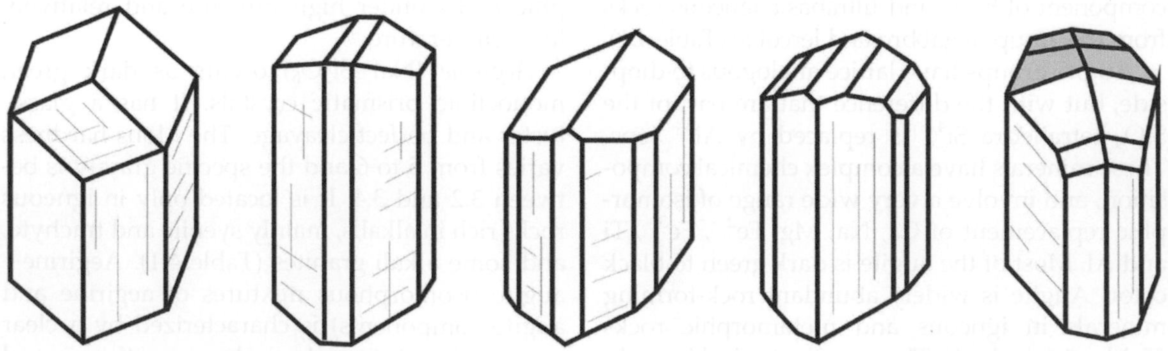

FIGURE 2.13 Typical prismatic crystals of amphiboles elongated direction of crystallographic axis *c*.

2.5.8.4.2.2. MONOCLINIC AMPHIBOLES Monoclinic amphiboles typically constitute complex isomorphic compounds with a wide possibility to replace a number of different ions, resulting in a very complex chemical composition. Specifically, Ca^{2+} ion can be isomorphic replacement with Na^+, K^+ and Fe^{2+}. Mg^{2+} ion with Fe^{2+}, Al^{3+} and Ti^{4+}, and Si^{4+} ion with Al^{3+}. Table 2.10 lists the general formula for some of the most important members of the isomorphic series of monoclinic amphiboles (for groups tremolite, actinolite, hornblende and alkali amphibole).

The color of amphiboles depends on their chemistry, particularly of iron. Tremolite, $Ca_2(Mg)_5(OH)_2Si_8O_{22}$, does not contain iron or it has only a very small portion, usually white. Actinolite, $Ca_2(Mg, Fe)_5(OH)_2Si_8O_{22}$, in which part of the magnesium is replaced with iron and the color is green. Basaltic hornblende, $Ca_2Na(Mg,Fe)_4(Al,Fe)(OH)_2(Si,Al)_8O_{22}$, which contains many isomorphic mixed trivalent iron and aluminum is black in color.

Tremolite and actinolite are calcium amphiboles forming a series of isomorphic mixed crystals. The minerals are in the form of long prismatic, radial needle and fibrous aggregates and known as actinolite asbestos. It occurs at relatively low temperatures in different ways:

1. Hydrothermal alteration of pyroxenes under the influence of steam and hot solution. This process of formation of actinolite in that way is known by name *uralite*.
2. A low degree of regional metamorphosis forming regular components of low degree of metamorphic schist, especially green schist.
3. Contact metamorphism, and are found in calcite and dolomite marble (Table 6.1).

Minerals actinolite–tremolite series are not stable, and can easily alter in the chlorites, tremolite and talc on the Earth's surface.

The group hornblende includes amphibole rich in trivalent iron and aluminum. A substantial portion of Si^{4+} is replaced with Al^{3+}, which requires the entry of Na^+ in the structure. Their chemical composition is very complex and can only show with general formulas of ferrohornblende and magnesiohornblende. There are ferrohornblende (common hornblende) and magnesium (basaltic) hornblende with specific reference to composition of divalent and trivalent iron and magnesium. Hornblendes, unlike most other amphiboles, are well-formed crystals with well-developed prism surfaces of the third and second positions (110, 101). In nature, there are only hornblendes in which the ratio of Mg:Fe is always greater than 4:6.

Ferrohornblende (common hornblende) is richer in ferrous iron, which is isomorphic replacement with Mg. Magnesiohornblende does not contain Fe^{2+}, but Fe^{3+} which is isomorphic is replaced with Al^{3+} (Table 2.10).

TABLE 2.10 The Most Important Petrogenic Minerals of Amphibole Group

Petrogenic Important Amphiboles		
Orthorhombic	**Monoclinic**	**Alkaline**
Anthophyllite $(Mg,Fe)_7(OH)_2Si_8O_{22}$	Tremolite $Ca_2(Mg)_5(OH)_2Si_8O_{22}$ Actinolite $Ca_2(Mg,Fe)_5(OH)_2Si_8O_{22}$ Ferrohornblende $Ca_2Fe_4^{2+}(Al,Fe^{3+})(OH)_2Si_7AlO_{22}$ Magnesiohornblende $Ca_2Mg_4(Al,Fe^{3+})(OH)_2Si_7AlO_{22}$	Glaucophane $Na_2Mg_3Al_2(OH)_2Si_8O_{22}$ Riebeckite Na–Fe amphibole with 15–30% Fe_2O_3 Arfvedsonite Na-amphibole with 5–10% Na_2O

Hornblende is formed by crystallization of magma, and the pneuma (aqueous solutions and vapors that are released from magma and lava). These minerals are found in almost all intrusive, extrusive and vein igneous rocks, and in particular diorite, granodiorites, andesite, gabbro, diabase and pegmatite (Table 4.1). In addition, hornblende occurs in regional and contact metamorphism. The common rocks are amphibolites, amphibolite schist and amphibolite gneiss and basic hornfels (metamorphites caused by contact metamorphosis). Hornblende often turns into clusters of chlorite, epidote, calcite and quartz under the influence of hydrothermal solution. Alkali amphiboles include monoclinic amphiboles, that contain a considerable amount of alkali elements (Na, K and Li), and only three are important petrogenic Na minerals, namely, glaucophane, riebeckite and arfvedsonite (Table 2.10).

Glaucophane ($Na_2Mg_3Al_2(OH)_2Si_8O_{22}$) is named from its typical blue color (in Greek, "glaucophane" means "blue appearing"). Glaucophane occurs at high pressures and low temperatures, and is an important mineral constituent of metamorphic rocks formed under conditions of low temperature and high pressure, such as glaucophane schists and some phyllites and mica schists (Table 6.1).

Riebeckite ($Na_2\ Fe_3^{2+}$, $Fe_2^{3+}(OH)_2Si_8O_{22}$) are mainly found in igneous rocks that crystallize from magma rich in sodium (alkali trachyte, syenite, rhyolite and granite—see Table 4.1).

Arfvedsonite ($Na\ Na_2\ Fe^{2+}{}_4$, $Fe^{3+}(OH)_2Si_8O_{22}$) the gray-black monoclinic alkali amphibole occurring from magma rich in sodium. It usually appears in very elongated prismatic fibrous or radial fibrous aggregates. It is a constituent of alkali syenite, clinkstone and their pegmatite (Table 4.1).

2.5.8.5. *Phyllosilicates—$[Si_{2n}O_{5n}]^{2n-}$*

The basic structure of the phyllosilicates is based on interconnected six-member rings of SiO_4^{-4} tetrahedra that extend outward in infinite

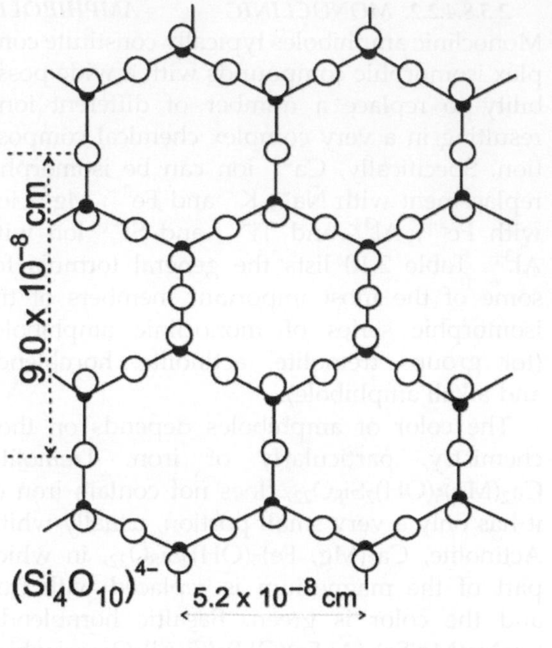

FIGURE 2.14 The structures of the phyllosilicates.

sheets. Three out of the four oxygens from each tetrahedron are shared with other tetrahedral as shown in Fig. 2.14.

The most important petrogenic minerals among phyllosilicates are group talc–pyrophyllite, mica, chlorite, vermiculite, smectite and kaolinite–serpentine (Table 2.11).

2.5.8.5.1. GROUP TALC–PYROPHYLLITE

Talc–pyrophyllite group comprises small sheets of soft white phyllosilicates which are greasy on touch. Talc and pyrophyllite cannot make mixed crystals because of large differences in ionic radii of magnesium and aluminum.

Talc ($Mg_3\ (OH)_2\ Si_4O_{10}$) crystallizes in monoclinic and triclinic system. It most often occurs as foliated to fibrous masses. It has a low hardness of 1in Mohs scale, which means it, can be scratched by a fingernail. The mineral is an essential ingredient of talc schists and chlorite schists (Table 6.1). Talc is a metamorphic mineral resulting from the metamorphism of magnesium

TABLE 2.11 The Most Important Petrogenic Minerals from the Group Phyllosilicates

Kaolin–Serpentine Group	
Kaolin Minerals Belongs to Clay Minerals	**Serpentine Minerals**
Kaolinite $Al_2(OH)_4Si_2O_5$	Lizardite $Mg_3(OH)_4Si_2O_5$
Dickite $Al_2(OH)_4Si_2O_5$	Chrysotile $Mg_3(OH)_4Si_2O_5$
Nacrite $Al_2(OH)_4Si_2O_5$	Antigorite $(Mg,Fe)_3(OH)_4Si_2O_5$
Talc–Pyrophyllite Group	**Vermiculite Group Belongs to Clay Minerals**
Talc $Mg_3(OH)_2Si_4O_{10}$ Pyrophyllite $Al_2(OH)_2Si_4O_{10}$	$(Mg,Al,Fe^{2+})_3(Si,Al)_4O_{10}(OH)_2 \cdot 4H_20$
Chlorite Group	**Smectite Group Belongs to Clay Minerals**
Includes hydrosilicates which make mixed crystals of complex chemical composition whose general formula is:	Includes dioctaedric aluminum mica series montmorillonite–beidellite and iron mica nontronite
$M_{4-6}\ T_4\ O_{10}(OH,O)_8$	Montmorillonite–beidellite $Na_{0.5}Al_2(Si_{3.5},Al_{0.5})O_{10}(OH)_2 \cdot nH_2O$
$M = Al,Fe^{3+},Fe^{2+},Li,Mg,Mn^{2+},Cr$, Ni and Zn $T = Si,Al,Fe^{3+}$,Be and B	Nontronite = Fe–smectite $Na_{0.5}Fe_2(Al,Si)_4(OH)_2 \cdot nH_2O$
MICA GROUP	
Muscovite $KAl_2(OH)_2AlSi_3O_{10}$	Celadonite $KFE^{3+}(Mg,Fe^{2+})(OH)_2Si_4O_{10}$
Celadonite $KFe^{3+}(Mg,Fe^{2+})(OH)_2Si_4O_{10}$	Paragonite $NaAl_2(OH)_2AlSi_3O_{10}$
BIOTITE PHOLOPITE GROUP	
Biotite $K(Mg,Fe)_3(OH)_2AlSi_3O_{10}$	Phologopite $KMg_3(OH)_2AlSi_3O_{10}$
Illite $K_{0.65}Al_2(OH)_2Al_{0.65}Si_{3.35}O_{10}$	Glauconite is the name of series with mixed layer rich in iron mica

minerals such as serpentine, pyroxene, amphibole, olivine, in the presence of carbon dioxide and water. The main uses of talc are in the manufacture of powder, the production of soap and paper, and as a refractory material.

Pyrophyllite ($Al_2(OH)_2Si_4O_{10}$) crystallizes in monoclinic system and occurs as crystalline folia and compact dense masses, and occasionally a fan aggregates of greenish, yellowish or light bluish color. It is a frequent mineral ingredient of crystalline schists that are rich in aluminum and some of bauxite. It occurs in regional metamorphism, and by the action of acidic hydrothermal solutions on the rocks rich in aluminum.

2.5.8.5.2. THE MICA GROUP

Micas have been one of the most widespread mineral group in the lithosphere. The group contains minerals that are key ingredients of many igneous rocks (Table 4.1) and metamorphic rocks (Table 6.1). The primary minerals from the mica group, especially muscovite, are regular constituents of clastic sedimentary rocks, particularly sand and sandstone (Section 5.5.3.1). Some types of crystalline schist (mica schists and phyllites) are composed almost entirely of mica. In gneisses, micas are regular and very important ingredients (Table 6.1). All micas crystallize in monoclinic system and are characterized by perfect cleavage into thin slices, which often have a hexagonal outline. The microcrystalline mica flakes are known under the name of sericite.

Muscovite ($KAl_2(OH)_2 AlSi_3O_{10}$) also known as common mica whose name originates from Muscovy-glass, a name formerly used for the mineral because of its use as windows in Russia. It does not contain iron, so it is colorless mica, and highly resistant to weathering. Muscovite is formed by crystallization of magma from the hot gases, vapors (pneuma) and hot solution. Muscovite is the most common mica, found in granites, pegmatites, gneisses, and schists, and as a contact metamorphic rock.

Celadon ($KFe^{3+}(Mg, Fe^{2+})(OH)_2 Si_4O_{10}$) is very small, only visible by electron microscopy crystal length of only a few micrometers, green to bluish green color. It appears in volcanic, especially altered rocks as filling veins, vesicles, and voids of different origin, usually together with chlorite, calcite, and zeolites. It may also occur as a product of changes of and olivine. It is very similar to glauconite with which it is often replaced, but glauconite is found in sedimentary rather than volcanic rocks.

Paragonite ($NaAl_2(OH)_2 AlSi_3O_{10}$) also known as sodium mica, and is very similar to muscovite by appearance. It is typical mineral formed at low and medium temperatures and a regular and important component of crystalline schist, phyllite and mica in particularly paragonite schist (Table 6.1).

Series biotite: biotite and phlogopite are the only important petrogenic minerals.

Biotite ($K(Mg,Fe)_3(OH)_2 AlSi_3O_{10}$) is iron–magnesium mica, dark brown to black color. The black color is due to the variations of the contents of Fe^{2+} which is isomorphic replacement with Mg^{2+} and Fe^{3+}, which is substituted by Al^{3+}. The ratio of Mg/Fe in biotite in igneous rocks decreases from basic to acid, and biotite in the volcanic rocks typically contain more Fe^{3+} than Fe^{2+}. Iron-rich biotite is entirely black and known as lepidomelane. Biotite is generally found in the form of small sheets of perfect cleavage on (001), hardness of 2–3, and their relative density is greater as the iron content gets higher (2.7–3.4). Biotite is general constituent of many igneous rocks, especially the pegmatite, granite, tonalite, granodiorite, diorite and syenite (Table 4.1). Biotite is also forms part of crystalline schist from the group of mica-schist and gneiss. It is significantly less resistant to chemical weathering of muscovite and therefore rarely found fresh in clastic sedimentary rocks. The color rapidly changes from black to bronze-brown during weathering and become vermiculite. Biotite changes into chlorites and in special conditions to muscovite under the influence of hydrothermal effects.

Phlogopite ($KMg_3(OH)_2 AlSi_3O_{10}$) is the magnesium mica in composition and structure. It is equivalent to biotite with no iron, and similar crystallographic properties as biotite. Phlogopite is brownish red, dark brown, yellowish brown, green and white in color. It is usually, with the OH– group and also contains some fluoride. Phlogopite occurs from crystallization of gases and vapors, and contact metamorphism. Phlogopite is found in the pegmatite and marbles.

Clay minerals form *illite* series (named after the state Illinois, USA that originate a number of samples analyzed as clay) belong to dioctahedral mica with lack of the layer cations. Part of K is replaced with OH− group in lattice similar to Muscovite. Therefore, the illite series is called as hydro-micas and illite hydro-muscovite. It often forms mixed-layer minerals with disordered illite or smectite layers. According to the current nomenclature, illite includes a series of complex mineral with general chemical formula: $\{K_{0.65}Al_2(OH)_2Al_{0.65}Si_{3.35}O_{10}\}$. Clay minerals of illite series occur in different environments, and are important components of many soils and marine, lacustrine and terrestrial clayey sediments, and matrix of graywacke sandstones (Sections 5.5.3.1 and 5.5.3.3). Mainly derived from the processes of surface rock weathering: illitization of feldspar and muscovite, and illitization kaolinite and smectite (Section 5.5.5).

Glauconite (from Greek "Glaukonos" = "blue-green") includes a series of green-mixed interlayer mica rich in iron, having complex and variable chemical composition. It contains variable amounts of isomorphic replacement of K^+, Fe^{2+}, Fe^{3+}, Mg^{2+} and Al^{3+} ions. Glauconite aggregates appear in the form of grain diameter of several micrometers to several millimeters and as cryptocrystalline coating. Such grains cluster rarely and contain only glauconite, and more often enclose interstratified mixed-layer glauconite/smectite minerals. These are minerals that provide green color to many of the rocks, for example, glauconite sandstones and green sandstones.

2.5.8.5.3. VERMICULITE GROUP

Vermiculite group $(Mg, Al, Fe^{2+})_3$ $(Si,Al)_4$ O_{10} $(OH)_2 \cdot nH_2O$, belongs to clay minerals, encompassing minerals in morphology similar to muscovite and biotite. It forms by replacing the K with some hydrated cations, usually, Mg^{2+}, a rare and Ca^{2+} and Na^+. Vermiculite loses adsorbed interlayer water when heated. Vermiculite is found in fine, rare and somewhat larger sheets, and in other clay minerals as particles <4 μm. It is formed by weathering or hydrothermal alteration of biotite or phlogopite. Vermiculite is essential ingredient of soil and clay sediments. The mineral can be transformed, easily and quickly, into smectite in surface weathering.

2.5.8.5.4. SMECTITE GROUP

Smectite group includes dioctahedral aluminum micas of montmorillonite–beidellite and nontronite series or Fe–smectite belonging to clay minerals. It has complex chemical composition. The structure is made up of three-layer package, one layer of Al octahedron sandwiched between two layers of SiO_4 tetrahedra. Water molecules are set between three-layer ions of Al, Mg and Fe. The package can expand or narrow due to these ions, and manifests itself by changing of the volume or strong swelling. The connection between these two packages is weak. Smectite crystallizes in monoclinic system in the form of small sheets (<1 μm) and can be explored only by the electron microscope (Fig. 2.15) and X-ray powder diffraction and chemical analysis with the help of electronic microsonde.

Smectite minerals are widespread in soils and clay sediments. The strong absorption of fluids and some cations are widely used for wastewater treatment, textile industry, production of cosmetics and medicine, petroleum industry for removal of organic liquid and gaseous impurities, addition of drilling mud, as the insulating buffer layer in preparing the impermeable barrier in landfills and as adsorbent of harmful substances in medicine. On the geotechnical and

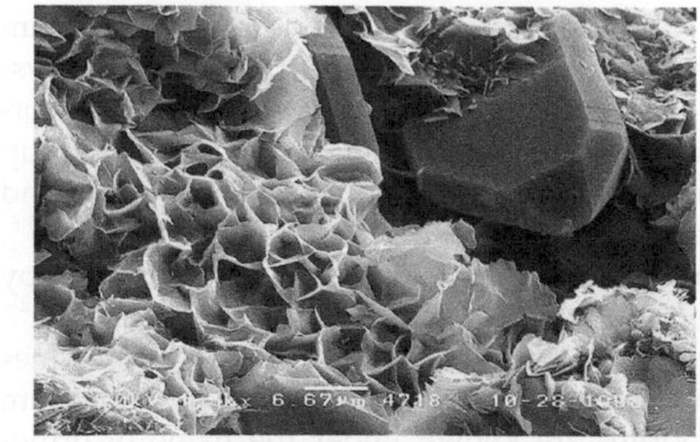

FIGURE 2.15 Sheets clusters and plate crystal of smectite visible by electron (SEM) microscopes.

construction-technological standpoint, smectite minerals have unfavorable characteristics by increasing the volume of the rock mass causing collapse, cracking or even complete collapse of construction.

Montmorillonite (per site montmorillonite) is the main representative of montmorillonite–beidellite group with variable Mg content, and beidellite (per site Beidell in Colorado), member of the series with little or no Mg. Clay minerals montmorillonite–beidellite groups are important mineral constituents of soils, especially in tropical areas. It is created by the surface weathering of rocks, particularly acid tuffs, with the presence of solutions containing Na, Ca, Mg and ferrous iron.

The acidic volcanic glass is altered to smectite phyllosilicates, mainly montmorillonite–beidellite series under the conditions of low pH, which is characteristic of fresh and seawater. The alteration may be associated with an opal, cristobalite and zeolite. Bentonite clay is the product of weathering of acid vitreous volcanic tuffs and ash most often in the presence of water (Section 5.6.3).

Nontronite or Fe smectite is an authigenic ingredient of clay and mud, usually above the basalt on the seafloor, and also can be created by hydrothermal processes of weathering basalt and ultramafic rocks (Sections 4.3.1.3 and 4.3.1.4).

2.5.8.5.5. CHLORITE GROUP

Chlorite group includes hydrated magnesium and iron hydrosilicates which make mixed crystals of complex chemistry, whose general formula is given in Table 2.11. Chlorite group mostly contains Mg^{2+}, Al^{3+}, Fe^{2+}, Fe^{3+}, and some more chlorites Mn^{2+}, Cr^{3+}, Ni^{2+}, V^{3+}, Zn^{2+} and Li^{+}. Part of Al^{3+} can be replaced by Fe^{3+} and/or Cr^{3+}, and part of Mg^{2+} and Fe^{2+} with Mn^{2+} and Ni^{2+}. Part of Si^{4+} sometimes can be replaced by Fe^{3+}, B^{3+} or Be^{2+}. Magnesium chlorites are known under the name of orthochlorites and iron chlorite as leptochlorites. Chlorite is commonly found in small sheets, and can have pale green color depending on the content of iron. Certain types of chlorite, and their chemistry, can safely be determined only using X-ray diffraction combined with chemical analysis. Chlorites are formed by regional metamorphic processes at lower temperatures, and are regular components of low-grade metamorphic schist, particularly chlorite and green schist (Table 6.1). Chlorites are regular secondary ingredients of igneous rocks, resulting from hydrothermal modifications at temperatures ranging between 50 and 400 °C from primary Fe–Mg minerals, mainly mica, pyroxene, amphibole, garnet and olivine. These minerals are common in sedimentary rocks, but usually in small amount. It can quickly turn into vermiculite, smectite mixed layered minerals with disordered layers of chlorite–vermiculite and chlorite–smectite by process of weathering.

2.5.8.5.6. KAOLIN–SERPENTINE GROUP

A group of kaolin–serpentine phyllosilicates includes two subgroups: group of dioctahedral kaolin minerals and group trioctahedral serpentine minerals (Table 2.11).

Kaolin group implies real clay group of minerals. These are pure aluminum silicate with hydroxyl groups that includes kaolinite, dickite and nacrite. All these three minerals have the same chemical composition of $\{Al_2(OH)_4Si_2O_5\}$, and are mostly located in very small particles (<2 μm), and rarely visible as macroscopical crystals. Kaolinite has the only petrogenic significance and is the most abundant mineral in this group. Dickite is much less prevalent, and nacrite is very rare. Kaolinite crystallizes in triclinic, and dickite and nacritein monoclinic system.

Kaolinite forms small sheet like crystals or clusters of white earth (clay particles with a diameter <2 μm). Plate kaolinite minerals are visible only by electron microscopy. It becomes plastic in contact with water. It is produced by the chemical weathering of aluminum silicate minerals like feldspar in igneous and metamorphic rock under the influence of CO_2. Kaolinite is an important mineral constituent of many clastic sediments, especially, clay, sand and pelite sediments, marl and clay (Sections 5.5.3 and 5.5.4). The complete pure white kaolinite clay is known as *kaolin* and is highly valued mineral resources for getting porcelain (Fig. 2.16). Kaolin minerals are important raw material for production of refractory materials, cement, paints, rubber, plastics, as a filler in paper production and less pure clay for making pottery and bricks.

The serpentine groups, $Mg_3\,(OH)_4\,Si_2O_5$, are usually known under the common name of three serpentine minerals: lizardite, chrysotile and antigorite (Table 2.11). The minerals are typically found in microcrystalline sheet and fibrous aggregates. Sheet aggregates are called antigorite, and fibrous type as chrysotile. The characteristic color of antigorite ranges between pale, gray or gray-green, but it changes to dark green to almost black, if part of magnesium isomorphic is replaced with iron.

Lizardite is the most abundant and most important petrogenic serpentine mineral, color of green as apples, and commonly found in association with chrysotile.

Serpentine forms as a large rock mass mainly by pneumatolytic-hydrothermal and hydrothermal processes of silicate minerals that contain aluminum, especially, olivine and orthopyroxenes. Special type of such rocks that contain only serpentine is called serpentinite. Its origin belongs to the metamorphic rocks formed in hydrothermal metamorphism.

Antigorite forms by metamorphism of amphiboles, and are often found as constituents of amphibolite and amphibolite schist.

Serpentinite rocks are used for making ornaments and as a valued dimension stone (especially churches) due to the beautiful green color, relatively low hardness and homogeneity. The other economic importance of serpentine is for obtaining asbestos that serves as thermal insulation and electromaterial resistant to wear. However, its use is reduced to a minimum today due to the carcinogenic action.

2.5.8.6. Tectosilicates

Tectosilicates ($[Al_xSi_yO_{2(x+y)}]^{x-}$) structure is composed of interconnected tetrahedrons going outward in all directions forming an intricate framework (Fig. 2.17). All the oxygens are shared

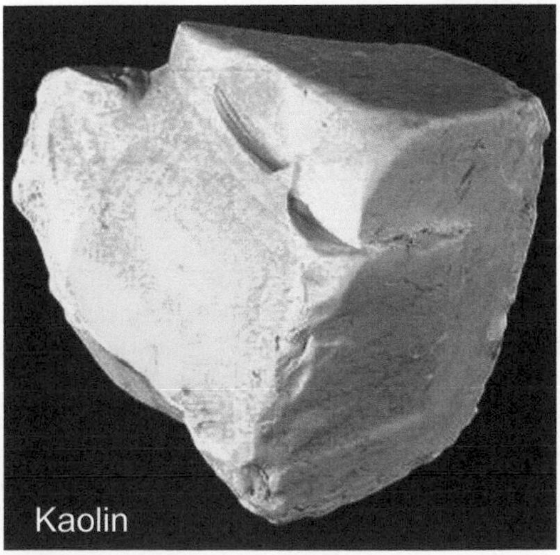

FIGURE 2.16 Kaolin—rock composed of micron fine particles of white kaolinite.

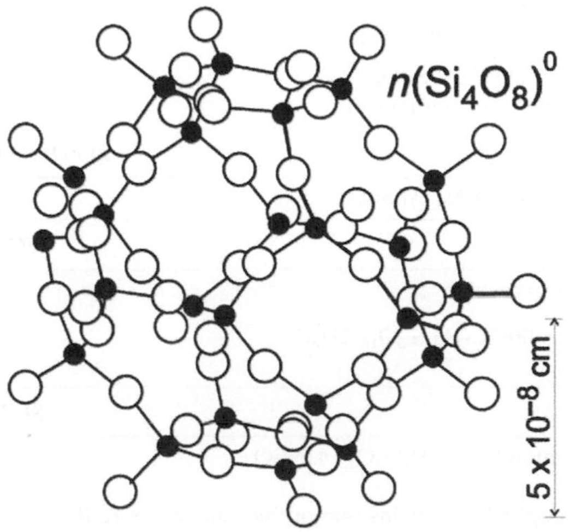

FIGURE 2.17 Tectosilicates framework.

with other tetrahedrons in this subclass. In the near-pure state of only silicon and oxygen, the prime mineral is quartz (SiO_2). Aluminum ion can easily substitute for the silicon ion in the tetrahedrons. In other subclasses, this occurs to a limited extent but in the tectosilicates it is a major basis of the varying structures. While the tetrahedron is nearly the same with an aluminum at its center, the charge is now a negative five (−5) instead of the normal negative four (−4). Since the charge in a crystal must be balanced, additional cations are needed in the structure and this is the main reason for the great variations within this subclass (Table 2.12).

TABLE 2.12 The Most Important Petrogenic Minerals from the Group Tectosilicates

FELDSPAR GROUP	
ALKALINE FELDSPARS	
Orthoclase $KAlSi_3O_8$ Medium-temperature monoclinic K−feldspar	Sanidine $(K,Na)AlSi_3O_8$ High-temperature monoclinic K−Na feldspar
Microcline $KAlSi_3O_8$ Low-temperature trilinic K−feldspar	Anorthoclase $(Na,K)AlSi_3O_8$ High-temperature triclinic Na−K feldspar
PLAGIOCLASE ***Isomorphic series of Albite (Ab) - Anorthite (An)***	
Acid or Na−Plagioclase	Neutral or Na/Ca−Plagioclase
Albite $NaAlSi_3O_8$ (Ab) 0−10% an component Oligoclase 10−30% an component	Andesine 30−50% an component
BASE OR CA−PLAGIOCLASE	
Labrador 50−70% an component	Bytownite 70−90% an component
Anorthite $CaAl_2Si_2O_8$ 90−100% an component	
FELDSPATHOIDES GROUP	
Nepheline $KNa_3Al_4Si_4O_{16}$	Leucite $KAlSi_2O_6$
ZEOLITE GROUP	
Fibrous zeolite	Cubic zeolites
Natrolite $Na_2Al_2Si_3O_{10} \cdot 2H_2O$	Analcime $NaAlSi_2O_6 \cdot 2H_2O$ Phillipsite contains isomorphic admixtures K, Na, Ca, and $6H_2O$
SLIP ZEOLITES	
Laumonitite $CaAl_2Si_4O_{12} \cdot 4.5H_2O$	Heulandite contains isomorphic admixtures K, Ba, Na, Sr, Ca and $12H_2O$
Clinoptilolite contains isomorphic admixtures K, Ba, Na, Sr, Ca, Mg, Fe^{2+} and $12H_2O$	

2.5.8.6.1. FELDSPAR GROUP

Feldspar group is petrogenic most important assemblage of silicate minerals, as it covers almost 58% of the Earth's crust (Table 2.3). The proportion of feldspar is extremely high in igneous, sedimentary and metamorphic rocks. The chemical compositions of feldspar group represent the aluminosilicates of potassium (Or-component), sodium (Ab-component) and calcium (An-component). It often forms isomorphic mixture of sodium and calcium components, i.e. plagioclase (Table 2.12). Potassium and sodium component form isomorphic mixture only in igneous rocks that crystallize at high temperatures and the product is known as *alkali feldspar*. This compound is unstable at low temperatures and divides on orthoclase and sanidine (Sa). There is only a small part of the K replacement by Na in orthoclase. The share of $NaAlSi_3O_8$ component usually is about 30% and can reach up to 65% in sanidine at high temperature. The isomorphic compounds of Na—feldspar (Ab) and K—Na feldspar sanidine (Sa), is known as anorthoclase.

The crystallographic characteristics of feldspar are divided into monoclinic and triclinic system.

Monoclinic feldspars that crystallize in the monoclinic system (orthoclase and sanidine) have a cleavage on plane side (010) and base (001) pinacoid and cleavage cracks intersect at right angles and are called *orthoclase feldspar* (from the Greek. "Ortho" means "vertically and klasis" cleavage).

Triclinic feldspars that crystallize in the triclinic system (microcline, anorthoclase and plagioclase) also have a lateral cleavage planes (010) and base (001) pinacoid but their cleavage cracks intersect at a sharp angle of 85—86° (Fig. 2.18). Therefore, the Na/Ca isomorphic series albite—anortite named plagioclase (from Gk. Plagios—slope and klasis—cleavage).

2.5.8.6.1.1. ALKALI FELDSPARS The alkali feldspars include monoclinic feldspars (orthoclase and sanidine) and triclinic feldspars (microcline and anorthoclase). At high temperatures, it is possible to form mixed crystals of isomorphic replacement of Or-component and Ab-component. Slow cooling at lower temperatures leads to separation and mutual intergrowths characteristic modes of these two components. K—feldspar albite intergrowths known as pertite and albite intergrowths K—feldspar as antipertite. The hardness is from 6 to 6.5 and the relative density of 2.55—2.63. The color is usually white, and sometimes changes from pale pink to reddish due to admixtures of iron (especially microcline). In microcline, K can be in small quantities of isomorphic replacement with Pb^{2+} and changes to green color and is known as amazonite.

Orthoclase ($KAlSi_3O_8$) is the monoclinic medium-temperature K—feldspar. It occurs in different ways, usually, by crystallization of magma. It is an important ingredient of all acidic (granite, granodiorite) to neutral (syenite) igneous rocks (Table 4.1). It often occurs in the pegmatite-stage crystallization of magma and is the essential ingredient of pegmatite. It can also occur by hydrothermal process and contains little natrium components and is known as *adular*. Orthoclase feldspar may arise from regional metamorphic processes, and is a common ingredient in various crystalline schists (Table 6.1). Orthoclase changes to kaolinite (Section 5.2.1.2) and sericite with the process of kaolinization and sericitization, respectively.

Sanidine ($(K,Na)AlSi_3O_8$) is the monoclinic high-temperature alkaline K—Na feldspar, which usually contains about 30%, but sometimes up to 62% sodium (Ab) component. It is found only in young volcanic discharges or (volcanic) rocks (rhyolite, trachyte and dacite). It forms by the crystallization of lava at high temperatures and its rapid cooling. Sanidine crystallizes orthoclase during slow cooling of lava. Sanidine, as a high-temperature alkali feldspar, is not stable in rocks on the surface or shallow under the surface of the Earth,

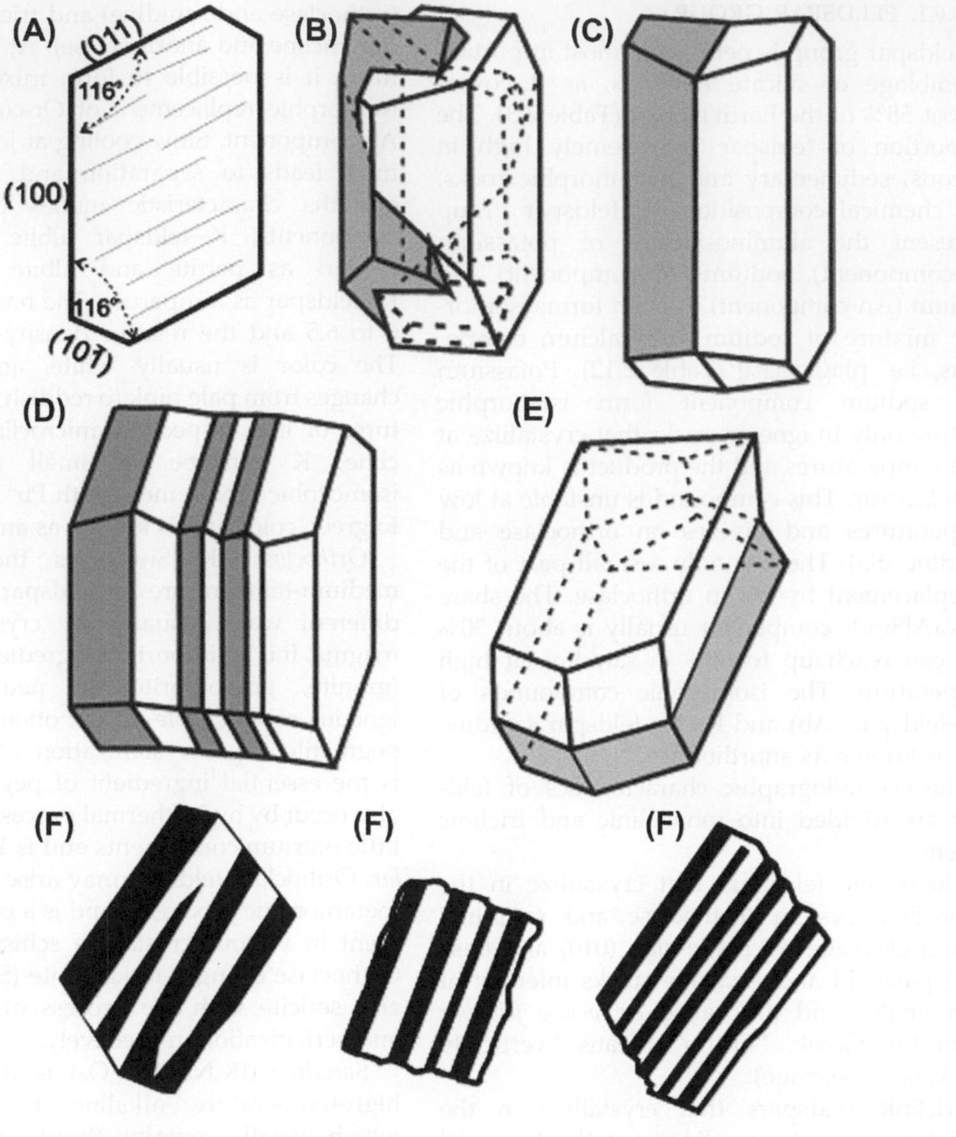

FIGURE 2.18 Characteristic crystal forms of feldspar: (A) cross-section through the crystal of triclinic feldspars; (B) orthoclase crystal twinning—Carlsbad law; (C) plagioclase crystal twinning—Carlsbad law; (D) plagioclase crystal twinning—Albite law; (E) orthoclase crystal twinning—Braveno law; and (F) cross-section of triclinic feldspars with polysynthetic twins.

and gradually recrystallize in orthoclase over time.

Microcline ($KAlSi_3O_8$) is the triclinic low-temperature K—feldspar stable at temperatures lower than 500 °C. It is usually formed by recrystallization from feldspar, and sometimes by direct crystallization from magma and hydrothermal processes. Microcline typically displays albite and pericline twining. This combination leads to a grid pattern (Fig. 2.19).

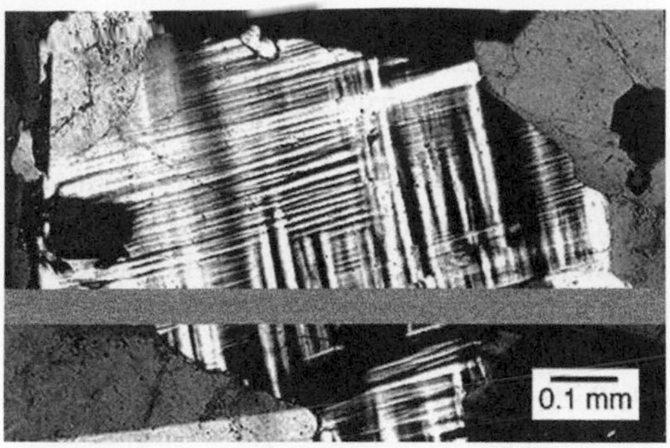

FIGURE 2.19 Photograph of thin section microcline showing the grid structure of twining under cross-polarized light.

Microcline is an essential component of many rocks, especially granite, syenite, pegmatite (Table 4.1) and gneisses (Table 6.1). In granite, syenite and pegmatites, i.e. acidic, neutral and core igneous rocks, microcline is commonly found together with feldspar. If the rocks are geologically very old, microcline cannot be found with feldspar because eventually orthoclase is recrystallized in microcline. Microcline is a regular and essential ingredient of sandstone, especially feldspar arenaceous rocks and feldspar graywacke (Sections 5.5.3.2 and 5.5.3.3).

Anorthoclase ($(Na,K)AlSi_3O_8$) is a crystalline solid solution in the alkali feldspar series, in which the sodium–aluminum silicate member exists in larger proportion. It typically consists of 10–36% of $KAlSi_3O_8$ and 64–90% of $NaAlSi_3O_8$.

2.5.8.6.1.2. PLAGIOCLASE FELDSPAR SERIES *Plagioclases* are triclinic feldspars that form a complete isomorphic compounds which are the final members of the Na–plagioclase albite $NaAlSi_3O_8$ (Ab) and Ca–plagioclase anortite $CaAl_2Si_2O_8$ (An) (Table 2.13). In albite one of the four Si^{4+} ions is isomorphic substituted with one Al^{3+} ions, and one free (−) valence is related to Na^+ ion. In anortite, two of the four Si^{4+} ions are isomorphic replaced with two Al^{3+} ions, while the remaining two (−) valence neutralize one Ca^{2+} ion. The acidity and basicity of silicates define with the amount of SiO_2, Na–plagioclase with three Si^{4+} ions have more silicon than Ca–plagioclase with two Si^{4+} ions. Na–plagioclases belong to acid plagioclase and Ca–plagioclases belong to basic plagioclase (Table 2.12). In fact, pure albite contains 68.7% SiO_2, and a pure anortite only 43.2% SiO_2.

Plagioclase is usually found in the form of granular aggregates in kaolinite, while well-formed crystals are rare. Special features of plagioclase crystallization occur from the magma and lava. It has a tendency to format polysynthetic twinning (Fig. 2.18(F)) and zonal structure due to the crystallization sequence from more basic to the acidic crystals twinning. Plagioclase zone formation, especially those in the volcanic rocks is reflected in the fact that from the center to the edge of the crystals formed plagioclase are of acid composition, i.e. with less An-component. It can be clearly observed as a zonal darkening of the grain due to changes in the optical properties of this phenomenon.

Plagioclase occurs at high and low temperatures, where high temperature creates disordered and low-temperature forms ordered crystal lattice. Plagioclase forms at high temperatures are significantly less stable than plagioclase forms at lower temperatures. Plagioclase, particularly high-temperature type, is relatively

TABLE 2.13 Plagioclase Minerals and Their Compositions

Plagioclase Minerals and Their Compositions		
Mineral	**% $NaAlSi_3O_8$ (%Ab)**	**% $CaAl_2Si_2O_8$ (%An)**
Albite	100–90	0–10
Oligoclase	90–70	10–30
Andesine	70–50	30–50
Labradorite	50–30	50–70
Bytownite	30–10	70–90
Anorthite	10–0	90–100

easily influenced by water and weathering. It generates kaolinite and sericite, and under certain conditions metamorphosed in clusters of zoisite, epidote, albite, quartz and actinolite known as sosirite. High-temperature basic plagioclases of volcanics (basalts and diabase) are replaced in hydrothermal processes with low-temperature acid plagioclase—albite (albitization). Such rocks are known as spilites (Table 4.1).

Plagioclase minerals are extremely widespread and abundant. It is an essential or important constituent of many igneous, metamorphic and sedimentary rocks (acidic, neutral and basic, pegmatite igneous rocks, amphibolite schist and gneiss, feldspar sandstone, siltstone and shale (Tables 4.1 and 6.1)).

2.5.8.6.2. GROUP FELDSPATHOIDS

The feldspathoids are a group of tectosilicates and alkali alum-silicate minerals which resemble feldspar, but have a different structure and much poor in silica content and alkali-rich elements like sodium, potassium and lithium. Feldspathoids occur in rare and unusual types of igneous rocks. The main minerals of the feldspathoids group are nepheline and leucite (Table 2.12).

Nepheline ($KNa_3Al_4Si_4O_{16}$) (along with the variety known as eleolite) is the most widespread of all feldspathoids. It crystallizes in the hexagonal system. It usually has the form of a short, six-sided prism terminated by the basal plane. It is found in compact, granular aggregates, and can be white, yellow, gray, green, or even reddish color (in the eleolite variety). The hardness is 5.5−6, and the specific gravity between 2.56 and 2.66. It is often translucent with a greasy luster. It is characteristic of alkali rocks as nepheline syenites and gneisses, alkali gabbros, in sodium-rich hypabyssal rocks, tuffs and lavas, and pegmatites, as a product of sodium metasomatism. Nepheline has economic importance as a raw material in chemical industry, leather tanning, and manufacture of glass, ceramics and paints.

Leucite ($KAlSi_2O_6$) crystallizes in cubic system, and usually has well-developed free, colorless, white or pale-gray crystals in the form of cubic icositetrahedra. It is a high-temperature mineral which crystallizes in alkalis lava rich and poor in silicon. It is never found together with quartz. It is often associated with nepheline and alkali feldspar, alkali pyroxene and analcime. It can be found in young volcanic rocks rich in K and poor in SiO_2, and in intrusive rocks only located in alkali syenite. It is a regular ingredient of younger alkaline effusive rocks, for example, in lava of Vezuves, clinkstones, trachyte and tuff (Table 4.1). It is an unstable mineral that quickly destroys into the clay minerals on Earth's surface.

2.5.8.6.3. ZEOLITES GROUP

The zeolites includes hydrated alumosilicates of alkali (Na and K) and earth-alkaline (Ca, Ba, and Sr) elements. The group is represented by large number of minerals of different chemical composition, but with similar properties. The basic feature of their chemical composition is the water content, which is in adsorption and poorly connected to the grid (zeolite water). Such water zeolites are losing when heated, but water is easily readmitted in its lattice. Zeolite crystallizes in different morphological forms, in different crystal systems: cubic, orthorhombic monoclinic and hexagonal. All, however, have very similar properties: usually colorless or gray due to impurities, the relative density of 2.1−2.4 and weakly resistant to chemical weathering.

Zeolites arise from the secretion of aqueous solutions. It is often found in crevices and cavities of younger igneous rocks. Zeolites are common constituents of soil, where it occurs as colloidal weathering products of various minerals. Petrogenic most important zeolites are given in Table 2.12.

Natrolite is very common fibrous zeolite, end-member of isomorphic series of Na−Ca zeolites. It crystallizes in the orthorhombic system and occurs as needle or radial crystals, particularly in basalt.

Laumontite, heulandite and *clinoptilolite* are monoclinic zeolite. It usually crops up as prismatic (laumontite), plate or wedge shape (heulandite and clinoptilolite). Laumontite occurs in igneous rocks, typically as a product of hydrothermal exchange, and in sedimentary rocks as conversion of plagioclase.

Analcime and *phillipsite* are cubic zeolites, exists in collective community with other zeolites. It occurs in cavities of effusive rocks (especially in the basalt).

FURTHER READING

Reading of text books authored by Pirsson,[33] Dana[9] and Rösle[38] will be an intelligent approach to step into the theme of *Mineralogy and Petrology*. The fundamental concepts of mineralogy and petrology are explained by Gaines et al.[21] and Klein et al.[27]

CHAPTER

3

Basic Petrology

OUTLINE

3.1. Introduction 81

3.2. Structure of the Earth 83

3.3. Classification of Rocks 85

3.3.1. *Igneous Rocks* 85

3.3.2. *Sedimentary Rocks* 85

3.3.3. *Metamorphic Rocks* 86

3.4. Origin of Earth and Theory of Plate Tectonics 86

3.4.1. *Origin of the Earth* 86

3.4.1.1. The Protoplanet Hypothesis 87

3.4.1.2. The Nebular Hypothesis 88

3.4.1.3. Age of the Earth 88

3.4.2. *Plate Tectonics* 89

Further Reading 91

3.1. INTRODUCTION

Petrology (from Greek: "Petra"—"rock" and "logos"—"knowledge") is the study of rocks, their occurrences, composition, origin and evolution. This research also focuses on the study of minerals and meteorites (Fig. 3.1) as model to unravel the interiors of planetary bodies. Petrography deals with the detailed description and classification of rocks, whereas petrology focuses primarily on the rock formation, or petrogenesis. A petrological description includes definition of the unit in which the rock occurs, its attitude and structure, its mineralogy and chemical composition, and conclusions regarding its

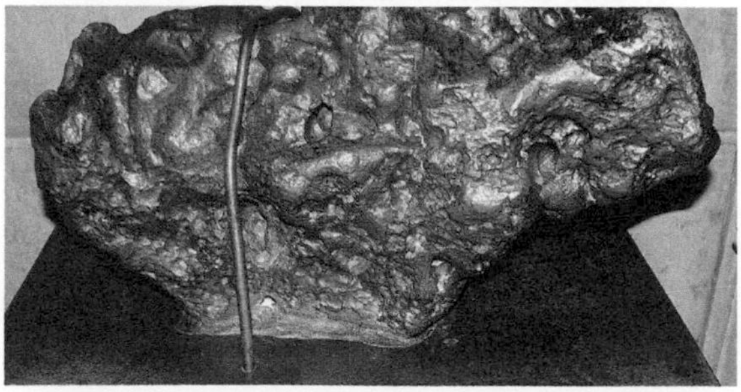

FIGURE 3.1 Iron meteorite of 179 kg by weight: fragment of a huge massive object that crashed into the Arizona desert about 50,000 years ago and created the Barringer lunar meteor impact circular crater of 1280 m diameter. *Photo source and courtesy: Griffith Observatory, Los Angeles, CA.*

Introduction to Mineralogy and Petrology
http://dx.doi.org/10.1016/B978-0-12-408133-8.00003-1

origin. The task of petrologists is to carry out research and study rocks, independent of geological bodies, which are integral parts of the lithosphere and are clearly different from their surroundings.

Meteorite, small to extremely large size, is a natural object originating in the outer space that falls on Earth creating great surface impact. Most meteorites are derived from small celestial bodies as well as produced by impacts of asteroids from the solar system. Meteorites are composed of silicate minerals and/or metallic iron–nickel. The structure of Igneous Complex at Sudbury Mining District, Canada is formed as the result of a meteorite (1850 Ma age) impact that produced a 150–280 km multiring crater, containing 2–5-km-thick sheet of andesite melt. The immiscible sulfide liquid differentiated into Ni–Platinum Group of Element dominated contact deposits by crystallization. There are 100+ deposits/mines having a total resource, including past production, of 1648 million tonnes at 1% Ni, 1% Cu, and 1 g/t Pd + Pt.

The primary and most significant processes to be focused are the following:

1. Tectonic movements of rock masses.
2. Volcanic eruptions and injection of magma into the lithosphere.
3. Physical, chemical and biological weathering and deposition in the surface areas of rocky crust and in the hydrosphere and atmosphere.
4. Mutual chemical reactions and biological processes in aqueous solutions.
5. Metamorphic changes due to increasing pressure and temperature at greater depths of covering.
6. Melting, migration, recrystallization, degassing and similar events on rocks.

Petrology is essentially a fundamental part of geology. Rocks, as mineral aggregates, are composed of certain minerals, so petrology is closely associated with the mineralogy.

Determination of mineral constituents and chemical composition of rocks is necessary to know and distinguish minerals, and also for resolving the origin of rocks. It is also necessary to have a good knowledge of the origin of minerals (mineral genesis). The researchers in this area conduct field- and laboratory-based experimental modeling supported by advanced computational tools to read the records and understand the potential of high-temperature and high-pressure processes.

Petrology is closely associated with chemistry, especially mineral chemistry and geochemistry, for the purposes of studying the complex chemical reactions and processes that lead to crystallization, or conversion of minerals and rocks, as well as studying the share of chemical elements in minerals, rocks, lithosphere, hydrosphere and atmosphere.

Petrology experiments and draws conclusive records about evolution and constitution of the Earth beneath its rocky crust. It is primarily based on volcanic eruptions, shape and composition of igneous bodies that have reached Earth's surface by tectonic processes and erosion. The fundamental research of meteorites in solar system provides additional support to these studies. The geophysical responses of contrast in seismic velocity propagating through certain parts of the rocky crust and underneath have also been acclaimed significant importance. It has been understood that the Earth has different laminate structures distinguished by either chemical or their rheological properties. This is based on remarkable geophysical response of primary seismic wave, particularly establishing the existence of two major and a number of less pronounced discontinuity obtained. The two major planes of discontinuity are established between

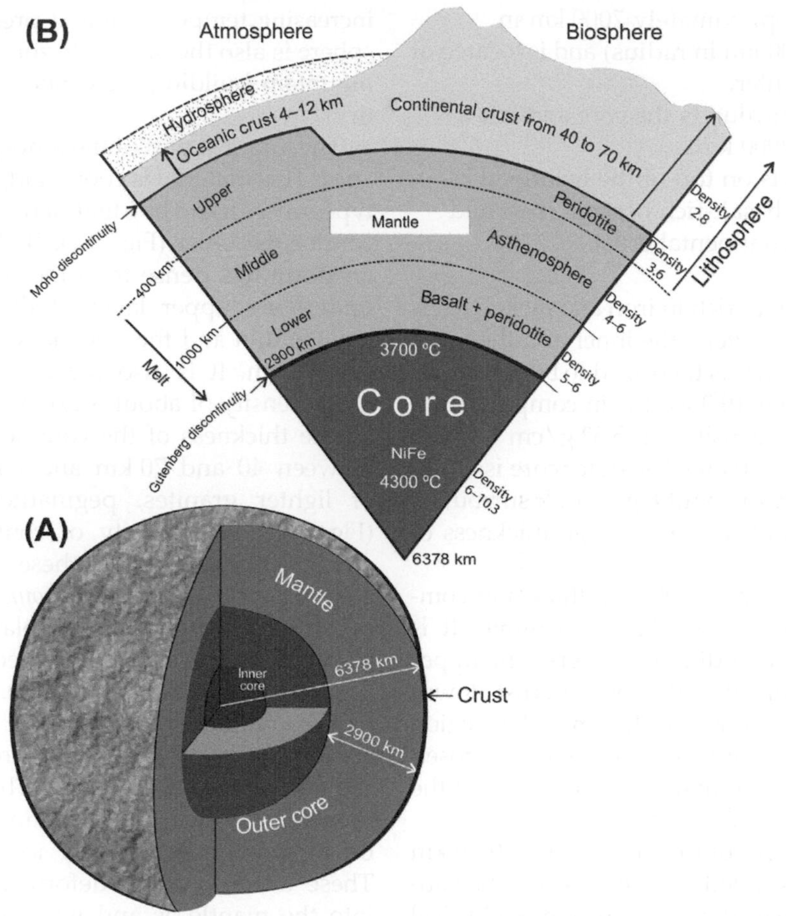

FIGURE 3.2 (A) Schematic diagram of the Earth's structure representing a three-dimensional perspective and (B) a sectional view portraying from central core to outer surface.

crust–mantle and mantle–core boundary. The minor discontinuities are set up between upper, middle and lower mantle as well as outer and inner core (Fig. 3.2). Therefore, petrology is also associated with geophysics and has an important role in geophysical research of the lithosphere.

Rheology is the study of the flow of the mantle, normally in the liquid state, but often as sift solids or solids that react as plastic flow in contrast to deforming elastically to the applied force.

3.2. STRUCTURE OF THE EARTH

The Earth is an oblate spheroid. It is composed of a number of different layers in spherical shells as determined by deep drilling and seismic evidence (Fig. 3.2). These layers are the following:

1. The Earth can broadly be modeled as an outer solid silicate crust, a highly viscous mantle, a liquid outer core that is much less viscous than the mantle, and a solid inner core.

2. The core is approximately 7000 km in diameter (3500 km in radius) and is located at the Earth's center.
3. The mantle surrounds the core and has a thickness of 2900 km.
4. The crust floats on top of the mantle. It is composed of basalt-rich oceanic crust and granitic-rich continental crust.

The *core* is a layer rich in iron and nickel that is composed of two layers: the inner and the outer cores. The inner core is theorized to be solid with a density of about 10.3 g/cm^3 in comparison to Earth's average density of 5.52 g/cm^3 and a radius of about 1220 km. The outer core is liquid and has a density of about 6 g/cm^3. It surrounds the inner core and has an average thickness of about 2250 km.

The *mantle* is almost 2900 km thick and comprises about 83% of the Earth's volume. It is composed of several different layers. The upper mantle exists from the base of the crust downward to a depth of about 400 km. This region of the Earth's interior is thought to be composed of peridotite, an ultramafic rock made up of the minerals olivine and pyroxene.

The middle layer of the mantle, 400–1000 km below surface, is called the *asthenosphere*. Scientific studies suggest that this layer has physical properties that are different from the rest of the upper mantle. The rocks in this upper portion of the mantle are more rigid and brittle because of cooler temperatures and lower pressures. The lower mantle stands below the upper mantle and extends from 1000 to 2900 km below the Earth's surface. This layer is hot and plastic. The higher pressure in this layer causes the formation of minerals that are different from those of the upper mantle.

The *lithosphere* is a layer that includes the crust and the upper portion of the mantle (Fig. 3.2). This layer is about 400 km thick and has the ability to glide over the rest of the upper mantle. The deeper portions of the lithosphere are capable of plastic flow over geologic time because of increasing temperature and pressure. The lithosphere is also the favorable zone of earthquakes, mountain building, volcanoes, and continental drift.

The topmost part of the lithosphere consists of *crust*. This material is cool, rigid, and brittle. Two types of crust can be identified: oceanic crust and continental crust (Fig. 3.2). Both these types of crust are less dense than the rock found in the underlying upper layer of the mantle. Ocean crust is thin and the thickness varies between 4 and 12 km. It is also composed of basalt and has a density of about 3 g/cm^3.

The thickness of the continental crust varies between 40 and 70 km and composed mainly of lighter granites, pegmatites and gneisses (Fig. 3.2). The density of continental crust is about 2.8 g/cm^3. Both these crust types are composed of numerous *tectonic plates* that float on top of the mantle. These plates move slowly across the asthenosphere caused by the convection currents within the mantle.

The continental and oceanic crusts have one common property. These tectonic plates have the ability to rise and sink. This phenomenon, known as *isostasy*, occurs because the crust floats on top of the mantle-like ice cubes in water. These tectonic plates deform and sink deeper into the mantle as and when the Earth's crust gains weight due to mountain building or glaciations. The crust becomes more buoyant and floats higher in the mantle if the weight is removed.

The *Mohorovičić discontinuity* (Moho) is the line between the Earth's crust and the mantle. It separates oceanic crust and continental crust from the mantle. The Mohorovičić discontinuity named after Andrija Mohorovičić, a Croatian geophysicist, who has established it. The Mohorovičić discontinuity is 5–10 km (3–6 miles) below the ocean floor and 20–90 km (10–60 miles) beneath the continents.

The *Gutenberg discontinuity*, named after German scientist Bruno Gutenberg, is located at 2900 km depth beneath the Earth's surface.

The boundary is observed by the applications of seismic waves. This discontinuity is due to the differences between the acoustic impedances of the solid mantle and the molten outer core.

Isostasy ("isos" is "equal" and "stásis" means "standstill") is the state of gravitational equilibrium between the lithosphere and asthenosphere such that the tectonic plates "float" at an elevation which depends on their thickness and density to explain the different topographic heights on Earth surface (Fig. 3.3). In the event of any dynamic change in isostasy, the plates collide or move causing Earthquake, Tsunami and related natural hazards and calamities.

Atmosphere is a layer of gases surrounding the Earth by gravity, distributed from surface upward as Troposphere, Stratosphere with ozone layer, Mesosphere and Ionosphere. It protects life forms by absorbing ultraviolet solar radiation and greenhouse effect (water vapor, carbon dioxide, methane and ozone). Air is the part of atmosphere used for breathing and photosynthesis.

Biosphere is the universal sum of total ecosystem or the zone of life for plants, animals and microbes on the Earth's crust controlled by natural self-regulating system.

Hydrosphere is the physical distribution of the combined mass of water found under, on, and over the surface of the Earth.

FIGURE 3.3 The Himalayan snow-capped peaks ranging between 10,000 and 15,000 ft (3000 and 4500 m) high above mean sea level (MSL) in the background and "Deodar (*Cedrus deodara*) and Chilgoza pines (*Pinus gerardiana*)" in the foreground, viewed from Kalpa town in Himachal Pradesh, India. The central core of the mountain range consists of intrusive granite rising as pointed high peaks within metaphyllites. The Himalaya Mountain is still young and rising by the force of the impacting Indian plates under the Tibetan plates and that makes the area earthquake prone.

3.3. CLASSIFICATION OF ROCKS

The rocks of Earth's crust are divided into three main groups (Fig. 3.4) according to the manner of their origin.

3.3.1. Igneous Rocks

Igneous ("ignis" means "fire" in Latin and eruptive) rocks (Fig. 3.5) are the primary rocks that are originally formed in and/or on the Earth. They occur in two main ways:

1. Underground: direct cooling, crystallization and solidification of rocks inside the Earth's crust from molten rock mass (magma). The type is intrusive (plutonic).
2. On surface: crystallization of the lava, i.e. magma poured onto the surface of the Earth, seabed or shallow under surface of the Earth. The type is extrusive (volcanic).

3.3.2. Sedimentary Rocks

Sedimentary (depositional) rocks are formed in the sea, fresh water or on land by precipitation, deposition and sedimentation processes (Fig. 3.6) of the following:

1. Solid waste material of physical and chemical weathering of rocks formed earlier.
2. Organogenic or fossil remains and other biochemical and chemical products extracted from the water.

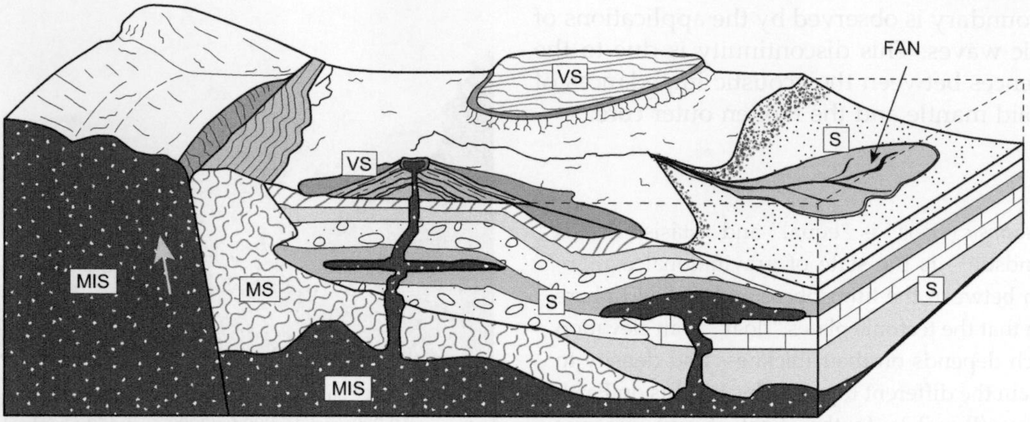

FIGURE 3.4 Conceptual diagram depicting the mode of formation and three fundamental genetic types of rocks that make up the Earth's crust: Igneous, Sedimentary and Metamorphic.

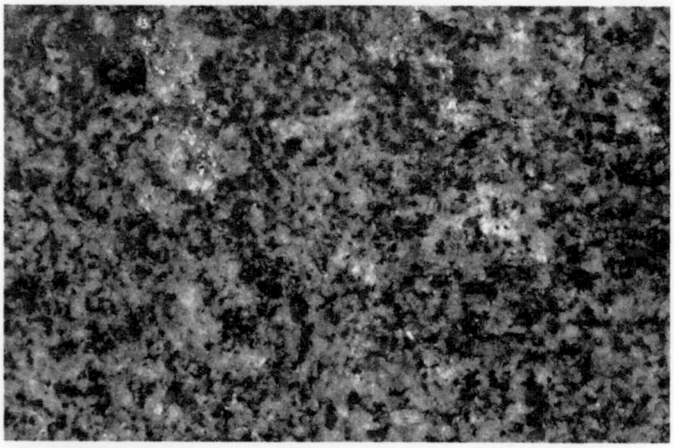

FIGURE 3.5 A typical light-color fine- to medium-grained igneous rock (granite) containing interlocking minerals of quartz (white) and feldspar (light rosy or pink) with minor grains (black) of actinolitic hornblende, biotite and chlorite. *Source: Prof. A.B. Roy.*

3.3.3. Metamorphic Rocks

Metamorphic (transformed) rocks (Fig. 3.7) are formed by metamorphism of preexisting igneous, sedimentary and metamorphic rocks with changes in crystal form (texture and structure) in solid state. The mineral transformation occurs under increased temperature (+150 to 200 °C) and pressure (+1500 bars) at greater depth of covering or at the contacts with magma in the Earth's rocky crust. The metamorphic process accelerates with the introduction of chemically active fluids.

Clarke[7], an US geochemist, calculated the lithosphere to a depth of 16 km. He opined that the lithosphere is consisting of 95% of igneous rocks and 5% of sedimentary rocks. The metamorphic rock components are included into the igneous or sedimentary group depending from which the original rocks are metamorphosed. He also observed that from a total of 5% of sedimentary rocks, around 4% are shales, only about 0.75% sandstones and remaining 0.25% limestone.

3.4. ORIGIN OF EARTH AND THEORY OF PLATE TECTONICS

3.4.1. Origin of the Earth

There are many different hypotheses on the origin of the solar system, including the Earth

FIGURE 3.6 A classical example of sedimentary rock formation of shale (gray) and limestone (yellowish brown) with sharp contact at the snow capped summit of Jungfrau, one of the main peak of Bernese Alps, Switzerland. Photo from top of Europe at 11,782 ft or 3571 m above mean sea level, September, 2009. The position of the in situ rock at high altitude is due to mountain building process of Alps.

as a planet of this system. The most famous and accepted among them are the following:

3.4.1.1. The Protoplanet Hypothesis

The protoplanet hypothesis suggests that a great cloud of gas and dust of at least 10,000 million kilometers in diameter rotated slowly in space about 5,000 million years ago. As time passed, the cloud shrank under the pull of its own gravitation or was made to collapse by the explosion of a passing star. Most of the cloud's material gathered around its own center. Its shrinking made it rotate faster, like a spinning whirlpool. The compression of its material made its interior so hot that a powerful reaction, hydrogen fusion, began and the core of the cloud

FIGURE 3.7 A typical metamorphic rock composed of biotite, quartz, feldspar gneiss displaying strong fluxion banding with numerous white porphyroblasts and porphyoclasts of feldspar showing varying degrees of flattening into the fabric. This implies blastesis during the mylonitization of sediments.

blazed into a newborn *Sun*. About 10% of the material in the cloud formed a great plate-like disk surrounding the Sun far into space. Friction within the disk caused most of its mass to collect in a number of huge whirlpools or eddies. These eddies shrank into more compact masses called protoplanets and later formed planets and moons. Some uncollected material remains even today as comets, meteoroids, and asteroids'.

3.4.1.2. *The Nebular Hypothesis*

The nebular hypothesis is the most widely accepted model explaining the formation and evolution of the Solar System. It was first proposed in 1734 by Emanuel Swedenborg, a Swedish scientist with occupation as mining engineer, anatomist and astronomer. The hypothesis was originally applied only to our own Solar System. This method of planetary system formation is now thought to be at work throughout the universe. The nebular hypothesis postulates that the stars form in massive and dense clouds of molecular hydrogen—giant molecular clouds. They are gravitationally unstable, and matter coalesces to smaller and denser clumps within, which then proceed to collapse and form stars. Star formation is a complex process, which always produces a gaseous protoplanetary disk around the young star. This may give birth to planets in certain circumstances, which are not well known. Thus the formation of planetary systems is thought to be a natural result of star formation. A Sun-like star usually takes around 100 million years to form.

The protoplanetary disk is an accretion disk which continues to feed the central star. The disk is initially very hot and cools later in what are known as the "T Tauri Star (TTS)" stage by possible formation of small dust grains made of rocks and ices. The grains may eventually coagulate into kilometer-sized planetesimals. Planetesimals are solid objects thought to exist in protoplanetary disks and in debris disks. A protoplanetary disk is a rotating circumstellar disk of dense gas surrounding a young newly formed star, i.e. a TTS. If the disk is massive enough, the runaway accretions begin resulting in the rapid—100,000–300,000 years—formation of Moon- to Mars-sized planetary embryos. The planetary embryos undergo through a stage of violent mergers, producing a few terrestrial planets near the star. The last stage takes around 100 million–1,000 million years.

Star is a massive and luminous sphere of vast plasma held together by gravitational forces. Sun is the nearest star to the planet Earth and is the source of most of the energy on the planet. Stars are innumerable in number and can be seen glowing and twinkling far away in the night. Stars are grouped together forming constellations.

A *planet* is an astronomical or celestial object orbiting a star. Planet is massive enough to rotate in its own axis by its own gravity.

The Solar System consists of the Sun (Star) and its planetary system of eight, their moons formed 4,600 million years ago from the collapse of a giant cloud. The eight planets from nearest to the Sun outwards are Mercury, Venus, Earth, Mars (rocks and metals), Jupiter, Saturn (hydrogen and helium), Uranus, and Neptune (water–ammonia and methane). All planets rotate in almost circular orbits that lie within a nearly flat disk called the *ecliptic plane*.

Star, planets and solar system are originated from the same giant massive parent cloud and dust and complimentary to each other.

3.4.1.3. *Age of the Earth*

Some of the oldest surface felsic rocks on Earth had been found in the Canadian Shield, Australia and Africa with age varying between 2,500 and 3,800 million years. The oldest rock from Nuvvuagittuq greenstone belt on the coast of Hudson Bay in northern Quebec was dated as

3,800–4,280 million years at McGill University. In 1999, the oldest-known rock of the Acasta Gneiss of the Slave craton in northwestern Canada was dated to 4,031 ± 3 million years. The age of the Earth is estimated as 4,540 million years based on evidence from radiometric age dating of meteorite materials. This has been corroborated by the age dating of the oldest-known rocks of granulites gneiss–sedimentary siliciclastic–mafic/ultramafic sequence (zircon crystal) from Jack Hills, Western Australia as 4,404 million years. Basaltic rock samples, collected from the Moon surface and from the highlands during 1993 space mission, have been measured by radiometric dating techniques and age reported as 3,160 and 4,500 million years old, respectively.

The recent estimate by the astrophysicist as well as dating of meteorite can be summarized that the upper limit of the Solar System including the Earth is 4,567,000,000 years or 4,567 million years.

3.4.2. Plate Tectonics

The upper part of the Earth is composed of a solid rock mass and divided into continental crust, oceanic crust and upper mantle or the lithosphere as clearly illustrated in Fig. 3.2. The lithosphere is underlain below by a melt or asthenosphere. Since the formation of the Earth, more than 4,500 million years ago, the surface and its interior are constantly undergoing rebuilding processes that create, release and transfer heat energy, and the process of cooling of parts of its surface. While the Sun heats the atmosphere and the very surface of the Earth, the primary heat sources for the Earth's interior are radioactive processes which release a very high-temperature. Thus the emerged heat is transferred from the interior to the Earth's surface and in the lithosphere leads to remelting of rocks. The melting of rocks and minerals in the lithosphere is endothermic reaction, i.e. absorption of heat. The crystallization of minerals is exothermic reaction, i.e. heat release process.

The lithosphere portrays the Earth's evolution history which is constantly changing—it was changing yesterday—it is changing today and will continue to change in the future. It is assumed that initially the lithosphere was a uniform continuous mass (Supercontinent) during late Paleozoic era and known as *Pangaea*. The lithosphere was gradually separated and rifted into several parts, and then in the Carboniferous and Permian era (about the 350–250 million of years) recollected. The separation repeated again during the Triassic period (before ~250–200 millions of years) to current configuration of the component continents consisting of multiple faults separated plates of different thickness and extension. However, from the Triassic to the present, there is a global tendency of reapproaching these plates. The term *plate* includes some solid parts of the oceanic or continental crust which are apart from each other with large horizontal (transform) faults, mountain chains, oceanic arcs, oceanic ridges and trenches. There exist seven such primary plates on the Earth today with roughly defined boundaries: the Pacific Plate, the North American Plate, the South American Plate, the African Plate, the Eurasian Plate, the Indo-Australian plate and the Antarctic plate. There is equal number of smaller secondary plates on Earth: Arabian Plate, Caribbean Plate, Cocos plate, Indian Plate, Juan de Fuca Plate, Nazca Plate, Philippine Plate and Scotia plate. In addition, there are several small tertiary plates that are grouped with major primary plates, without having distinct identity. Some related definitions are the following:

1. *Transform faults* are the horizontal displacements, or spaces, between plates and also the main place of earthquakes that have shallow epicenter under the surface of the Earth.
2. *Oceanic ridges* stand as boundaries between the divergent plates movement. In that process, magma is injected between plates or poured on the ocean floor forming growth of oceanic crust. The mid-oceanic ridges and

growth of the oceanic crust represent in the form of effusive volcanic rocks. In this way, for example, Middle-Atlantic ridge at the bottom of the Atlantic Ocean was originated. In the lithosphere, this growth of oceanic crust, however, compensates with convergent plate movement (subduction) or underscores one plate under another, which leads to narrowing of areas of the ocean or even disappearance of the ocean.

3. *Subduction zone* is a place where the Earth's crust is broken down and consumes part of the oceanic crust or oceanic plate, which underscores (subduction) under a continent or island arc (Fig. 3.8). Subduction zone is slope surface tilted in the direction of subduction, along which the main focus of earthquake takes place and along which remelting of rocks occur.

Plate movement, mobility of mountain ranges and oceanic ridges are recognized under the common names such as plate tectonics or global tectonics. Plate tectonics and global tectonics explain almost all the geological phenomena on large scale, particularly closely

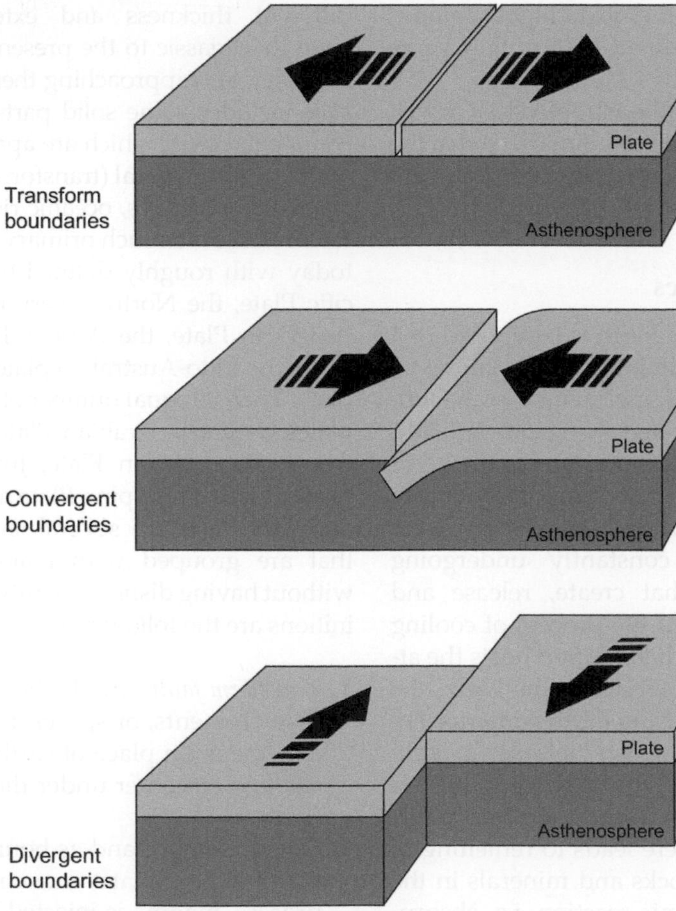

FIGURE 3.8 Illustrations of the three types of plate boundaries such as transform (top), convergent (middle) and divergent (bottom) associated with the relative process of plate tectonics.

associated with earthquakes, faults, volcanic areas, the origin of mountain chains (orogeny), oceanic arcs, oceanic ridges and the deep ocean trenches or furrows.

Orogenetic movements, i.e. orogeny or processes of forming mountain chains in the Earth's crust, are a direct consequence of plate tectonics and subduction. There are two basic types of orogeny: "collisional" and "noncollisional".

1. *Collisional orogeny* includes long-term underscore or subduction zone where it has consumed and melted much of the oceanic crust and oceanic plates. This leads to a mutual approach of two continents, or even up to their clash or the clash of the continent and the island arc. In this type of orogeny, clastic flysch sediments do not occur, and overthrust structures have only one direction. Subduction zone reaches deep into the lithosphere and along with it the magma inject from very deep source (from the asthenosphere), forming ophiolite igneous rocks. The ophiolite groups and assemblages are the community of basic, ultrabasic/ultramafic intrusives (gabbro, peridotite, and pyroxenite) and effusive igneous rocks that originate from oceanic crust (Sections 4.3.1.3, 4.3.1.4 and 4.3.2.3).
2. *Noncollisional orogeny* is associated with the converging trends by pinch of the plates and subduction. The orogeny takes place over part of the oceanic crust and the oceanic plate is subducted under the island arc, as, for example, the case in the present oceanic trenches in the western Pacific. The andesite volcanism is strong above the subduction zone and the basaltic volcanism is most active on the ocean bottom. Metamorphic zones exist on both sides of the subduction zone and the resulting overthrust structures are the consequence of pulling in both directions. The uplift causes deposition of clastic flysch sediments on both sides of the mountain areas.

Ophiolite is the thrust sheets of ancient oceanic crust and upper part of mantle rocks that has been uplifted and exposed above sea level and often emplaced on top of the continental lithosphere. Ophiolite is composed of green colored altered spilite (fine-grained oceanic basalt), pillow lava, serpentinites, gabbros and chert. It occurs close to the oceanic ridges, orogenic belt, within mountain belts of Alps and Himalayas, documenting the existence of former ocean basins that have now been absorbed by thrusts, subduction zone and plate tectonics.

The main volcanoes on Earth are located in areas of divergent and convergent plate movement, and the most significant places of formation of igneous (volcanic) rocks are now divergent plate margins, especially middle oceanic ridges, which makes annually over 20 km^3 of basalt rocks.

FURTHER READING

The Principles of Petrology-An Introduction to the Science of Rocks by G.W. Tyrrell[56] is worth reading for the beginners in this subject. Blatt et al.[2] will be a good reference for overall petrology. *The interior of the Earth* by Bott[5] is informative.

CHAPTER

4

Igneous Rocks

OUTLINE

4.1. Origin of Igneous Rocks 94
- 4.1.1. *Properties of Magma and Lava* 94
- 4.1.2. *Bowen's Reaction Series* 95
- 4.1.3. *Cooling of Magma after Crystallization* 96

4.2. Classification of Igneous Rocks 98
- 4.2.1. *Forms of Appearance and Structure of the Intrusive (Plutonic) Igneous Rocks* 100
  - 4.2.1.1. Forms of Intrusive (Plutonic) Igneous Rocks 100
  - 4.2.1.2. Textures of Intrusive (Plutonic) Igneous Rocks 101
  - 4.2.1.3. Shapes and Structures of Veins Igneous Rocks 102
- 4.2.2. *Forms of Appearance and Structure of the Extrusive (Volcanic) Igneous Rocks* 102
  - 4.2.2.1. Forms of Extrusive Igneous Rocks 102
  - 4.2.2.2. Textures of Extrusive Igneous Rocks 103

4.3. Main Group of Igneous Rocks and Their Composition 104
- 4.3.1. *Mineral Composition of Intrusive Igneous Rocks* 104
  - 4.3.1.1. Felsic Intrusive Igneous Rocks 105
  - 4.3.1.2. Intermediate Intrusive Igneous Rocks 108
  - 4.3.1.3. Mafic Intrusive Igneous Rocks 109
  - 4.3.1.4. Ultrabasic and Ultramafic Intrusive Igneous Rocks 113
- 4.3.2. *Mineral Composition of Extrusive Igneous Rocks* 116
  - 4.3.2.1. Felsic Extrusive Igneous Rocks 116
  - 4.3.2.2. Intermediate Extrusive Igneous Rocks 117
  - 4.3.2.3. Mafic Extrusive Igneous Rocks 118
- 4.3.3. *Veins Igneous Rocks* 120

Further Reading 120

Introduction to Mineralogy and Petrology
http://dx.doi.org/10.1016/B978-0-12-408133-8.00004-3

4.1. ORIGIN OF IGNEOUS ROCKS

The origin of the solar system and particularly the planet Earth including its internal structure (crust, mantle and core) is discussed in the previous chapter. Igneous rocks and its sedimentary and metamorphic complements constitute the entire Earth's crust. Therefore, systematic study of the igneous rocks, i.e. igneous petrology, is a fundamental necessity to understand the geological science.

Igneous rocks are the natural products of cooling, crystallization and solidification of extremely hot mobile molten material (magma) originated from the deepest parts of the Earth. This process of formation of igneous rocks is the earliest mechanism of rock formation and accountable for the growth and evolution of the present day solid Earth. The mode of formation can be either intrusive (plutonic) or extrusive (volcanic).

Intrusive igneous rocks are formed by cooling, crystallization and solidification of magma within the Earth's crust surrounded by preexisting country rocks. These rocks are generally medium to coarse grained. The rocks may be extremely coarse (pegmatite) and easily identifiable. The rocks are designated, according to the shape, size and relationship with the existing formation, as *abyssal* (deep seated), *hypabyssal* (near surface), *batholiths* (large felsic/intermediate massive plutonic), *stocks* (massive plutons), *laccoliths* (concordant plutonic sheets between sedimentary layers), *sills* (concordant tabular plutonic sheets within volcanic/sedimentary/metamorphic rocks) and *dykes* (plutonic sheets cut discordantly across existing rocks) (Fig. 4.5).

Extrusive igneous rocks are formed at the crust's surface as a result of the partial melting of rocks within the mantle and crust. The molten rocks, with or without suspended crystals and gas bubbles, erupt outside the crust due to lower density and spread as lava. Volcanic eruptions into air and ocean are termed as *subaerial* and *submarine*, respectively. The rocks cool and solidify very quickly and are fine grained in general. The mid-oceanic ridges (basalt) are example of submarine volcanic activity.

The igneous rocks include exceptionally large verities depending on the source, composition and types of parent magma, nature of emplacement, cooling, crystallization and finally solidification. The essential characteristic features of igneous rocks, in comparison to sedimentary and metamorphic counter parts, are the complete absence of fossils and distinctive internal texture and structure of the same. The first character is due to the amazing source material of extremely hot molten magma from deep inside the Earth. The second feature is the result of slow or fast cooling, crystallization and solidification of the magma.

4.1.1. Properties of Magma and Lava

Igneous rocks are formed in cooling, crystallization and solidification of minerals from magma inside the Earth or crystallization of lava ejected from volcanoes on Earth's surface or on the seafloor (Fig. 3.2).

Magma (Greek: magma means hot, molten mass) is the name for the molten mass in the Earth's interior that penetrates the lithosphere. If one visits inside the Earth's surface or on the seabed, he will come across a molten shiny mass, having temperature between 700 and 1200 °C, and the same is called the *lava* (Latin: lavare means flow).

Igneous rocks, those are formed by slow and gradual cooling and crystallization of minerals from magma inside the Earth, i.e. deeper below the surface, are called intrusive (*plutonic*) igneous rocks (from the Latin "intrudere" meaning to "break" or "Pluto" representing "God of the underworld"). The igneous rocks exist on the surface of the Earth today due to tectonic movements to rise near or at the very surface of the Earth or with strong erosion of the existing rocks that covered the surface.

Igneous rocks, those are formed by relatively rapid cooling, crystallization of lava from volcanoes on Earth's surface or on the seafloor, and are called effusive (*volcanic*) igneous rocks (from the Latin "effusio" means "discharge").

There are transitional types between these two main groups and are formed by cooling and crystallization of magma, lava and hot solution (*hydrothermal*), gases and vapors (*pneuma*) introduced into the cracks and cavities of rocks.

The rocks formed in accumulation and lithifaction of clasts and volcanic ash that originated from explosive volcanic eruptions is called *pyroclastics* or *pyroclastic rocks* (from Greek "pyros" meaning "fire" and "klastos" meaning "broken off"). The most of volcano-clastic fragments deposited after the transfer of pyroclastic flows, air and water, while in the sea, lakes or rivers mixes with nonvolcanic sediment material. The pyroclastic rocks resemble characteristics of clastic sedimentary rocks. These rocks are usually included into the study along with the clastic sedimentary rocks (Chapter 5.6).

Each of these major genetic groups of igneous rocks is characterized by a particular shape, appearance and characteristic structures and textures by which geologists, especially petrologists, can recognize and identify their mode of origin.

The chemical composition of magma and lava is very complex and the magma or lava from different places can be very different. However, the most important chemical elements in any magma and lava are oxygen, silicon, aluminum, iron, calcium, sodium, potassium, magnesium and titanium. The magma and lava also contain many other elements, as well as different amounts of water vapor mixed with easily volatile components, i.e. gases and vapors, such as hydrogen sulfide (HS), hydrogen fluoride (HF), hydrogen chloride (HCl), carbon dioxide (CO_2), sulfur dioxide (SO_2), hydrogen, nitrogen and sulfur.

The chemical composition of the magma or lava, and particularly gases and steam, with both cooling and crystallization play a significant role as it defines mineral communities together and crystallize to form different kinds of rocks, determine the viscosity of magma and facilitate its penetration through the rocks and cracks in the stone. The gases and vapors have a decisive role in the formation of ore deposits of magmatic type. The origin of many of these ore deposits relates to the last stage of crystallization of magma, i.e. pneumatolytic and hydrothermal stage.

The viscosity of magma depends primarily on its chemical composition and temperature. The felsic magmas are rich in silica and are more viscous than basic magma, which is poor in silica. This cause changes the speed of lava flow and form of occurrences of volcanic rocks. The basic lava with poor viscosity flows much faster than the acid magmas and spills in the form of volcanic plates. The acid lava with high viscosity has almost no flow, but resembles as thick malleable paste from volcanic craters.

4.1.2. Bowen's Reaction Series

Bowen's reaction series (Fig. 4.1) is the work of Norman L. Bowen, a researcher of petrologist at Geophysical Laboratory, Carnegie Institution of Washington. He explained through his revolutionized experimental petrology the understanding of discriminating mineral crystallization. He could illuminate the reason for certain types of minerals tend to be found together, while others never associate jointly. He crushed and grinded original igneous rocks along with mixtures of chemicals that could make up igneous rocks and experimented with their melting. He would heat the powered material at 1600 °C or more until it completely melt. The melt is cooled to a target temperature, for example 1400 °C. He would hold it at that temperature for long enough (minutes, hours or days) to allow crystal formation and there after quick cooling the material by throwing into a bucket of water. The resulting crystallized minerals that formed in the process of melting and cooling are examined. Nonmineralized left-over material would be glass. He further observed that there are two sequences of minerals: the discontinuous reaction series and the continuous reaction series.

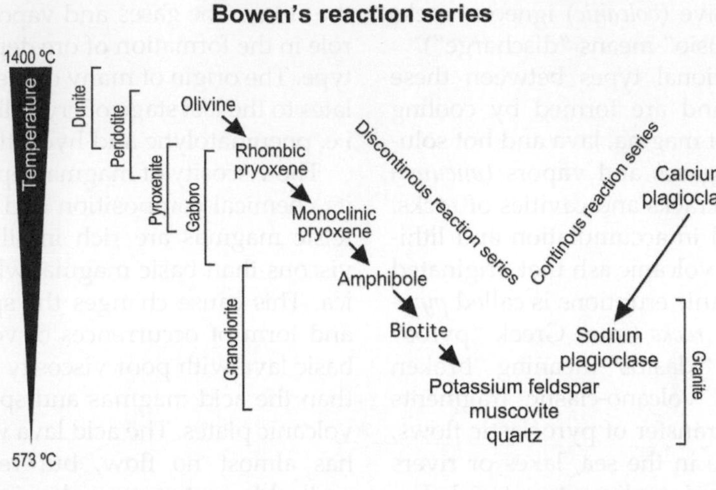

FIGURE 4.1 Bowen's reaction series depicting the sequence of crystallization of minerals in descending order of temperature due to cooling of the magma.

The *discontinuous reaction series* includes a group of mafic or iron–magnesium bearing minerals: olivine, pyroxene, amphibole, and biotite. These minerals react discontinuously to form the next mineral in the series. This means that in the igneous magmas, each mineral will change to the next mineral lower in the series as the temperature drops if there is enough silica in the melt. The silica content increases in mineral composition down the Bowen's reaction series.

The *continuous reaction series*, on the right side of the Bowen's reaction series, represents the enrichment of calcium → sodium → potassium in plagioclases feldspar with decreasing temperature. In the highest temperature, plagioclase has only calcium (Ca) and in the lowest temperature, only sodium (Na). In between, these ions mix in a continuous series from 100% Ca and 0% Na to 0% Ca and 100% Na at the lowest temperature.

4.1.3. Cooling of Magma after Crystallization

The rock mass is still relatively high in temperature after the crystallization of magma or lava (the foundation of intrusive and extrusive igneous rock). The rock mass and its mineral components, like most other substances in nature, reduce its volume and cracks (Figs 4.2 and 4.3) when cooled to ambient temperature. Such originally compact rock mass over time breaks and separates into pieces of various sizes and shapes such as plates, three-sided, four-sided, five-sided or six-sided prisms, cube, sphere, or completely irregular bodies. This joining of a cooling mass, both intrusive and extrusive, manifests remarkably distinguished "columnar structure" in igneous rocks. It is most commonly displayed in basalt (Fig. 4.4). This cracking and separation must strictly be distinguished from cracks and crushing of rocks caused by tectonic movements. This breaking of rocks is especially significant feature of igneous rocks. It is characteristic only for such rocks, and plays a decisive role in breaking and processing stone.

In equal intensity of cooling of large areas (which takes place faster in shallow than in the deeper parts of the igneous rock mass), rocks are separated in thinner or thicker plates due to differential shrinkage. Rock mass cracks for

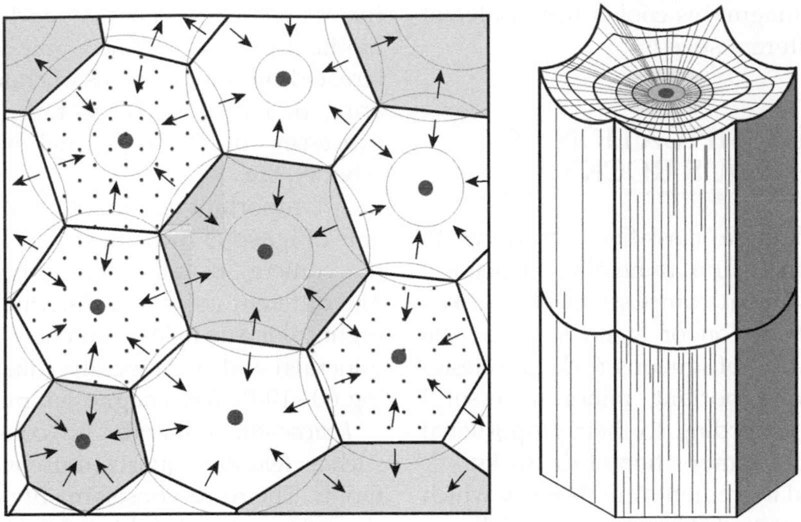

FIGURE 4.2 Conceptual development of six-sided prismatic cracks and joints (left) during cooling and crystallization of magma and over time breaks and separates into distinct hexagonal prism (right).

FIGURE 4.3 Plan view of the polygonal cracks and joints in basalt developed during cooling and crystallization of magma at Albert Hill, Mumbai, India.

FIGURE 4.4 Sectional view of the columnar structure (cracks and joints) in lava basalt flow developed during cooling and crystallization of magma, and separation over time at Albert Hill, Mumbai, India.

every regular flat surfaces that enclose the proper three-, four-, five- and six-sided prisms. As a result, the prismatic polygonal bodies appear with diameter of several centimeters to several decimeters, and the length of few decimeters to 10 m (Fig. 4.2).

The cracks can also be formed more or less regular cube or parallelepiped during slow cooling and slow crystallization of igneous magma. The rocks can be separated into spherical shapes in irregular cooling. It can be separated in sharply angular bodies of irregular shapes if the rock mass (magma) is cooled from different directions at different speeds.

4.2. CLASSIFICATION OF IGNEOUS ROCKS

Igneous rocks are classified according to mode of formation and mineralogy (chemical composition). Mineral composition plays a key role in the distribution of igneous rocks. The mineral ingredients of igneous rocks are designated as major, important, minor (accessory) and secondary, according to their proportional significance in the composition of the rocks.

Major mineral ingredients are those by which the rocks are classified. These are essential minerals for the rock, and which makes them different from others. For example, quartz, potash feldspar and biotite are essential components of granite and without any of those minerals, the rocks would no longer be designated as granite.

Important mineral constituents of a rock are those by which a special name is assigned to the rock, as for instance, the olivine in gabbro, nepheline in syenite, etc. Gabbro contains plagioclase and pyroxene, and with them may, but need not contain olivine. If gabbro contains olivine, then it is olivine gabbro and olivine is its important ingredient.

Minor (accessory) mineral ingredients are not important or relevant to the rock in which they are associated. The amount is small, typically <1% and it may but need not be the essential ingredients of the rocks. For example, zircon and rutile are minor minerals in granite.

Secondary minerals do not occur during the formation of the parent rock, rather later introduced or substituted during the weathering process or changes in the primary or original mineral constituents of the rock. The most common secondary minerals are kaolinite (created by the processes of change and chemical weathering of feldspar), chlorite (created by the processes of change and weathering of biotite, pyroxene and amphibole), sericite (created by the processes of change and weathering of feldspar), serpentine (created by the processes of hydrothermal modification of olivine), etc.

An important feature of the mineral constituents of igneous rocks is their color as a result of the content, usually isomorphic inserted iron. The different colors are described as leucocratic (light color due to low content of ferromagnesian minerals) and melanocratic (dark color containing 60–100% ferromagnesian minerals).

Leucocratic minerals are colorless or white, such as feldspar, quartz, muscovite and feldspathoids. The rocks consist mainly of them and are characterized by bright and light shades of gray, such as the granites and granodiorites.

Melanocratic or ferromagnesian minerals are green, dark green or completely black color due to greater or lesser amounts of isomorphic iron (especially Fe^{2+}) (Sections 2.5.8.1 and 2.5.8.4.1). This group of minerals includes olivine, pyroxenes, amphibole, biotite, etc. The rocks are mostly composed of ferromagnesian minerals showing dark green to black colors, such as gabbro, dunite, peridotite and pyroxenites (Table 4.1).

Chemical composition of rocks is determined and expressed with oxide content of main chemical elements, i.e. the content of SiO_2, FeO, Fe_2O_3, Al_2O_3, CaO, MgO, K_2O, Na_2O, MnO, P_2O_5 and TiO_2. The content of silicon dioxide (SiO_2) in the rock is one of the most significant chemical characteristics of igneous rocks. The amount of SiO_2, which varies from 35% to 80%, defines the rock as "acid". The terms "acid" and "basic" do not apply on the hydrogen ion concentration (pH—used by chemists), but only on the chemistry of rocks and the proportion of SiO_2 with respect to the total oxide content of the above-mentioned chemical elements.

"Acid" igneous rocks, i.e. the amount of silicon dioxide (SiO_2) in their chemical composition,

TABLE 4.1 Mineral Composition of Major Igneous Rocks

	Intrusive Rocks	Extrusive Rocks	Main Minerals
Felsic	Granite	Rhyolite	Quartz 20–40% K–feldspar > Na–plagioclase and mica
	Adamellite	Dellenite	Quartz, K–feldspar = Na–plagioclase
	Grandiorite	Dacite	Quartz 10–30%, Na–plagioclase + Na–Ca plagioclase > K–feldspar, biotite and hornblende
	Tonalite and quartz diorite	Dacite	Quartz, Na–plagioclase + Na–Ca–plagioclase > K–feldspar, biotite and hornflende
Intermediate	Monzite	Latite	Quartz, K–feldspar = Na–plagioclase biotite, hornblende and pyroxene
	Diorite	Andesite	Na–plagioclase and Na–Ca–plagioclase 60–80%, amphibole and pyroxene
	Syenite	Trachyte	K–feldspar 60–80%, Na–Ca plagioclase, hornblende, biotite, pyroxene and riebeckite
	Nephelene syenite	Phonolite	Nepheline, leucite, aegirine, K–feldspar, riebeckite, biotite, pyroxene and arfvedsonite
Mafic	Gabbro	Basalt diabase spilite	Ca–plagioclase (40–70%) pyroxene (augite, hypersthene), small quantities of hornblende and biotite, with or without olivine
	Norite	Basalt	Ca–plagioclase, pyroxene (hypersthene) with or without olivine
	Anorthosite		Ca–plagioclase (90–100%) with pyroxene, ilmenite, magnetite (0–10%) ± olivine
Ultramafic	Peridotite		Olivine, one or more pyroxene
	Dunite		Mostly Mg–olivine with little pyroxene
	Lherzolite		Olivine, bronchite, and dialage
	Serpentine		Serpentine derived from olivine
	Pyroxenite		Monoclinic pyroxene (augite, diopside, and dialage)

are a direct consequence of their mineral composition. It is higher as the rock contains more free quartz and/or more silicate minerals rich in silica. The igneous rock that contains more free quartz, more K-feldspar and Na-plagioclase; within the SiO_4-tetrahedra isomorphic, one Si^{4+} ion is replaced with Al^{3+} ion. Such rock has in its chemical composition higher content of SiO_2, and a higher degree of "acidity" unlike the rock that does not contain quartz or Na–Ca and Ca plagioclase.

The best example of "acidity" is minerals from isomorphic series of plagioclase: in pure acid plagioclase, i.e. albite ($NaAlSi_3O_8$), in each SiO_4 tetrahedron, only one Si^{3+} ion is replaced with an Al^{3+} ion and in pure basic plagioclase, i.e. anortite ($CaAl_2Si_2O_8$) in all the SiO_4 tetrahedra is replaced by two Si^{4+} ions with two Al^{3+}

ions. The chemical composition of albite contains 68.68% SiO_2, and anortite only 43.16% SiO_2.

The igneous rocks are divided into four types according to the content of SiO_2 in the chemical composition:

1. Acid igneous rocks in general contain >63% SiO_2. Acid igneous rocks, with K feldspar, also contain acid plagioclase and mineral quartz.
2. Neutral igneous rocks usually contain ~52–63% SiO_2. Neutral igneous rocks contain neutral plagioclase and do not contain quartz.
3. Basic or Mafic igneous rocks, by and large, contain 45–52% SiO_2. Basic igneous rocks contain basic plagioclase and ferromagnesian minerals (pyroxene, amphibole and olivine), which are poor in silica.
4. Ultrabasic or Ultramafic igneous rocks normally contain <45% SiO_2. Ultramafic igneous rocks do not contain plagioclase, but contain only ferromagnesian minerals, i.e. minerals rich in iron and magnesium, and low in silica.

4.2.1. Forms of Appearance and Structure of the Intrusive (Plutonic) Igneous Rocks

4.2.1.1. Forms of Intrusive (Plutonic) Igneous Rocks

It is established that most of the intrusive igneous rocks are formed by cooling and crystallization of magma at depths of 1.5–20 km.

Slow cooling of magma, deep in the lithosphere, under the surface of the Earth, created a huge body of igneous intrusive rocks of irregular shape, whose propagation is several thousand kilometers with an unknown base in depth. Such massive intrusive bodies are called *batholiths* (Fig. 4.5). There are often smaller or larger enclaves, *xenoliths* of surrounding rocks at the edges of batholith, which are incorporated in the magma and partially altered or completely metamorphosed under the influence of high-temperature fluids from the magma.

Stocks are smaller irregular bodies with 10 km in maximum dimension, and are associated with the batholiths.

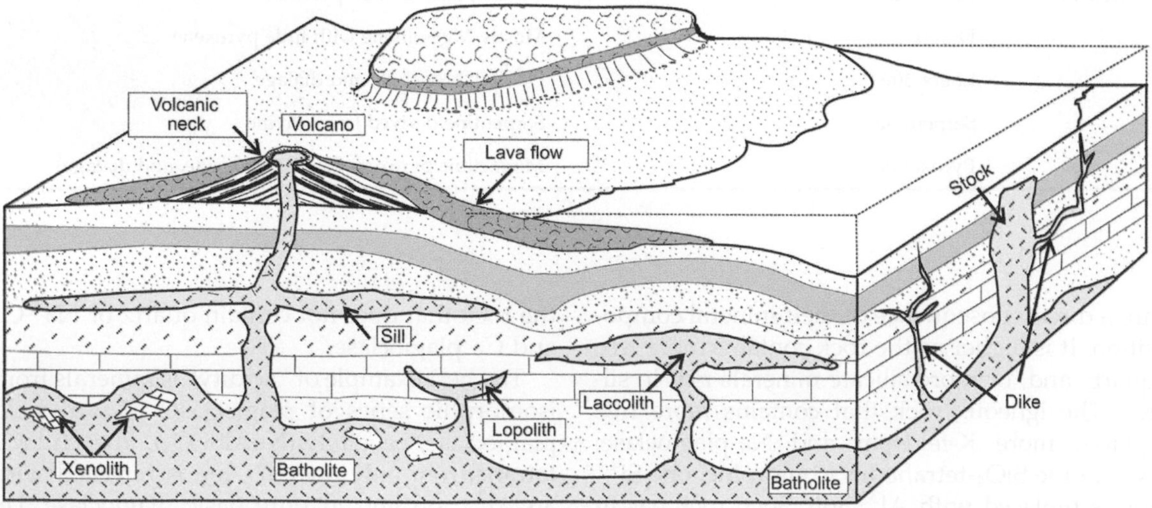

FIGURE 4.5 Conceptual diagram showing the major forms of igneous rocks such as batholiths, lopolith, xenolith, laccolith, sill, dyke, stock, volcano, volcanic neck and lava flow.

Round and irregular intrusive body of larger size is known as *massive*. The batholith, stock and massive, occurs by crystallization in the depths of lithosphere, can reach on the Earth's surface by variety of tectonic movements, erosion and denudation processes. The batholiths, stocks and massive of granodiorite, diorite, peridotite, gabbro and granite are often found on the Earth's surface or at shallow depth.

Magma, at movement and penetration through the lithosphere, can be injected into the surrounding sedimentary rock layers, and thereby raising the layers above it, so that creates a smaller igneous body. The newly created body has shape like dome or mushroom and is well known as *laccolith*. The length of laccolith usually does not exceed a few hundred meters to several kilometers, similar to the *lopolith*, which is smaller, lenticular in shape with a depressed central region (Fig. 4.5).

4.2.1.2. Textures of Intrusive (Plutonic) Igneous Rocks

The intrusive (plutonic) and extrusive (volcanic) igneous rocks have mutually and clearly different structure. The texture of the igneous rocks involves the size, relationship, arrangement and shape of certain mineral constituents of rock, and it depends on the speed and degree of crystallization of magma, lava, pneuma (gases and vapors) and the hot solution.

A large intrusive body (batholite) takes many hundreds of thousands or even millions of years due to slow cooling and slow crystallization. As a consequence, the large intrusive rocks are characterized by a high degree of crystallinity. Most mineral ingredients in slow cooling of magma and complete crystallization take the form of minor or major crystalline grains (Fig. 4.6). This is unlike to amorphous mass often formed under rapid cooling of lava at the surface.

The intrusive rocks contain smaller or larger mineral grains, i.e. the mineral ingredients are all fully crystallized. Therefore, the intrusive rocks are principally holocrystalline. The rocks show a granular texture (Fig. 4.6). The characteristic texture of the individual grains, all or most mineral constituent, has equal size of crystalline grains (Fig. 4.6(A)). The grain size and degree of crystallinity of mineral grains are a direct consequence of the cooling rate, size, viscosity and chemistry of magma and magmatic body.

The rock can be designated, according to the size of crystals and grains as the following:

1. Macrocrystalline (crystals visible to the naked eye)

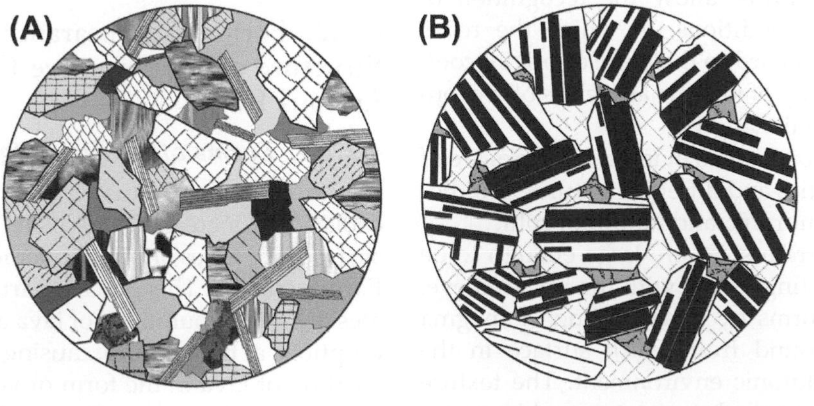

FIGURE 4.6 Typical textures of intrusive igneous rocks: (A) granite with hypidiomorphic (greater proportion of subhedral crystal forms) and characteristic granular texture, and (B) gabbro texture showing large plagioclase embedded in fine matrix of ferromagnesian minerals.

2. Microcrystalline (crystals visible to the microscope)
3. Cryptocrystalline (crystals visible only in large microscopic increments).

Given the form of minerals, crystal in rocks can develop in its following forms:

1. Ideal, i.e. idiomorphic or euhedral crystalline forms.
2. Only partially proper, i.e. hipidiomorphic or subhedral forms.
3. Completely improper, i.e. alotriomorphic or xenomorphic forms.

A special type of macrocrystalline grain is *porphyritic* texture that is characterized by extremely coarse-grained K-feldspar (phenocrysts) in relation to other macrocrystalline ingredients (Figs 4.18, 4.25 and 4.40).

The degree of crystallinity is higher as the cooling of magma is slower, and forms of crystals indicate the environment in which the crystals are formed.

Idiomorphic crystals form under conditions of slow cooling of slightly viscous magma with enough space for the growth of each mineral.

Alotriomorphic crystals form when their growth has lack of space, because at the same time in a small space crystallizes a number of mineral ingredients.

All these features allow the recognition of igneous rocks conditions, in which the rocks occurred with regard to the place of origin, cooling rate, viscosity of magma as well as other conditions of crystallization.

The shapes of crystals in intrusive rocks are of different varieties and different textures: hipidiomorphic, idiomorphic and alotriomorphic.

Phaneritic grain sizes are large enough to be visible and distinguished with the unaided eye. This texture forms by slow cooling of magma deep underground from Earth surface in the intrusive or plutonic environment. The texture may also be applied to metamorphic rocks. Examples of phaneritic igneous rocks are diorite, gabbro and granite.

Aphanite or aphanitic grain sizes are essentially so fine that their component mineral crystals are not detectable by the unaided eye. This texture results from rapid cooling of igneous volcanic or extrusive surface and shallow-surface environment. Aphanites are often porphyritic having large crystals embedded in fine-grained groundmass such as andesite, basalt, dacite, and rhyolite.

4.2.1.3. Shapes and Structures of Veins Igneous Rocks

At the end of magmatic crystallization, i.e. pegmatite and pneumatolytic stage of crystallization, often magma penetrates into cracks in surrounding rocks of stony crust and crystallize in the form of thin plates (dykes or sills) (Fig. 4.5). These are igneous rocks known as the veins rocks (Table 4.1). If the magma in pegmatite or pneumatolytic phase of crystallization is pushed parallel in-between layers, it forms igneous body with the shape of saucer. It is known by the name sill or concordant intrusive sheet (Fig. 4.5).

An important textural feature of the veins rock (aplite, pegmatites and lamprophyre, Table 4.1) is holocrystalline and microcrystalline in aplite and lamprophyre, macrocrystalline in pegmatites and often with some giant crystal of diameter up to several meters (Section 4.3.3).

4.2.2. Forms of Appearance and Structure of the Extrusive (Volcanic) Igneous Rocks

4.2.2.1. Forms of Extrusive Igneous Rocks

The magma gradually, but relatively fast, cools with increasing viscosity due to loss of steam and gases in its movement toward the Earth's surface. This process particularly accelerates in sudden outbursts of lava and in explosive eruptions at the volcano, causing a sudden solidification of lava in the form of volcanic glass.

Extrusive igneous rocks usually occur in the form of cup, plate and basin of volcanic lava (Fig. 4.5).

Volcanic necks are product of an old volcanic rim composed of several layers (outflow) of solidified lava. This occurs when the acidic, highly viscous and therefore poorly mobile lava solidifies around volcanic crater forming a conical hill or dome. Volcanic plate or lava cover presents body of large propagation and relatively small thickness, formed in spout or outbursts of low viscous voluble lava on a large area around the volcano. Basin (flow) of lava is formed by cooling lava flows like a fiery river poured down the slopes of the volcano (Fig. 4.5).

Volcanic rock also forms by outpourings of lava on the seabed typically within the volcanic mass. The pulsating pouring of lava and mixing with seawater create spherical or cushion shape-structure and known as "pillow lavas".

4.2.2.2. Textures of Extrusive Igneous Rocks

The effusive lava cools rapidly after the eruption on the surface or ocean floor. The initial high-temperature crystallization of some minerals includes olivine, pyroxene, Ca/Na plagioclase, sanidine, leucite, nepheline and cristobalite. These minerals in volcanic rocks are represented by properly developed crystals of phenocrysts embedded in vitreous, microcrystalline groundmass resulting from a sudden solidification of rest of the magma after eruption. This is the basic texture feature of the porphyritic igneous rocks (Fig. 4.7).

The characteristic textural feature of the intrusive rocks is essentially holocrystalline which means that all the mineral ingredients are crystallized with the most equal size crystals. In contrast, the extrusive rocks are characterized by two generations of mineral: initial crystallization of phenocrysts and later fine-grained or glassy matrix. Both the groups of minerals differ in size.

Porphyritic texture is characterized by single large crystals or phenocrysts and fine-grained matrix, which is partly glassy containing tiny crystals (Fig. 4.7).

Glassy or vitreous texture is created by sudden cooling and solidification of lava on Earth's surface in the form of amorphous volcanic glass with or without a few tiny crystals or crystallites of different embryos, sometimes dendritic forms. The examples of glassy or vitreous volcanic rocks are obsidian and pumice.

Special types of textures are diabase or ophite and intersertal (similar to intergranular) texture. Diabase or ophite texture is typical of the basic veins rocks. Diabase and ophite (altered diabase)

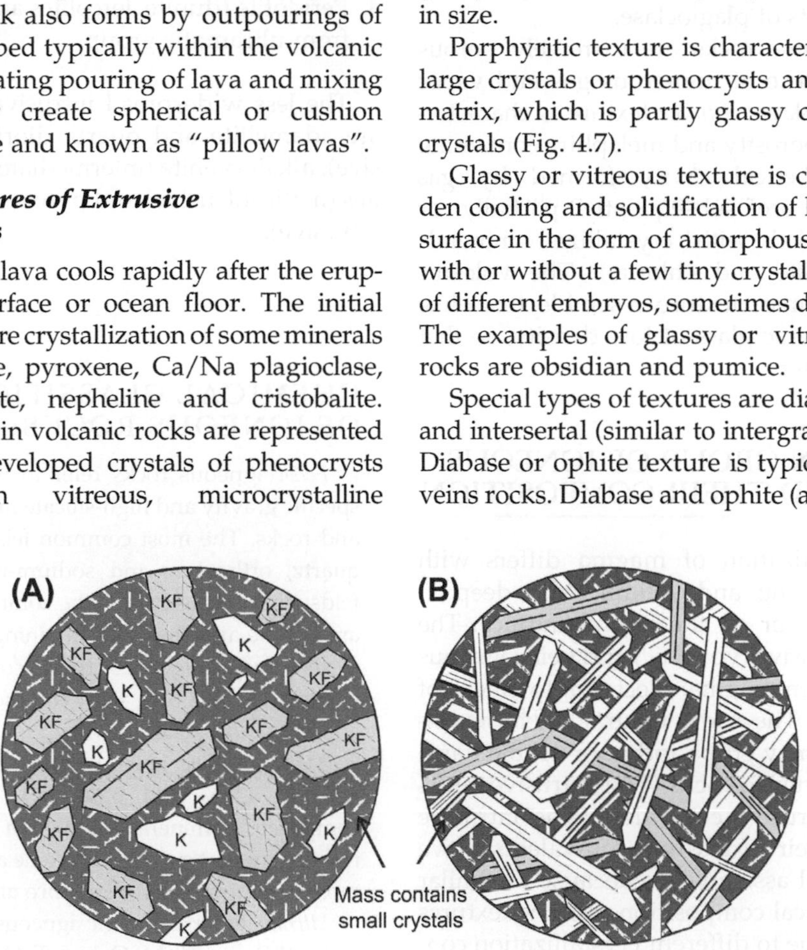

FIGURE 4.7 Typical textures of extrusive igneous rocks: (A) porphyritic texture distinguished by single large crystals (phenocrysts) embedded in partly glassy fine-grained matrix as in granite, (B) intersertal texture having stick-like plagioclase floating in fine-grained matrix as in spilite (oceanic basalt).

are characterized by unoriented stick-like plagioclase in their interstices irregular grains of augite or diopside.

Intersertal texture is most common in spilite that instead of augite or diopside in interstices of stick-like plagioclase containing glassy primary mass (Fig. 4.7(B)). The rocks which have diabase and intersertal texture are especially tough, with high compressive strength and high resistance to impact and abrasion. The texture becomes hyalopilitic if the basic glassy mass distinctly prevails over the tiny needle-like phenocrysts of plagioclase.

Many volcanic rocks are extremely porous with the presence of numerous gas and vapor bubbles in the lava. *Pumice* texture is characterized by great porosity and melaphire containing numerous spherical cavity formed by gas bubbles, that are subsequently partially or completely filled with crystallized minerals (calcite, prehnite and chlorite). The volcanic rocks with pumice texture resemble foam-like silica-rich volcanic glass of low density, so that it floats on water.

4.3. MAIN GROUP OF IGNEOUS ROCKS AND THEIR COMPOSITION

The crystallization of magma differs with respect to cooling and solidification deep in the lithosphere or on the earth's surface. The intrusive, extrusive (volcanic) and veins igneous rocks form from magma injected into cracks of rocks (Fig. 4.5). The acidic, neutral and basic volcanic magma may erupt and pour on the Earth surface as extrusive rocks. The intrusive rocks have their extrusive equivalent. The intrusive rocks and their extrusive equivalents have similar mineral assemblages because of similar primary chemical composition. But the textures vary widely due to different crystallization condition. Ultramafic magmas that are characteristic of the deep sea have their extrusive equivalent.

4.3.1. Mineral Composition of Intrusive Igneous Rocks

The most common and best-known intrusive igneous rocks (Table 4.1) are the following:

1. Granites and granodiorites from the felsic group.
2. Diorite and syenite from the intermediate group.
3. Gabbro, norite and anorthosite from the mafic group.
4. Peridotite (dunite, lercolites and serpentinites) from ultramafic group.

The less widespread intrusive igneous rocks are adamellite and quartz diorite (felsic intrusive), alkali syenite (intermediate intrusive), norites (mafic intrusive) and pyroxenites (ultramafic intrusive).

CHEMICAL CLASSIFICATION OF IGNEOUS ROCKS:

Felsic igneous rocks refer to light-color, low-specific gravity and high-silicate minerals, magma and rocks. The most common felsic minerals are quartz, orthoclase and sodium-rich plagioclase feldspar and muscovite. The common felsic rocks are granite and rhyolite containing +63% SiO_2.

Intermediate igneous rocks contain SiO_2 between 52% and 63% with common examples of andesite and dacite.

Mafic or basic igneous rocks have low silica between (45% and 52% SiO_2) and typically composed of minerals with high iron and magnesium content such as pyroxene and olivine. The most common rocks are gabbro and basalt.

Ultramafic or ultrabasic igneous rocks contain <45% SiO_2, >18% MgO, high FeO, low potassium and generally +90% mafic minerals. The common rocks are dunite, peridotite and pyroxenite.

4.3.1.1. *Felsic Intrusive Igneous Rocks*

Granites and *granitoids family* viz. alkali-feldspar granites, granites, quartz-monzonite, diorite/quartz-diorite/granodiorite, syenite and tonalities, are the most abundant rocks that constitute the upper crust of the continental areas. Granites are generally formed as "batholiths" at great depth and when exposed by erosion or other tectonic activity, these rocks occupy huge areas of the Earth's surface, often as series of domes (Fig. 4.8) and valleys. The central cores of major mountain ranges consist of intrusive igneous rocks, usually granites.

Granite landform changes slowly by physical and chemical weathering. A typical style of weathering produces smoothly curved irregular to rounded shapes of boulders. These boulders of granite often sit on smooth bare rock surfaces giving a mystic landscape (Fig. 4.9).

Granites have many contrast colors ranging between white, gray, black, and pink to red. Granites are intrusive felsic rocks (from the Latin "granum" means "grain") usually hypidiomorphic (greater proportion of subhedral crystal forms) and typically granular texture (Fig. 4.6(A)). The grain size varies between fine (Figs 4.10 and 4.11), medium and coarse. The rare and coarse porphyritic (large crystals or phenocryst floating in a fine-grained groundmass) texture is presented by extremely large crystals of K-feldspar, compared to other minerals. Granite that contains large K-feldspar grains of spherical shape, pink or reddish color is known as *porphyritic granite* and rapakivi (large rounded crystals of orthoclase/oligoclase feldspar) granite (Figs 4.12 and 4.13). The granites are usually fine to coarse grained, but occasionally as large lens shape enclaves (phenocryst) of mineral grains or mineral aggregates or older rocks embedded in fine-grained granitic groundmass (Fig. 4.14). These phenocrysts are partially ganitized with mineral aggregates of feldspar, quartz, biotite and amphiboles.

The most essential mineral constituents of granite are 20−40% quartz, 50−80% K-feldspar (orthoclase and/or microcline and pertite), Na-plagioclase and micas, mainly biotite and rare muscovite (Table 4.1). Pertite is an intergrowth

FIGURE 4.8 View of granite monolith "Half-Dome", 2693 m elevation, from Glassier Point, Yosemite National Park, East California and is a part of Sierra Nevada Mountain Range. The impression from the valley floor implies that this is a round dome which has lost its northwest half in an illusion. *Source: Soumi.*

FIGURE 4.9 Smooth irregular granite boulders (top), product of typical weathering, present a scenic landscape near ancient city of Aswan, southeastern Egypt rising through the blue water of River Nile.

FIGURE 4.10 Fine-grained granite composed of quartz, potassium/plagioclase feldspar and biotite ± amphibole from Aswan granite quarry, south-central Egypt.

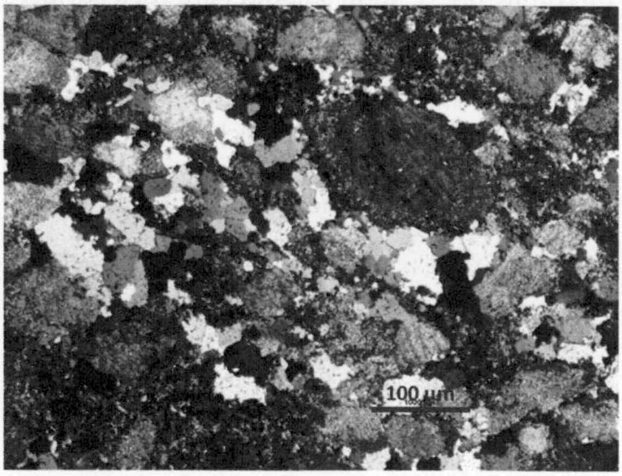

FIGURE 4.11 Photomicrograph of thin section of subhedral fine-grained aplitic variety of granite composed of feldspar (white), quartz (sky blue) and ilmenite (opaque). *Source: Prof. Arijit Ray.*

of albite or oligoclase with a microcline/orthoclase host. The granite may even contain small amounts of hornblende and augite with the main ingredients and as accessory ingredients of apatite, ilmenite, hematite, rutile, zircon, and tourmaline. Granites contain little ferromagnesian minerals (biotite, hornblende, and augite). Granites are largely leucocratic rocks, usually pale gray or pink, depending on the color of feldspars, for example, presence of pink microcline granite looks pink. It occurs mostly in the form of huge batholiths, stock, but rarely laccolith (Fig. 4.5).

Feldspar dominates the granite composition and is easily recognized by its appearance, color and cleavages. Quartz is typically anhedral

FIGURE 4.12 Coarse-grained porphyritic granite composed of quartz, potassium/plagioclase feldspar and biotite ± amphibole from Aswan granite quarry, south-central Egypt. The coarse feldspar grains are in the initial stage of linear alignment as in gneissic texture.

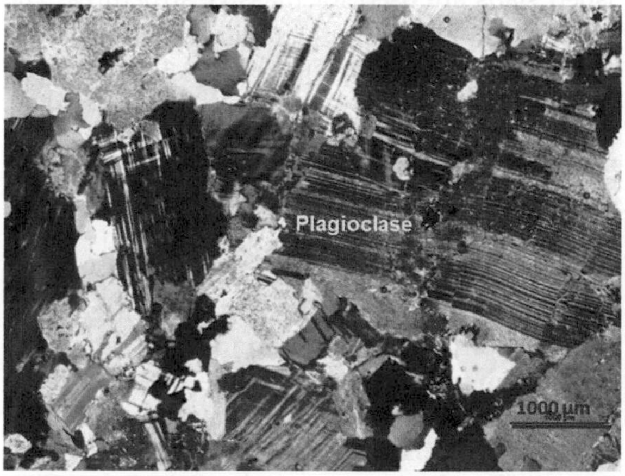

FIGURE 4.13 Photomicrograph of thin section of porphyritic granite showing deformed plagioclase lamellae in uncontaminated granite. *Source: Prof. Arijit Ray.*

(ill-formed crystals), and occurs as filling the interstices between the other minerals masking its own characteristic crystal shape. Quartz is generally colorless to smoky and identified by its glassiness, hardness, lack of cleavage, and conchoidal fracture. Biotite, muscovite and

FIGURE 4.14 Large lensoidal enclave (phenocryst) of older rocks embedded in fine-grained granitic groundmass. The phenocryst, in turn, is partially ganitized with mineral aggregates of potassium feldspar, quartz, biotite and amphiboles, Aswan quarry, south-central Egypt.

hornblende are distinguished by flaky black and silver color, and black/dark-green grains/prisms characteristics.

The builders and architects are often wrongly designates all the grain intrusive igneous rocks, especially the granodiorite, diorite and gabbro by the name "granite". The granite, granodiorite, diorite and gabbro are petrologically defined fundamentally with mineral composition as shown in Table 4.1. The common uses of granite are as building and decorative stones, tiles, kitchen counter, ancient and modern sculptures (Fig. 4.15), engineering, curling and rock climbing.

Adamellite (named after the town Adamello in Tyrol) is a felsic intrusive igneous rock with hipidiomorphic texture similar to *quartz monzonite* (monzonite with some quartz). It is medium- to coarse-grained rock with color varying typically between white, gray, pink, brown and bronze. Adamellite is composed of approximately equal proportion of orthoclase and plagioclase feldspars, significant amount of quartz, biotite and/or amphibole. It differs from granite and granodiorite that contain equal amount of

FIGURE 4.15 The Pillar of Pompey is monolithic red granite (from Aswan) column of 26.85 m high built in 297 AD commemorating the victory of Roman Emperor Diocletian over an Alexandrian revolt, Egypt.

quartz, K–feldspar, Na–plagioclase and Na–Ca–plagioclase. Adamellite rock seldom host for gold and silver deposits, primarily used as building stone for monuments (The Mormon temple, Salt Lake City, Utah), and aid in mountaineering.

Granodiorite differs to granite by containing less quartz (10–30%), more Na–plagioclase and K–feldspar. Na–plagioclase is approximately twice the K-feldspar in content. It also contains ferromagnesian minerals (biotite, hornblende, and augite). Granodiorite is usually of light gray color having large phaneritic crystal due to slow cooling. The rock is most often used as crushed stone for road building and occasionally as ornamental stone.

Tonalite (named by pass Tonale, Adamello massif in Tirol) is a felsic igneous plutonic rock with phaneritic texture and special variety of granodiorite in turn to diorite. It is composed of quartz, biotite and plagioclase (andesine or oligoclase). The share of K–feldspar, hornblende and pyroxenes is very small and as accessory minerals. Granodiorites and tonalities are found independently or together with the granite in the batholiths and stock.

Felsic intrusive rocks of granite, adamellite, granodiorites and tonalities are widely used in construction, as the crushed rock is particularly of high quality. The fine grain varieties and polished slabs are used for massive structures. The porphyritic granites and rapakivi granites that contain large pink microcline are specially appreciated.

4.3.1.2. Intermediate Intrusive Igneous Rocks

Monzonite is an intermediate igneous intrusive rock composed of approximately equal amounts of K–feldspars and Na–plagioclase with minor amount of quartz (<5%) and ferromagnesian minerals (hornblende, biotite and pyroxene). The rock seldom hosts gold and silver deposits, and uses as building stone for monuments (The Mormon temple, Salt Lake City, Utah), and aids in mountaineering.

Diorite is an intermediate intrusive igneous phanerites (large grain size) presenting hypidiomorphic (granular) to allotriomorphic (very large crystallographically continuous crystals) texture with about 60–80% of the Na–plagioclase, oligoclase and Na–Ca–plagioclase and little or no quartz. It contains much more ferromagnesian minerals than granite and granodiorite. The ferromagnesian minerals

FIGURE 4.16 Gray syenite contains predominantly of alkaline feldspar (60–80%) and 20–40% hornblende and biotite (black) with no or only negligible amount of quartz.

from the group of amphiboles (hornblende) and biotite, and usually pyroxenes, present a gray to dark gray color to diorite with bluish, greenish and brownish tinge. Diorite represents a transitional type between the granodiorite intrusive rocks and gabbro according to the mineral composition (Table 4.1). The different varieties of diorite are determined by color, coarseness of grain, and mineral composition. Diorite is usually located on the edges of granite batholith or in the form of smaller massif. The common uses are as aggregate, fill in construction and road industries, cut and polished dimension stone for building facings and foyers, statue and vase made during ancient Inca, Mayan and Egyptian civilization.

Syenite is a coarse-grained intermediate intrusive igneous rock with pandiomorphic (euhedral crystals of same size) and hypidiomorphic (subhedral crystals of equal size) texture. The color varies between white and gray or reddish. The rock contains predominantly of K–feldspar (60–80%, white, red or pink orthoclase) and 20–40% hornblende, biotite and pyroxene (Fig. 4.16). The mineral composition and the texture resemble granite with only difference that it does not contain quartz or a negligible quantity. It contains much more ferromagnesian minerals biotite and hornblende, and rare pyroxene. The specific gravity varies with the constituent minerals and their proportion ranging between 2.6 and 2.8. Syenites are rare rocks that appear in the small forms, usually on the edges of the granite massif.

Alkali syenites are rich in alkaline minerals, consisting of feldspathoids (nepheline and leucite [$K(AlSi_2O_6)$]), alkali amphibole (riebeckite and arfvedsonite), alkali pyroxene (aegirine and aegirine–augite) and K–feldspars. It does not contain quartz, and poor/no plagioclase. The name alkali or nepheline syenite is assigned because of its chemical composition containing substantial amounts of alkali oxides Na_2O and K_2O. Alkali syenite or nepheline and leucite syenites are more common in nature than the normal syenite. Syenite and alkali syenite crystallized from alkali-rich and silica-poor magma and are therefore not presented in Bowen's series of crystallization. Syenites possess better fire-resistant qualities and are suitable for dimension stone for building facings, foyers and aggregate in road industries.

4.3.1.3. *Mafic Intrusive Igneous Rocks*

Gabbro is mafic, intrusive, coarse-grained rock with allotriomorphic texture. Gabbros contain mainly ferromagnesian minerals and plagioclase, the amount of ferromagnesian minerals equaling or exceeding that of the plagioclase. Gabbros are plutonic rocks formed by cooling and crystallization of molten magma trapped under the Earth's surface and chemically equivalent to extrusive basalt. The ferromagnesian minerals are pyroxene (diopside or diallage, augite and hypersthene), hornblende, and olivine, occurring either together or singly. The pyroxene in gabbros is mostly clinopyroxene (diopside and augite) with or without small amounts of orthopyroxene (hypersthenes). The feldspar in gabbros is chiefly calcic plagioclase, generally 50–60% labradorite [$(Ca, Na)(Al, Si)_4O_8$], and also plagioclase composition of

FIGURE 4.17 Gabbro is dark gray to greenish black color plutonic rock and chemically equivalent to volcanic basalt. The rock mainly contains Ca–plagioclase and ferromagnesian minerals such as pyroxene (augite and hypersthene or diallage) ± olivine. *Source: Prof. Arijit Ray.*

FIGURE 4.18 Photomicrograph showing large deformed plagioclase phenocryst embedded in finer matrix of ferromagnesian minerals in gabbro. *Source: Prof. Arijit Ray.*

bytownite to anortite. Gabbros are mostly dark colored, ranging between dark gray and greenish black because of the high proportion of ferromagnesian minerals (Fig. 4.17). The rocks are at the turn of diorite in peridotite and pyroxenes according to the mineral and chemical composition (Fig. 4.1). Gabbro with olivine is called olivine gabbro. If olivine gabbro does not contain pyroxene and is primarily composed of calcium plagioclase and olivine, it is known as *troctolite*. The type of gabbro that contains predominantly orthorhombic pyroxene (hypersthene) and the Ca–plagioclase is known as *norite*. It may also contain some olivine, and then it is olivine norite.

Plagioclases can undergo processes of saussuritization with the interaction of hot solutions extensively change into a dense, compact mixture of zoisite, epidote, albite, quartz, muscovite and actinolite to form "sosirite". In similar conditions, pyroxenes in uralitization processes can be modified in dense clusters of actinolite minerals known as "uralite".

Gabbro texture (Figs 4.6(B) and 4.18) is formed by simultaneous long crystallization of bright (leucocratic) and dark ferromagnesian mineral ingredients, so deeply related to each other, that gabbros are extremely solid and tough rocks. Gabbros appear as a densely homogeneous rock often fairly the same texture and composition throughout the rock mass. Gabbros are greatly valued in the construction industries as dimension stone, especially for sculptures and tombstones (black granite) because of its high strength, toughness, dark color and excellent polishing capabilities.

Dolerite and *Diabase* are both mafic igneous rocks having same mineralogical composition, but differ in formation. The colors are frequently dark gray, black and green. Dolerite is medium-grained (Fig. 4.19) intrusive equivalent of volcanic basalt or plutonic gabbro, and usually occurs as dykes (Fig. 4.20), sill and plugs. Dolerite is heavy with specific gravity ranging between 2.9 and 3.3. Dolerite dykes are often exposed to the surface and exhibit as walls in straight line. Diabase is a subvolcanic rock equivalent to volcanic basalt or plutonic gabbro. Diabase is subsurface volcanic rock formed by injecting

FIGURE 4.19 Dolerite is a medium-grained mafic intrusive igneous rock composed primarily of plagioclase set in a finer matrix of clinopyroxene $\pm$ olivine, magnetite and ilmenite. *Source: Prof. Arijit Ray.*

FIGURE 4.20 Massive fine-grained dolerite (dark gray color in the center) intruded in dolostone (buff color on either side) as dyke cutting at steep angle. Photograph has been taken from the underground zinc lead silver mine at Zawar Group, India.

gabbroic magma or lava as shallow dykes and sills under the surface of the Earth. Diabase is typically fine grained having chilled margin. The main ingredients of dolerite and diabase are mafic lath-shaped plagioclase of about 60% (labradorite and rarely bytownite) set in a finer matrix of clinopyroxene (typically 20–30% augite) and $\pm$olivine (up to 10% in olivine

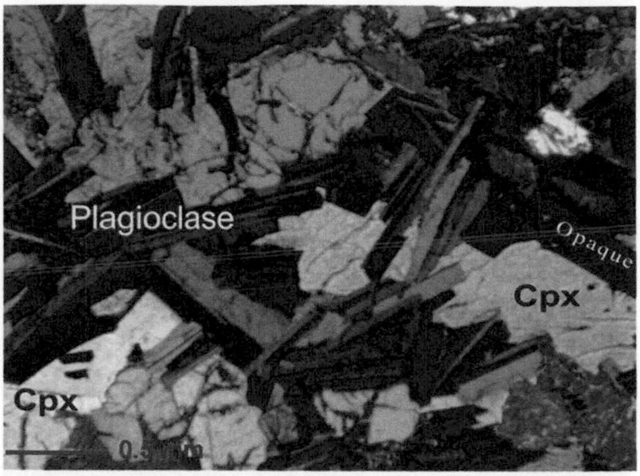

FIGURE 4.21 Photomicrograph of thin section showing intersertal and intergranular texture of lath-shaped plagioclase of (labradorite, rarely bytownite) set in a finer matrix of clinopyroxene in dolerite dyke. *Source: Prof. Arijit Ray.*

diabase), magnetite and ilmenite (Fig. 4.21). The accessory minerals are chlorine, uralite and calcite. The rocks usually display intersertal and intergranular texture. The coarse-grained diabase with pyroxene specifically alter to uralite (uralite diabase) and plagioclase from the labradorite and oligoclase type, known as "ophite," which are characterized by a special structure known as "ophite structure". Dolerite and diabase rocks are used as crushed stone in road making, concrete mixture in rough masonry, and block paving and ornamental stone in monumental purposes.

Norite is a mafic intrusive igneous rock with color ranging between light to dark gray and brown. The rock is indistinguishable from gabbro, other than type of pyroxene under microscope. The rock is composed of Ca-rich plagioclasc (labradorite), Mg-rich orthopyroxene/hypersthene (enstatite) and olivine. The rock occurs in close association of mafic gabbro and ultramafic layered intrusion igneous complex, e.g. Bushveld (South Africa) and Stillwater (Montana, USA) with large platinum group of deposits, and layered igneous complex with large deposits of chromite at Sukinda and

FIGURE 4.22 Field photograph of coarse-grained light-color Norite from footwall of open-pit chromite mine at Boula-Nausahi layered igneous complex, Orissa, India.

FIGURE 4.23 Photograph of drill core (Norite) cutting across the mafic and ultramafic layered igneous intrusive complex, being explored for chromium and platinum group of minerals at Boula-Nausahi, Orissa, India.

Nausahi (Figs 4.22 and 4.23), India The common usages are ornamental facing, paving, graveyard headstone at funerary rites and kitchen countertops.

Anorthosite is usually a coarse-grained intrusive igneous rock with color varies between white, yellowish to brown, shades of gray, blush and smoky pigment. The rock is characterized by the predominance of plagioclase feldspar (90–100%), generally labradorite, and remaining mafic components of pyroxene, magnetite, ilmenite (0–10%) ± olivine (Fig. 4.24). The fine-grained, nearly monomineral composition and light color anorthosite resembles both marble and quartzite in hand specimen. If the quantity of pyroxene increases in anorthosite, the rock

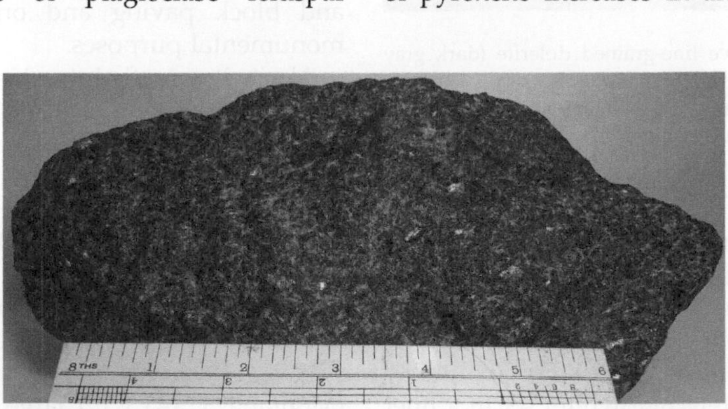

FIGURE 4.24 Anorthosites is typically coarse-grained dark-color rock and composed primarily of plagioclase (labradorite) with minor amount of pyroxene, magnetite and ilmenite ± olivine. *Source: Prof. Arijit Ray.*

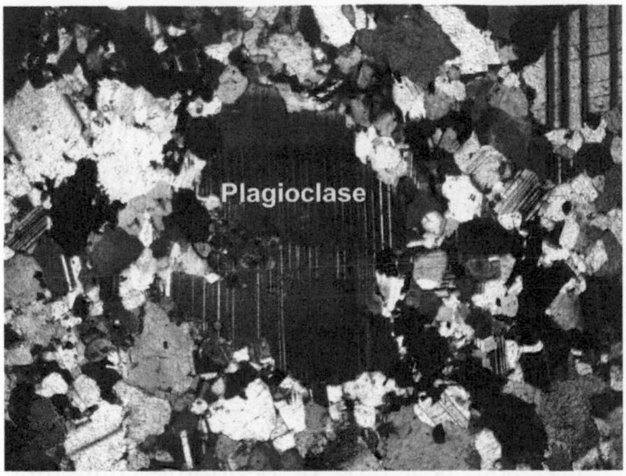

FIGURE 4.25 Photomicrograph of thin section showing large phenocryst of deformed plagioclase (labrodorite) in fine-grained pyroxene and plagioclase-rich groundmass in massif Anorthosite. *Source: Prof Arijit Ray.*

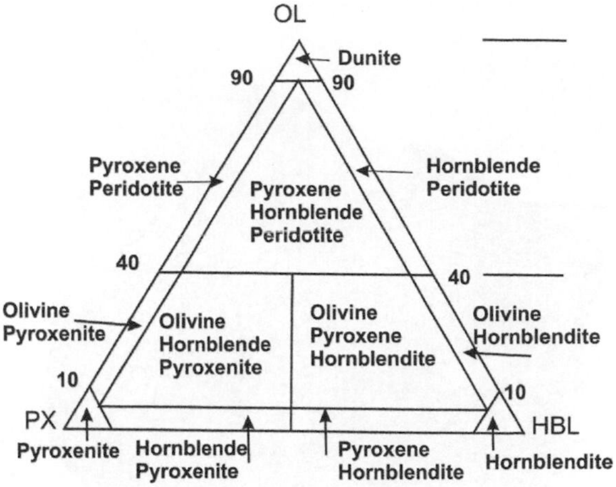

FIGURE 4.26 Modal plot for ultramafic members indicating classification and nomenclature of hornblende bearing peridotites and pyroxinites. OL, olivine; PX, pyroxene; HBL, hornblende. *Source: Modified after Bose[4].*

grades into gabbro and vice-a-versa. The rock can be identified with certainty under microscope with predominance of feldspar (labradorite) and texture (Fig. 4.25). It can be distinguished by hardness and presence of well-developed cleavages from quartzite and marble, respectively. The principal modes of occurrence of anorthosite are either (1) as large independent intrusive mass or (2) as layers with variable thickness, as members of banded or layered gabbro lopoliths. Anorthosite can be a source for hosting titanium, aluminum, gemstones, building material and scientific research of similar composition of Moon, Mars, Venus and meteorites.

4.3.1.4. Ultrabasic and Ultramafic Intrusive Igneous Rocks

Ultrabasic or Ultramafic igneous rocks contain <45% SiO_2, >18% MgO, high FeO, and low potassium. The group of rocks, generally dark colored, with high (+90%) magnesium and iron bearing mafic minerals. The mode of occurrences of ultramafic rocks are commonly intrusive (dunite, peridotite and pyroxenite) and rarely as extrusive. The rocks occur as large layered intrusive complex hosting chromium, nickel, platinum and palladium ± massive sulfides. The categorization of ultramafic group of rocks can be explained with "classification diagram based on modal percentages of minerals like olivine, pyroxene and hornblende" (Fig. 4.26).

Peridotite is the general name for the ultrabasic or ultramafic intrusive rocks, dark green to black in color, dense and coarse-grained texture, often as layered igneous complex. It is composed of ferromagnesian minerals (>40%), high proportion of magnesium-rich olivine, both clinopyroxenes and orthopyroxenes, hornblende, and <45% silica (Figs 4.27 and 4.28). Regular secondary mineral ingredients are chromite, magnetite, nickel, copper and platinum group of metals. The rocks are composed of entirely single mineral or in combination of various proportions. The components are branded on the basis of the minerals present such as peridotite, kimberlite, lherzolite, harzburgite, hornblendite, dunite, and pyroxenite. Peridotite is the most dominant constituent of the upper part of the Earth's mantle. A special variety of the peridotite is kimberlite, composed predominantly of olivine,

FIGURE 4.27 Peridotite is a dense coarse-grained dark green to black color intrusive rock, often layered, composed primarily of ferromagnesian minerals (magnesium rich olivine and pyroxene) and less of silica. Regular secondary mineral are chromite, magnetite, nickel, copper and platinum group. *Source: Prof. Arijit Ray.*

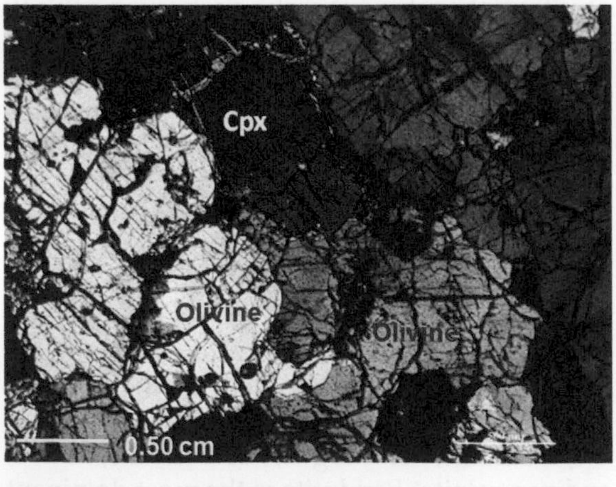

FIGURE 4.28 Photomicrograph of peridotite rock shows intercumulus texture between olivine and clinopyroxene (Cpx) resulting from the settling of a crystallizing magma. *Source: Prof. Arijit Ray.*

phlogopite, orthopyroxene and clinopyroxene, in which a secondary valuable gem component is diamonds (Section 2.5.1). Peridotites are formed due to the low stability of olivine and are very susceptible to changes, i.e. serpentinization of olivine in the fibers and/or sheet clusters of serpentine and monoclinic pyroxene (diallage) in uralite at Urals region. In this way creates a new rock "serpentinite" (Table 4.1). Layered intrusive variety is most suitable host rock of chromium, nickel, copper and platinum–palladium ore bodies, and glassy green type as gem and ornamental stones.

Lherzolite is a type of peridotite containing idiomorphic developed olivine and equal share of orthorhombic pyroxene bronzite (with irregular grains) and monoclinic pyroxene diallage.

Harzburgite is a type of peridotite with no or very little monoclinic pyroxene and consisting only of olivine and orthorhombic pyroxene bronzite.

Dunite is a special type of peridotite family consisting almost entirely of magnesium-rich olivine (+90%) and very small amounts of chromite, pyroxene and pyrope. Dunite is an igneous plutonic rock of ultramafic composition with coarse-grained granular or phaneritic texture and often massive or layered. The color is usually light to dark green with pearly or greasy look (Figs 4.29 and 4.30). Dunite is the olivine-rich end member of the peridotite group of mantle-derived magma/rock.

One variety of dunite is the end product of differential cooling, crystallization and solidification of hot molten ultramafic magma processed in a huge chamber within the Earth and develop layered igneous complex. The composition of layered igneous complex is often shared by large presence of chromite ± nickel, copper and platinum group of ore deposits (Fig. 4.31). Finely grounded dunite used as sequesters of CO_2 and mitigate global climate change, source of MgO as flux in metallurgical blast furnace, refractory and foundry applications, filtering media and filler.

FIGURE 4.29 Dunite is typically coarse grain with light to dark green color rock and consists almost entirely of magnesium rich olivine with minor amount of chromite, pyroxene and pyrope. *Source: Prof. Arijit Ray.*

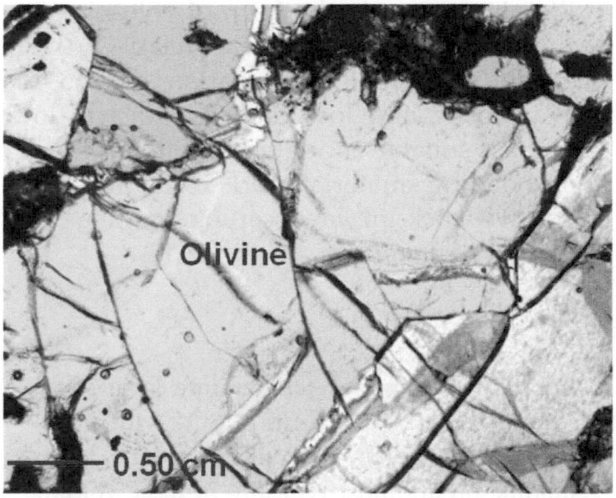

FIGURE 4.30 Photomicrograph of thin section of dunite composed entirely of olivine. *Source: Prof. Arijit Ray.*

Pyroxenites are the ultramafic intrusive igneous rocks composed essentially of pyroxene group of minerals, such as, augite and diopside, hypersthene, diallage, bronzite and enstatite. The absence of feldspar and olivine makes it different from gabbro-norite and dunite,

FIGURE 4.31 Dumite, composed of alternate layers of olivine and chromite forming a part of the Sukinda layered Igneous Complex, represent + 90% of chromite resources in India. The group of mines with production capacity of ~4 Mt/a is the second largest in the World after South Africa.

respectively. The accessory minerals are chromite, magnetite, garnet, rutile and scapolite. The rock is dense and coarse grained (Fig. 4.32) with dark green, gray and brown color. Pyroxenites are classified into clinopyroxenites, orthopyroxenites, and the websterites which contain both pyroxenes. Pyroxenites occur either as cumulates at the base of the intrusive chamber or as thin layers within peridotites and/or xenoliths in basalt. Pyroxenites are source of MgO as flux in metallurgical blast furnace, refractory and foundry applications, filtering media and filler, building materials and sculptures and often host deposits of Cr–Ni–Cu–Platinum group of minerals.

Ophiolites or the ophiolite complex is the general name for the community of mafic and ultramafic intrusives (gabbro, peridotite, and pyroxenite), and extrusive igneous rocks (spilite–keratophyre–basalt) that originate from the oceanic crust and mantle, and greywacke sandstone, shales and siliceous sedimentary rocks—chert.

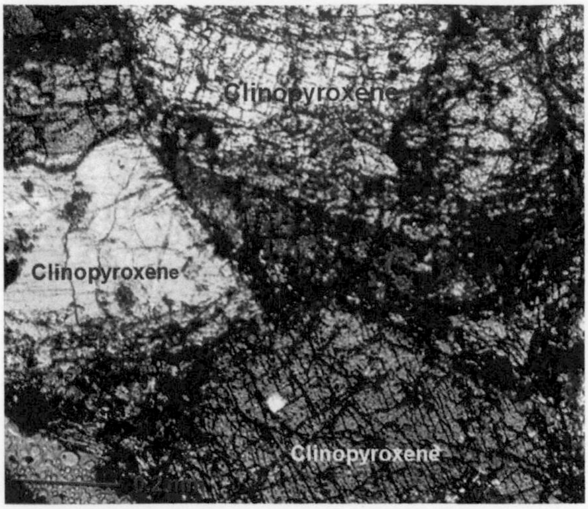

FIGURE 4.32 Photomicrograph of thin section of coarse-grained pyroxenite composed entirely by clinopyroxene. *Source: Prof. Arijit Ray.*

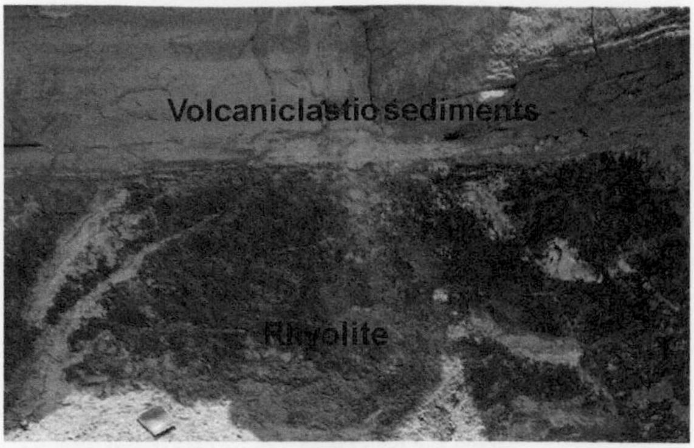

FIGURE 4.33 Field photograph of Rhyolite (bottom) and volcaniclastic sediments (top) at Khnaiguiyah Zn–Cu deposit forms a part of Shalahib formations, Saudi Arabia.

4.3.2. Mineral Composition of Extrusive Igneous Rocks

The most common extrusive igneous rocks are rhyolite and dacite (felsic), andesite and trachyte (intermediate), basalt and diabase (mafic) and spilite (plagioclase-rich rocks occur in changes and albitization of basalt).

4.3.2.1. Felsic Extrusive Igneous Rocks

Rhyolite is extrusive equivalent of granite magma. It is composed predominantly of quartz, K–feldspar and biotite. It may have any texture from glassy, aphanitic, porphyritic, and by the orientation of small crystals reflecting the lava flow. There is distinct porphyritic texture characterized by sanidine and rare quartz, plagioclase and biotite.

The various types of rhyolites are of gray, bluish gray or pink color with vitreous texture and individual spherical aggregates of feldspar and a number of concentrically arranged and spiral cracks, known under the name of perlite. The gray, black or pink, porous volcanic glass without phenocrysts are called *obsidian*, and very porous volcanic glass and full of unrelated gas bubbles are light and floats on the water, known under the name of pumice, as shown in Section 5.6.1.

Rhyolite occurs in the form of volcanic plate and lava basin (Fig. 4.33) with relatively large thickness and small propagation due to the high viscosity and low capacity of lava flow. Rhyolite is suitable as aggregate, fill-in construction, building material and road industries, decorative rock in landscaping, cutting tool, abrasive and jewelry.

Sanidine is the high-temperature form of potassium feldspar (K,Na) $(Si,Al)_4Os$ with monoclinic crystal system and vitreous-pearly luster. Sanidine occurs most typically in felsic volcanic rocks such as rhyolite, trachyte and obsidian.

Dacite is an extrusive equivalent of granodiorite magma, along with quartz, more Na–plagioclase than K–feldspar and more ferromagnesian minerals (biotite, amphibole, and augite) than rhyolite. Dacite is gray to dark gray color. It has a distinct porphyritic texture

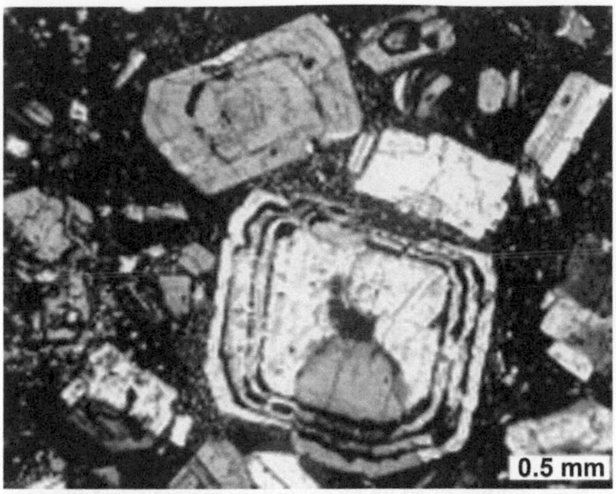

FIGURE 4.34 Dacite with sanidine (high-temperature form of potassium feldspar $(K,Na)(Si,Al)_4O_8$) phenocrysts in sanidine dacite.

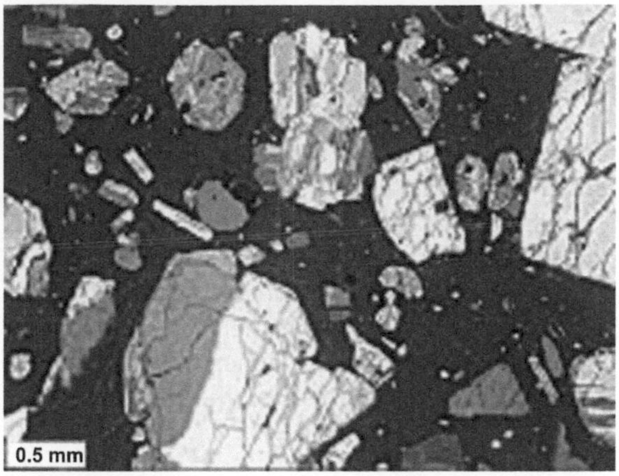

FIGURE 4.35 Porphyritic texture of trachyte with sanidine phenocrysts.

with large felsic plagioclase (oligoclase), improperly damaged by quartz and rare sanidine in glassy groundmass. A special type of dacite is sanidine dacite (Fig. 4.34) containing large sanidine phenocrysts, and in the mineral composition close to dellenite, i.e. a transitional extrusive type from rhyolite to dacite (Table 4.1).

Dellenite is the extrusive equivalent of adamellite and quartz monzonite, with 5–20% quartz, equal amounts of K–feldspar and Na–plagioclase with Na–Ca–plagioclase. Dacite is a relatively rarely represented extrusive rock. Dacite is suitable as aggregate, fill-in construction, building material and road industries, decorative rock in landscaping, cutting tool, abrasive and jewelry.

4.3.2.2. Intermediate Extrusive Igneous Rocks

Latite is the extrusive equivalent of monzonite (Table 4.1), characterized by the porphyry texture. Latites are in fact andesite rich in K–feldspar, and contains more than 10% K–feldspar (sanidine and orthoclase). It has a lower amount of K–feldspar compared to the total amount of feldspar in andesite. In some latites, there are smaller amounts of leucite.

Andesite is the extrusive equivalent of diorite magma characterized by holocrystalline and often porphyritic texture. It contains grains of Na–Ca–plagioclase and hornblende in holocrystalline groundmass. Andesite is named after "Andes," the longest continental mountain range in the World, where large volume of lava with such mineral composition exists. Andesite is most widespread extrusive rocks after basalt. The hornblende andesites usually contain ferromagnesian minerals (biotite and pyroxene). Hydrothermally altered andesite (and dacite) is known as *porphyrite* (Table 4.1). Andesite is suitable mainly for naturally slip-resistant tiles, bricks, water or landscape gardens, aggregates, and fill-in construction.

Trachyte is the extrusive equivalent of syenite magma. The typical porphyry texture, known as *trachyte texture*, is also characterized by sanidine grains, and sometimes Na–plagioclase in the groundmass. It is composed of parallel arranged tiny rod-like crystal sanidine within the glassy matrix (Fig. 4.35). Geologically old trachyte that occurred before the tertiary is known by the old nomenclature as porphyry.

Phonolite is the extrusive volcanic equivalent of alkaline syenite magma. It is characterized by relatively high content (>10%) feldspathoids—nepheline and leucite. Phonolite is named after a strong echo from the blows of a hammer (in Greek "sounding stone"). Phonolite is a rare rock of intermediate chemical composition between felsic and mafic, with texture ranging from aphanites (fine-grain) to porphyritic (mixed fine- and coarse-grain). The highly porphyritic texture is characterized by large nepheline or leucite phenocrysts within a glassy or finely crystalline core mass. The rock is named as nepheline phonolite and leucite phonolite as per the predominance of phenocrysts component.

4.3.2.3. *Mafic Extrusive Igneous Rocks*

Basalts are common aphanitic igneous extrusive (volcanic) rocks. Basalts are composed of minute grains of plagioclase feldspar (generally labradorite), pyroxene, olivine, biotite, hornblende and <20% quartz. Nepheline or leucite may associate or proxy the feldspar giving rise to verities with special names. The ferromagnesian minerals are mainly amphibole and rarely biotite. Basalts are usually dark gray to black color.

Basalts are formed by the rapid cooling of basaltic lava, equivalent to gabbro-norite magma, from interior of the crust and exposed at or very close to the surface of Earth. These basalt flows are quite thick and extensive, in which gas cavities are almost absent. In case of thin and irregular lava flows, gas cavities are formed on the rock surface. The rock is called "vesicular" basalt when the gas cavities are empty. The majority of the gas cavities are filled up by secondary minerals (zeolites, calcite, quartz, or chalcedony) and amygdales are formed. The rock containing such filled-up gas cavities is called "amygdaloidal" basalt (Fig. 4.36).

FIGURE 4.36 Dark grayish black massive basalt showing surface cavities filled up by secondary minerals and the rock is designated as "amygdaloidal basalt". *Source: Prof. Arijit Ray.*

Outcrops of basaltic lava flow are easily susceptible to mechanical and chemical weathering by penetration of groundwater along the polygonal joints (Fig. 4.37) and fractures, loosening and decaying the rock layer by layer. The surface of weathering grows more and more rounded as the process progresses into blocks resembling spheroidal shape on a larger scale in plutonic rocks. The process is accelerated by insolation effect and repeated expansion (hot days) and contraction (cold nights) causing stresses that lead to the weakening of ties between the mineral, cracking and disintegration. It is also known as *onion skin* or *concentric weathering* (Fig. 4.37).

Basalts show, almost always, aphanitic or fine-grained mineral texture resulting from rapid cooling of volcanic magma on or close to surface of Earth. The component minerals are so fine that they are not identifiable by the unaided eye. The texture can sometimes be porphyritic containing the larger crystals formed prior to the eruption, that brought the lava to the surface, and embedded in a fine-grained matrix. Glomeroporphyritic is the extension to describe porphyritic texture in which phenocrysts of plagioclase and pyroxenes are clustered into aggregates and settled in groundmass due to surface tension (Fig. 4.38). Glomeroporphyritic textures are particularly common in basalt, andesites and dacites.

Tholeiitic basalts are the most common eruptive rocks produced by submarine volcanism from tholeiitic magma series, forming much of the ocean crust and mid-oceanic ridges. The tholeiitic magma is relatively rich in silica and

FIGURE 4.37 "Spheroidal" or "onion-skin" or "concentric weathering" caused by penetration of groundwater along the polygonal joints and fractures, loosening and decaying the rock layer by layer in basaltic lava flow over sustained periods. The process is accelerated by insolation effect by repeated expansion (hot days) and contraction (cold nights) causing stresses that lead to the weakening of ties between the mineral, cracking and disintegration at Albert Hill. Mumbai, India.

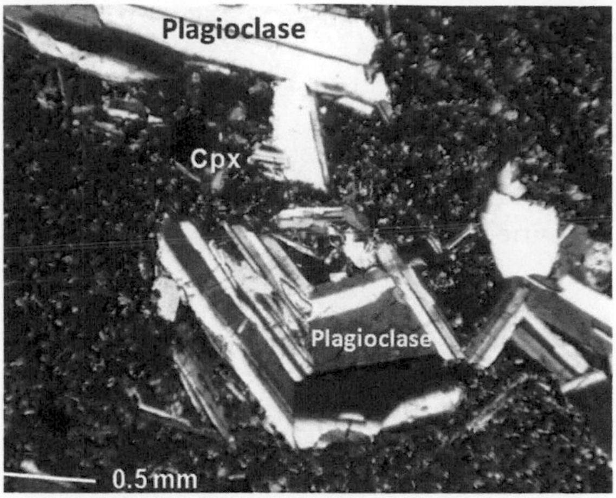

FIGURE 4.38 Photomicrograph of thin section showing glomeroporphyritic texture defined by plagioclase and clinopyroxene grains embedded in fine-grained groundmass of porphyritic basalt. *Source: Prof. Arijit Ray.*

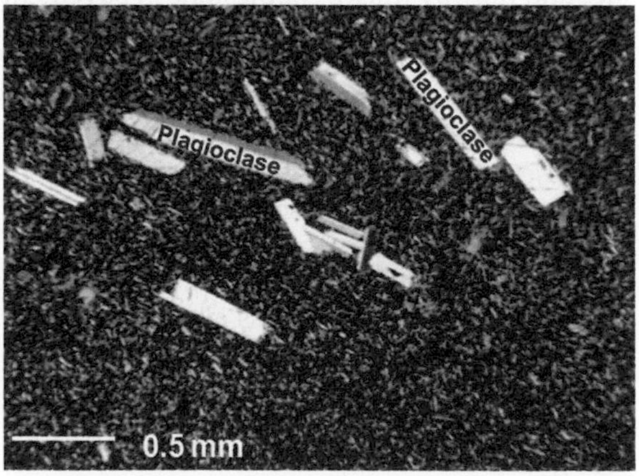

FIGURE 4.39 Photomicrograph of thin section showing plagioclase phenocrysts in tholeiitic basalt flow. *Source: Prof. Arijit Ray.*

poor in sodium. The rock is composed of clinopyroxene, hypersthenes and plagioclase with minor iron–titanium oxide, and ±olivine. Tholeiitic basalt often represents a fine, glassy groundmass consisting of fine-grained quartz and other main constituent minerals. Tholeiitic basalt has fine-porphyritic texture, which is characterized by pyroxene/plagioclase phenocrysts (Fig. 4.39) in fine glassy groundmass.

Alkali basalt is a fine-grained dark color volcanic rock composed of phenocrysts of olivine, titanium rich and iron oxides (Fig. 4.40). Alkali basalt is relatively poor in silica and rich in sodium. It is silica unsaturated and contains feldspathoids (nepheline, leucite), alkali feldspar and phlogopite (magnesium mica) in the groundmass. Alkali basalts are typically found on up-domed and rifted continental crust, and on oceanic islands such as Hawaii Island, in the North Pacific Ocean.

The basalt is compact, rough and tough, abundant and widely distributer rock. The most common uses are as construction materials (building

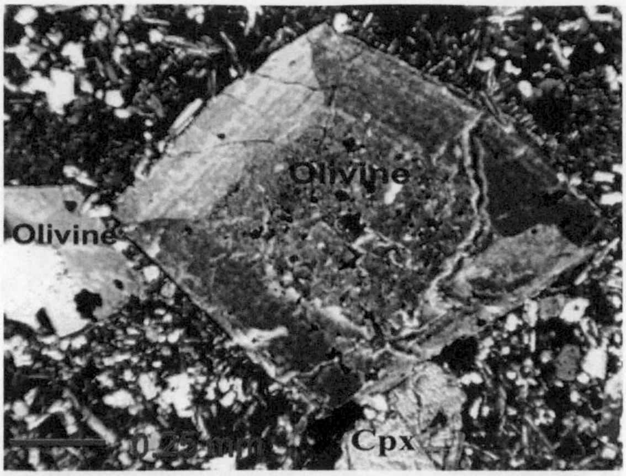

FIGURE 4.40 Photomicrograph of thin section showing zoning in olivine phenocryst settled in the fine-grained groundmass in alkali basalt. *Source: Prof. Arijit ray.*

blocks, flooring titles and aggregates, road surface and railway track), cobblestone in pavement (columnar variety), architecture, statues, and stone–wool fiber as excellent thermal insulator.

Spilite is the sodium-rich volcanic rock formed by turning basalt and/or diabase in albitization processes at low temperatures in the presence of CO_2 and water rich in sodium (e.g. seawater). So, albite in the spilite is not caused by crystallization of lava rather than secondary processes of Ca–plagioclase. The minerals albite and pyroxenes are transformed into green minerals such as chlorites, uralite and epidote with the conversion of Ca–plagioclase in the Na–plagioclase. The green color of spilites is the effect of the newly formed minerals. They have intersertal texture—typically microcrystalline. Spilite is usually found in submarine lava effusion in cushioned forms, i.e. as a "pillow-lava".

4.3.3. Veins Igneous Rocks

Veins igneous rocks, unlike other igneous rocks, never appear alone but are associated with intrusive and sometimes extrusive igneous rocks or as veins found in metamorphic and some sedimentary rocks (Fig. 4.5). They are classified as aplite, pegmatite and lamprophyre.

Aplites are the fine-grain equivalent of granite and composed only of quartz and alkali feldspar, and very small amounts of muscovite and biotite. They are mostly fine grain, white to light gray. They are found only in felsic intrusives like granites and granodiorites.

Pegmatites are veins rocks, white to pale gray or pink color and are composed of very large crystals of quartz, feldspar (feldspar, microcline, and rarely oligoclase) and muscovite. Pegmatites sometimes include the immense size of the crystals up to several meters long. In addition to these essential minerals, pegmatites regularly contain many other, often economically important, minerals such as beryl, monazite, uraninite, fluorite, molybdenite, apatite, wolframite, and many others. It occurs, most often, in the form of veins located in the felsic intrusives of granites and granodiorites, and in metamorphic rocks of gneisses.

Lamprophyre is the veins rock, dark gray to black in color. The rock consists of ferromagnesian minerals such as biotite, amphibole, pyroxenes, and rarely olivine. The share of leucocratic (white or colorless) minerals is small, typically found only a few tiny grains of feldspar. They are found in intrusive and extrusive rocks formed from syenite magma.

FURTHER READING

Igneous Petrology by Bose[4] and Taider et al.[46] will be an initial reading for beginners in the subject. *Principles of Igneous and Metamorphic Petrology* by Winter[63], *Igneous Petrogenesis* by Huges[26] and Wilson[62] will be an excellent reading in the subject with much greater details.

CHAPTER

5

Sedimentary Rocks

OUTLINE

5.1. Function, Significance, Classification and Transformation 122

5.2. Sedimentary Rock Formation 124
5.2.1. *Weathering* 124
5.2.1.1. Physical or Mechanical Weathering 124
5.2.1.2. Chemical Weathering 126
5.2.1.3. Biological Weathering 128
5.2.2. *Sediment Transport* 128
5.2.2.1. Fluvial Processes 128
5.2.2.2. Aeolian Processes 129
5.2.2.3. Glacial Processes 130
5.2.3. *Deposition* 130
5.2.4. *Lithification* 132

5.3. Texture and Structure of Sedimentary Rocks 133
5.3.1. *Bedding* 133
5.3.1.1. External Bedding 134
5.3.1.2. Internal Bedding 134
5.3.1.3. Upper Bedding Plane Structures 137
5.3.1.4. Lower Bedding Plane Structures 140
5.3.1.5. Forms Created by Underwater Slides and with the Destruction of the Layers 141
5.3.2. *Packing of Grains* 142

5.4. Classification of Sediments and Sedimentary Rocks 144

5.5. Clastic Sediments and Sedimentary Rocks 145
5.5.1. *Genesis and Classification of Clastic Sedimentary Rocks* 145
5.5.2. *Coarse-Grained Sediments—Rudaceous* 146
5.5.2.1. Intraformational Breccias and Conglomerates 147
5.5.2.2. Extraformational Breccias 149
5.5.2.3. Extraformation Conglomerates 152
5.5.3. *Medium Granular Clastic Sediments—Arenaceous Rocks* 153
5.5.3.1. The Composition and Distribution of Sandy Sediments 153
5.5.3.2. Arenite Sandstones or Arenaceous Rocks 155
5.5.3.3. Graywacke or Wackes 157
5.5.3.4. Mixed or Hybrid Sandstones 159
5.5.4. *Fine Granular Clastic Sediments—Pelite* 159
5.5.4.1. Classification of Pelitic Sediments 159
5.5.4.2. Marlstone 162

Introduction to Mineralogy and Petrology
http://dx.doi.org/10.1016/B978-0-12-408133-8.00005-5

5.5.4.3. Organic Matter in the Argillaceous Sediments 163
5.5.5. Diagenesis of Clastic Sediments *164*
5.5.5.1. Diagenetic Processes in Sandy Sediments 164
5.5.5.2. Diagenetic Processes in Clayey Sediments 168
5.5.6. Residual Sediments: Laterite, Kaolin, Bauxite and Terra Rossa *171*

5.6. Volcaniclastic Rock **174**
5.6.1. Definition and Origin of Volcaniclastic Sediments and Rocks *174*
5.6.2. Composition of Volcaniclastic Sediments and Rocks *177*
5.6.3. Alteration of Tuff *178*

5.7. Chemical and Biochemical Sedimentary Rocks **179**
5.7.1. Limestone *179*
5.7.1.1. Mineral Composition, Physical, Chemical and Biological Conditions for Foundation of Limestone 179
5.7.1.2. The Structural Components of Limestone 182
5.7.1.3. Limestone Classification 189
5.7.1.4. Limestone Diagenesis 195
5.7.2. Dolomites *199*
5.7.2.1. The Origin of Dolomite 200
5.7.2.2. Early Diagenetic Dolomites 201
5.7.2.3. Late-Diagenetic Dolomite 201
5.7.3. Evaporites *203*
5.7.3.1. Mineral Composition, Origin and Classification of Evaporites Rocks 203
5.7.3.2. Petrology and Diagenesis of Evaporite Sediments 205
5.7.4. Siliceous Sediments and Rocks *207*
5.7.4.1. Mineral Composition, Origin and Classification of Silicon Sediments and Sedimentary Rocks 207
5.7.4.2. Siliceous Sediments and Siliceous Rocks of Biogenic Foundation 208
5.7.4.3. Siliceous Sediments and Siliceous Rocks of Diagenesis Origin 210

Further Reading **212**

5.1. FUNCTION, SIGNIFICANCE, CLASSIFICATION AND TRANSFORMATION

Sedimentary rocks are formed by one or in combination of the complex physical, chemical, biological and geological diagenetic processes of sediments. The sediments are deposited on or near the Earth's surface at a temperature and pressure appropriate to these conditions. The rocks are formed under specific processes derived from other preexisting rocks of igneous, sedimentary, and metamorphic origin, and/or as products of life activities of organisms or chemical secretion.

The title of sedimentary rocks (from the Latin sedimentum residue) suggests that these rocks are formed by deposition of inorganic and organic, solid or excreted material from aqueous solutions. However, the sedimentary rocks also include rocks originated by diagenetic chemical processes of existing sediments and sedimentary rocks (e.g. late-diagenetic dolomite and anhydrite, and some diagenetic siliceous sediment). While the term "sediments" usually cover nonlithified, and in some cases, soft deposits, the entitled sedimentary rocks mainly include lithified deposits in the form of solid rock, and all sediments and sedimentary rocks along.

The basic requirement for the formation of sediment is the existence of underlying materials, which are primarily in the evolution of Earth's rocky crust, composed of igneous, metamorphic and older sedimentary rocks. The deposition process continues at the Earth's surface with its evolution, so that the size and thickness of sedimentary cover increases in the rocky crust. The rock configuration changes by tectonic movements (global tectonics), volcanism and erosion. The sedimentary rocks also include residues resulted by accumulation of volcanic materials after its transfer and deposition by wind, water or ice, i.e. pyroclastic rocks.

The sediments and sedimentary rocks are divided into two basic groups with respect to the type of physical, chemical, biochemical and geological processes. There are mixed sediments and sedimentary rocks between these two distinct and different groups (Table 5.2).

1. Clastic (exogenous) sediments and sedimentary rocks.
2. Chemical and biochemical (endogenous) sediments and sedimentary rocks.

Clastic sediments and sedimentary rocks are divided according to size of clasts, regardless of their origin, except those ejected from volcanoes, as follows:

1. Coarse grain or gravel (rudaceous)
2. Medium-grain or sandy (arenaceous)
3. Fine grain and argillaceous (pelite).

Sediments that are generated by deposition of clasts ejected during volcanic eruptions belong to a special group of clastic sediments and called as "volcanoclastic" or "volcaniclastic" sedimentary rocks.

Special group of clastic sediments include residual sediment (residue) remaining after an intense chemical weathering of rocks.

Chemical and biochemical sediments and sedimentary rocks include the following:

1. Carbonate, e.g. limestone.
2. Evaporate, e.g. halite and carnallite.
3. Siliceous sedimentary rocks, e.g. sandstone.

The thick file of sediments moves down to deeper depth due to overlying accumulation of fresh sediments. The package at depth will be under the influence of elevated temperatures and pressures, making some components unstable and transform into new stable ingredients. The same sediments gradually transform or metamorphose under increasing pressure and elevated temperature at great depth.

It is experimented that the diagenesis with significant metamorphic changes takes place at a depth of 4–5 km, pressures up to approximately 2530 bar and temperatures below 220 °C. The temperature increase, due to geothermal gradient of 1 °C/33 m depth, corroborates the temperature of 200–220 °C at Central European depths of about 6600–7260 m. It also supports that the temperature increase in the Earth's crust is not an exclusive function of the depth of the overlay, but due to the proximity of magma, volcanism and other thermodynamic factors. Therefore, the depth at which the metamorphism starts will be very different from place to place. It can be safe to assume that the sedimentary rocks gradually transform or metamorphose into metamorphic rocks at great depths of covering supplemented by the rise of pressure and temperature above 220 °C (Fig. 6.9).

Many of the mineral resources are sedimentary in origin. All mineral fuels, i.e. oil, natural gas, coal and oil shale, are confined only in sediments, except otherwise in some special cases. The oil and gas-filled pores in lithified sediments, coal and oil shale are sedimentary rocks. Many metallic and nonmetallic minerals are hosted by sedimentary rocks, e.g. majority of iron ore, and partly of manganese, copper, uranium and magnesium. In addition, many sedimentary rocks are directly used as raw material for obtaining cement (marl and limestone), glass (quartz sand), ceramic and porcelain (clay and kaolinite), bricks and tiles (clay), building materials like concrete (aggregates of limestone, dolomite or sandstone, gravel, and sand) or used as a technical, architectural and building stone (limestone, dolomite, and sandstone). All the mineral phosphate, nitrate and potassium fertilizers mineral, salts (halite and carnallite) as well as gypsum and anhydrite are of sedimentary origin. The sedimentary rocks are the excellent holders or the collectors of freshwater, which today, become the primary life essential of all humanity.

5.2. SEDIMENTARY ROCK FORMATION

Sedimentary rocks are, sediments, formed by the following precesses:

1. Sedimentation of solid residues (clasts) left over from weathering of older rocks (clastic sediment).
2. Biochemical and chemical secretion from aqueous solutions, as well as the deposition and accumulation of fossil evidence.
3. Skeletons and shells of organisms (chemical and biochemical sediments).

The formations of sedimentary rocks include the following:

1. Processes of physical and chemical weathering of older rocks.
2. Transfer or transport of materials in solid or dissolved state.
3. Deposition or sedimentation.
4. Complex processes of diagenesis and, significantly, the lithifaction.

5.2.1. Weathering

Weathering is the process of destruction of rocks on Earth's surface or shallow water due to:

1. physical or mechanical,
2. chemical, and
3. biochemical factors, or due to the effects of atmosphere, water, ice, climate and temperature changes, erosion, sunshine and life activity of organisms.

The first two, and the third factor of weathering, are closely related and interactive. The rocks are weathered out in two ways, by the action of water, especially water flows like a river, occasional torrents and storm tides and waves:

1. Mechanically because of the speed and power of water flows and water activities during the transmission of material.
2. Chemical dissolution due to the action of the weak carbon and humic acids.

5.2.1.1. Physical or Mechanical Weathering

The physical or mechanical wear and tear (deterioration and weathering) includes fragmentation and disintegration of existing minerals and rocks, without the formation of new minerals. The development is primarily caused by mechanical action of water, ice or wind, sunshine and frost. The process of grinding stones in finely dispersed particles is a basic element of physical weathering. This causes an increase in volume and decrease in density that facilitates and accelerates their chemical weathering because of the intensification of the oxidation and hydration of primary mineral constituents of the rocks. Products of the physical weathering are solid particles or clasts of different sizes:

1. Mud (0.004−0.063 mm)
2. Sand (0.063−2 mm)
3. Gravel (2−256 mm)
4. Boulder (>256 mm).

These materials may be transferred to a greater or lesser distances by water, wind or glaciers, or may remain in place.

Insolation (solar radiation energy) is the most important factor of physical weathering, especially in arid region (dry climate), i.e. in the deserts. The repeated changes of hot days and cold nights cause recurring expansion and contraction of certain mineral constituents of rock. The anisotropic properties of minerals are affected by various stresses causing to weakening of ties between the mineral, cracking and disintegration of rocks. This process is particularly intense in the surface areas of dark colored rocks, e.g. basalt (Fig. 4.37).

Absorption and *desorption* of water (hydration–dehydration) caused by large temperature differences that change the pressure of water vapor in the air and in the pores of rocks. It creates dissolution or extraction of minerals salts in the rock or enhances the strong hydration of some minerals (e.g. anhydrite in gypsum). The frequent volume change can result into complete destruction of rocks. The atmosphere is saturated with water vapor and decrease evaporation of water from the pores of rocks, specifically, at lower temperatures, typically at night. However, the evaporation of water from rocks increases at high temperatures during day and the pressure of water vapor in the air fall. These changes are caused by repeated dissolving and crystallization of salt. The rocks formed cavities during dissolving the salt, which further enhance its physical and chemical weathering. The crystallization of salts in the pores space makes destructive stresses due to increase in volume of the rocks. The hydration of anhydrite to gypsum increases the volume by 38%.

Freezing–thawing is at temperate climates in the high mountains of rocks saturated with water. This exerts high pressure and great stress due to increased volume of ice compared to the volume of water. Such stresses can destroy the hardest rocks. The intensity of destruction is subject to freezing of high-porosity rocks that are already tectonically fractured.

Erosion is the process of destruction of relating parts of the Earth's surface with accessibility of streams, ice and wind. The river erosion is strongest with torrential water flows. The water carries a large mass of rock debris. The quantity of material transfer will be more and the size of individual pieces will be larger at the highest speed of the water flow. The water swirling along with rock material will impose strong impact on the rocks at the bottom and sides of the river bed. The tear-off pieces collapse into the river and become part of its course. The rock materials crumble, fall-apart, and crush in such a transfer situation. Thereby, the large pieces gradually become sandy and the final product of these processes results in grain sizes of tiniest powder (silt).

Denudation is the name for the processes of erosion, leaching, stripping, and reducing the mainland due to removal of material from higher to lower areas like valleys, river valleys, lakes and seas with a permanent filling of low lands.

Glacial erosion is among the most devastating factors of physical weathering of rocks on Earth's surface. The ice and rock carried by glaciers acted as sandpaper which moves down the valley, smoothing and widening them, leaving U-shaped profiles for the valley cross-sections. There will be, finally, large steep-walled bowls called "glacial cirques" at the heads of these valleys. A sharp-sided ridge separates them when glaciers carve out valleys next to each other.

Abrasion wave activity is more intense as greater as their speed and strength. The abrasion is stronger with larger grain sizes, larger quantity of material transfer, and the weaker ground rock strength. Abrasion on the seacoasts depends on the strength of the waves, which can move the heavy stone blocks of several hundred tons in the stormy weather. The high waves hit the rocky shore and destroy rocks due to the power of water and waves. The hydraulic effects of water contained in the hollows and crevices of coastal rocks crack and break the massive boulders into pieces (Fig. 5.1). In the activity of waves and moving of rock material by mutual collision and friction of the fragments of the original angular remain form well-rounded grains of sand and gravel. The rocky debris becomes smaller all the while.

FIGURE 5.1 Strong action of wind and wave of Pacific Ocean crack, break and split the giant hard rock mass to smaller fragments of sand and silt along West Coast Highway to San Francisco. *Source: Soumi.*

FIGURE 5.2 Chemical weathering by surface oxidation process of rain and seawater on basalt. The dark gray/black color has changed to reddish forming open erosion cavities, Mumbai coastal area, India.

5.2.1.2. Chemical Weathering

The chemical weathering occurs mainly due to the action of water, carbonic acid and oxygen on the minerals and rocks, where they are chemically susceptible to changes. As a consequence, some disappear and some appear as "authigenic" minerals, which are stable under conditions prevailing in the Earth's surface or just below the surface. The chemical weathering in action of water and water as a mild carbonic acid is hydration. The same process with action of oxygen is oxidation.

Hydration is the process of receiving H^+ ions and the release of alkali elements (Na, K, Li, Ca, Mg, Sr) and silicon (Si). A great amount of alkali liberates during the chemical weathering of rocks, transfer by water in dissolved, mainly ionic, but partly in collide state over long distances in rivers, lakes and seas. The same water excretes new minerals in the favorable physical and chemical conditions. The weathering products of some rocks in the hydration and carbonation generate soils that are very different in different climatic zones.

Oxidation is a very significant factor of chemical weathering of rocks. The oxidation process changes the primary color, porosity, volume, and mineral composition of rocks (Fig. 5.2). The oxidation zone is the deepest in rocks of areas where the basic water is deep below the surface. The oxidation occurs mainly above the basic level of groundwater table in areas with steep relief and warm climate. The rainwater enriched with oxygen (which causes oxidation processes) penetrates into the depth of the pores of rocks. In the deeper layers, water gradually loses oxygen with more and more saturation of dissolved cations and anions, and loses the oxidizing effect. The effect of oxidation processes on color change of rocks can be seen best in fresh outdoor shoots of dark gray sediments along and around the tectonic cracks, crevices or open layer spaces. The transition in dark tan or reddish color can be seen along zones of circulation of oxygen-enriched rainwater. Such color change is the result of oxidation of Fe^{2+} to Fe^{3+} with the formation of goethite (yellow-brown color) or hematite (reddish color), and oxidation of organic matter.

The easily oxidized mineral constituents of rocks are sulfides, such as, pyrite (FeS_2), hematite (Fe_2O_3) or goethite (α-FeO,OH), and from other constituents of rocks and organic matter.

The chemical weathering depends on the climate. The heat, daytime and annual temperature fluctuations and humidity significantly accelerate the abrasion. The increase in temperature by 10 °C accelerates the flow of chemical reactions up to 2–2.5 times more. The chemical weathering is related to the Earth's surface, shallow area and underwater. The weathering process can be under the influence of the atmosphere, water and seawater. The main factor of chemical weathering of rocks is water that contains dissolved CO_2 and dissociates to the free $[H]^+$ and $[HCO_3]^+$ ions, representing a mild carbonic acid ($CO_2 + H_2O = H_2CO_3$). The content of free $[H]^+$ ion determines its chemical activity and the share of water acts neutrally (pH = 7), acidic (pH = 1–7) and alkaline (pH = 7–14).

The process of formation of new mineral directly depends on the pH of the water. Kaolinite is the most important mineral in the zone of weathering and occurs in acidic pH ~5. Montmorillonite occurs in the weak alkaline solution at pH >7. The water in the chemical weathering has a significant role in transferring large amounts of easily soluble anions and cations. The large amounts of silicic acid (H_4SiO_4) or Si ions that are released during the weathering of silicate minerals are equally significant.

If the pressure of CO_2 increases in the atmosphere, it will increase its solubility in the water creating an increase in acidity of water, i.e. its conversion into weak carbonic acid. Such water is chemically very aggressive toward many petrogenic minerals intensively destroying carbonates, feldspars and amphiboles.

The effect of water containing dissolved carbon dioxide on calcite is very fast and powerful process (Fig. 5.3). The soluble calcium hydrogen carbonate occurs in water, which can be illustrated by the following chemical reaction:

$$CaCO_3 + H_2O + CO_2 \longrightarrow Ca(HCO_3)_2$$

$$\underset{\text{Calcite}}{} \; \underset{\text{dissociation carbonic acid}}{[H]^+ + [HCO_3]^-} \longrightarrow \underset{\text{calcium hydrogen-carbonate}}{Ca(HCO_3)_2}$$

The effects of water on feldspars are leaching of alkalis, liberation of Si and combination with H_2O in the silicic acid to form kaolinite. It is a well-known process of kaolinitization:

$$\underset{\text{orthoclase}}{2KAlSi_3O_8 + 10H_2O + CO_2} \longrightarrow \underset{\text{kaolinite}}{Al_2Si_2O_5(OH)_4} + \underset{\text{silicic acid}}{4H_4SiO_4} + 2K^+ + (CO_3)^{2-}$$

Kaolinitization process is an example of chemical weathering, hydration and the leaching of mineral constituents of parent rock and generates new or authigenic minerals. The formation of new minerals is called "autigenesis". The most common authigenic minerals formed during the chemical weathering are clay minerals and the aluminum hydroxides. The minerals

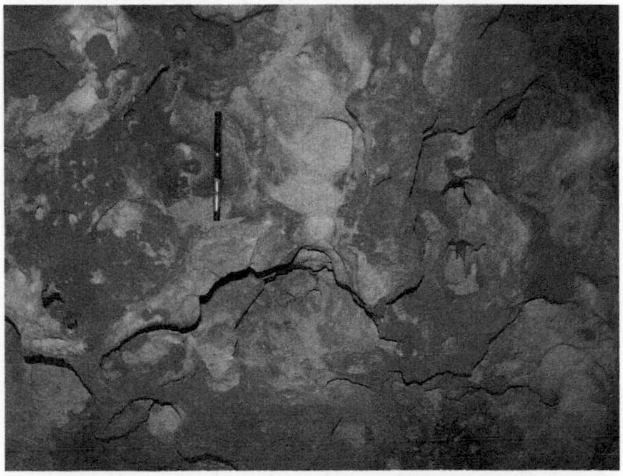

FIGURE 5.3 Chemical weathering of limestone in association of rain and seawater with erosion of surface leaving large open cavities, Mediterranean Sea at west bank of Alexandria, Egypt.

from kaolinite group generate during weathering of rocks with large precipitation. It forms weak acidic solution in the soil at pH of 5 and contains enough dissolved silicon in water in the form of silicic acid (H_4SiO_4).

5.2.1.3. *Biological Weathering*

Biological weathering takes place under the influence of life activities of organisms. The organic processes involve biological dissolution of rocks from bacterial activity, humic acids and bioerosion or destruction. The changes occur by the growth of roots, and drilling of organisms (shells, lichens, cyanobacteria, algae and the fungi) in the rock on which they cultivate. The bioerosion of carbonate rocks (limestone and dolomite) caused by cyanobacteria, lichens and fungi has particularly significant role. The large areas where these organisms live, and their prolonged activities during the geological period destroy significant amount of rock with the formation of massive quantity of very fine-grained carbonate detritus of limestone sludge (Section 5.7.1.2).

5.2.2. Sediment Transport

The transfer of detritus, i.e. solid materials or clasts remaining after the physical and chemical weathering, primarily takes place by water, and lesser part by wind and glaciers.

5.2.2.1. *Fluvial Processes*

The transfer of detritus by flowing water is the most natural way of transport and deposition of sedimentary rock. The flow of water can be orderly, laminar or turbulent. Laminar movement of water is gentle with certain parts of the fluid move in the parallel layers. The movement of detritus is also parallel to the flow of water without mixing. In turbulent or vortex movement, the main flow of water changes the speed and also the direction of flow. The turbulence movement of water carries large masses of fluid mixed with debris material due to the difference in the speeds and whirling motion. The detritus can be carried by dragging, suspension, sediment flow, underwater sliding and gravity flow. This is similar to rock transfer in air by landslides and avalanches.

The particles and grains will slide or roll on the bottom during the material transfer. The grains moves in short jumps during transmission, i.e. hitting in the bottom and bounce off the bottom back in the fluid. The detritus finally settle when the energy of water (or wind) is so much limited that it can no longer move. The transfer with the suspension is possible only if the intensity of turbulent water movement is greater than the speed of deposition of material by the action of gravity. The fine grain of clay-dusty (mud) or clay-sandy-dusty detritus transmits mainly in suspension. The flow of sediments is the movement of a mixture of unbound sediment and water. The underwater sliding includes sliding of poorly bound sediment down the slope on the bottom of the nearly flat or a little wavy surface. The sedimentary body originated by sliding (Section 5.3.1.5) in the lower and upper parts shows a strong deformation of primary inner and outer stratification or slump structure. The deformation in central part of such body can be weak and strong in the lower and upper part. There is always a clear angular discordance toward the basement and roof (Fig. 5.17).

Blurry or turbidity currents are flow of material mixed with water under the influence of gravity moving down the slightly inclined (1–3°), but long underwater slopes. The blurry or turbidity current of material forms due to the increase mixing of solid particles with water and differences in effects of gravity on large grains and small particles. The large grains (gravel) move forward with greater acceleration and accumulate in the bottom. The smaller grains (sand) lag increasingly behind in stream, and the smallest particles (dust, silt and clay) left behind suspended in the tail of currents. The suspension raises high above the bottom due to the turbulence of water. The coarse grains move faster up to 60 km/h compared to the smaller particles. The larger grains are increasingly separated from the small grains and accumulate in the frontal part of the flow and in its bottom. The finer materials that are lagging behind in the suspension are lifted above the bottom. In this way, the current or flow separates coarse material on the forehead, medium material in the middle and fine material in the tail end. The granulometric differentiation takes place horizontally and vertically (large grains in bottom, small in suspension above the bottom). The sediments of special structure, "turbidites," are common in flysch facies and are created from such distributed material within the turbidite current. These structures are deposited as sedimentary fan-shaped bodies (Figs 3.8 and 7.9) as shown in more detail in Section 7.2.5.

The detrite flows or debris flows are defined as more or less cohesive laminar flows of relatively dense sediment—fluid mixture of plastic types or clasts containing at least 4% clay component. The stability of the sediments is disturbed by extruding of fluids and clay, and thus initiates its movement down the slope. The debrite flows can be initiated by seismic shocks or can develop as a result of rapid accumulation of debris or formation of gases in sediments that cause local increase of pressures. When the gravity force is no longer stronger than the internal friction of sedimentary masses, or when there is no exceeded pore pressure, the flow suddenly stops or "freezes". The detrite flows can move down the slope angle $>1°$ with the speed of up to 20 cm/s.

The sediments formed by precipitation of detrite flows are called "debrites". The typical debrites are mainly composed of clasts of different sizes: coarse granular debris, with a diameter of several millimeters up to tens or even hundreds of meters, and medium granular to fine-grained muddy matrix in such a mutual ratio that the clasts have matrix support (clasts "floating" in fine matrix). Coarse blocks in debrite and debrite breccias are known under the name "olistolith" (Fig. 7.8; Section 7.2.5). The clasts that originate from the strong physical weathering or erosion of rocks outside participation area belong to extraclasts, and which originate from erosion of older sedimentary rocks inside the participation area called *intraclasts*.

5.2.2.2. Aeolian Processes

The wind can carry substantial amounts of material of small dimensions to long distances, especially in areas of bulk material with a dry arid climate or in deserts. The area is characterized by lack of moisture and vegetation. The main activity of the wind in the desert consists in puffing away, blowing up and transfer of sand and dust grains of a certain size depending on the intensity and wind speed. On the other hand, the grains are sorted from coarse to smallest particles in weakening of power and speed of wind. The regular wind usually does not carry sand-size grains far away and only the strongest winds can move larger sand grains by jumping on the ground and thereby transport sand in the direction of the wind and settles in the form of sand dune.

In contrast, the strong wind or air vortex currents can lift fine sand and dusts high off the ground and transfer to very long distances. The wind can pick up the smallest dust thousands of feet high in the air and carry it hundreds of miles far away from where it was raised. The wind, except in deserts, puffs sand and dust from river flood plains, upper delta plains and low coastal sea areas, especially low coastal areas above high tide level. All this material is redeposited closer or further from the original place in the form of special sediments and sedimentary bodies, which are known as *aeolianite*, aeolian dunes and mega-dunes or loess (Section 5.5.4.1.4).

5.2.2.3. Glacial Processes

The glaciers carry and transport large amount of materials by erosion and scraping of the sides and bedrock on which it descend in lower areas. Material transported by glaciers is not sorted because of its incorporation in an ice mass. There is no possibility of selection of detritus on grain size. Therefore, sediments deposited from glacier's transportation are extremely poorly sorted and contain smallest particles and up to several decimeter-diameter blocks, and even feet, as it is the case with diamictite (Section 5.5.2). The coarse and fine-grain detritus can be transported in the form of iceberg that floats on rivers, lakes or seas. Such material is deposited on the river, lake or sea bottom, after the gradual melting of icebergs and more often in their grain size. The mineralogical composition differs significantly from the usual lake or marine sediments.

The role of the glaciers in the transfer of material is limited only to areas with permanent ice and snow cover, particularly on high mountains with glaciers and moraines. A part of ice and snow melts during the summer and the rock debris accumulates in the form of moraines.

Moraines are composed of materials of different sizes and different types of sediments and sedimentary rocks (till, tillite, and diamictite). The sliding speed of glaciers is very different, depending on the angle of slopes, ice thickness, the width of the surface over which it moves, roughness and climate changes, particularly of temperature. In general, the speed of glaciers can be of only a few millimeters and up to several meters per day. The accumulated material can still carry with the water that originates from melting of ice in the forehead. These are fluvioglacial flows that have high energy and can strongly erode the surface.

5.2.3. Deposition

The deposition of material transferred by water, wind or ice begins to cease at a time and place when the power of water or wind or ice becomes too weak to continue moving all the materials. The glacier moves down on lower region and starts melting. The deposition processes are very complex and form different types of layer and shapes of sedimentary bodies (Section 5.3 and Chapter 7). There are three main ways of settling of material and sedimentary filling of space namely aggradation, progradation and retrogradation (Fig. 5.4).

Aggradation is the deposition process in which depositional area fills with vertical stacking of sediment from the thick layer of water, as for example the case of deposition in the deep water far away from shore. The new sediment settles just above the previously deposited material (Fig. 5.4(A)). Aggradation is simply the increase in land elevation step by step due to the deposition of new sediment in areas in which the

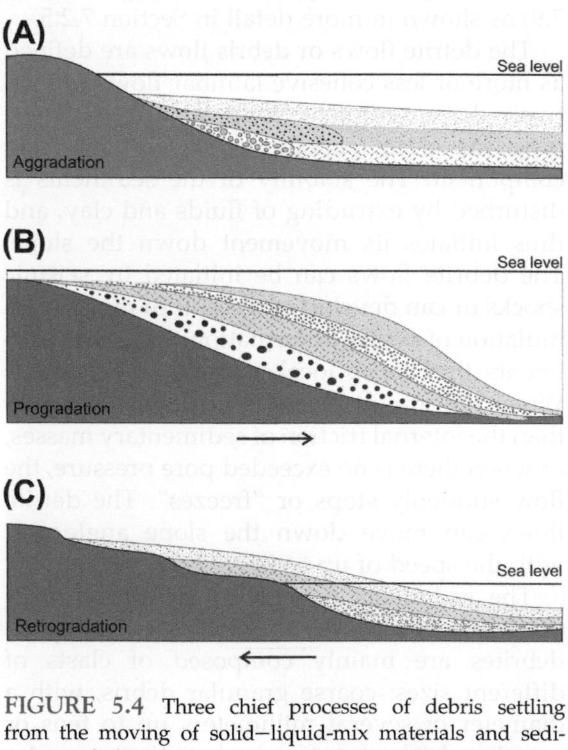

FIGURE 5.4 Three chief processes of debris settling from the moving of solid–liquid-mix materials and sedimentary feeling such as (A) Aggradation, (B) Progradation, and (C) Retrogradation.

supply of sediment is greater than the amount of material that the system is able to transport, often resulting subsidence. It typically includes lowland alluvial rivers, deltas and fans.

Progradation is the deposition process in which the depositional area fills most of its edges toward the center. The material deposits from the coast and moves progressively further toward the center of the depositional area (Fig. 5.4(B)). Progradation is the growth of river delta farther out into the sea over time, such that the volume of incoming sediments is greater than the volume of the delta that is lost through subsidence, sea level rise and/or erosion. The youngest sedimentary depositional units are usually not in its entire propagation deposited on the previously deposited units. The younger units settle on sediments deposited by aggradation rather than progradation, providing that portions of deposition by aggradation are significantly smaller than those formed in progradation. The consequence of progradation deposition is depositional sequence in which it does not always match the progradation sediment units. Progradation is the most prominent process of deposition in river deltas, turbidity fans, the deposition of fragments of skeletal organisms in coral and other reefs. The progress of coral reefs and other similar sediments deposits on the slope of the ridge. Progradation is very common in the tertiary deposits of Pannonian Basin, and has an important role in mutual relations between collector and isolator rocks in the oil and gas fields (Fig. 5.24).

Retrogradation is a process of deposition in which sedimentation area expands due to relative or global lifting of sea level, i.e. sinking of depositional area or rising of global sea level. Retrogradation is generally characteristic for the transgressive cycle during which the shallow-sea sediments precipitate farther and farther toward the mainland along with the influx of sea and coastline moving increasingly into the mainland (Fig. 5.4(C)). In this way over the initially deposited shallow marine sediments precipitate sediments of deeper and deeper waters. In other words, the process is the landward change in position of the front of river delta with time and occurs when the volume of the incoming sediment is less than the volume of the delta that is lost through subsidence, sea level rise and/or resulting erosion.

The gradual decline in the power of water precipitates the largest or larger particles first, and then all the finer particles. The major part of coarse river sediments deposit in alluvial fans (Section 7.2.2; Figs 7.1 and 7.2), deltas (Section 7.2.2; Figs 7.3 and 7.4), and the fine-grained sediments on the flood plains in floods, i.e. discharge from the river bed. A huge amount of sand and muddy sediments accumulates on the tidal plains and the sand deposits on the shallow sandbanks and sandy beaches (Section 7.2.3) and in the turbidity range (Section 7.2.5).

The turbidity current carries huge amounts of assorted sedimentary material down the slope and arrives on the flat bottom first. It, fairly and quickly, deposits majority of the large grains at that area from the head of turbidite flow within minutes or hours. All the sands and even finer grains are deposited in next few days and weeks, and the remaining tiny particles of clay from the tail of turbidite flow settle only a few hundred or even thousands of years duration. In such deposition of turbidite flow crop-up fan-shape sedimentary body or "turbidity fan" with proper vertical and lateral sorting of grains and particles, specific layered shapes and textural—structural features (Section 7.2.5). These sediments are known as "turbidites". Large quantities of coarse granular debris deposit by gravitational flows at the foot of steep cliffs or mountain ranges in the form of talus on land or debrites under the sea (Fig. 5.24; Section 7.2.4).

The alluvial fan, lake and marine deltas, tidal plains, shallow-sea sandbank, as well as turbidity fans and debrites are the most important sedimentary bodies with characteristics of oil and gas collectors. Large amounts of calcium carbonate can be deposited on the bottom of waterfalls and in freshwater lakes through the life processes of plants in the form of calcareous tuff (Section 5.7.1.3.2; Fig. 5.43, etc.). Materials created by direct excretion of minerals from the water, either inorganic or organic, or biogenic processes precipitate in the sea. Similarly, large quantity of skeletons and shell of organisms (Section 5.7.1) settle down in shallow and warm seas. In this way, carbonate platforms are formed with different shapes and types of carbonate sedimentary bodies, which can also be an important reservoir rocks for oil and gas (Section 7.3.1). The deeper seas/oceans are favored location for deposition of the finely grained pelitic sediments (Section 5.5.4), the carbonate mud (Section 5.7.1.1.2), and the silicon sediments (Section 5.7.4).

The sediments and sedimentary rocks receive special shapes and the textural–structural features as a result of different mechanisms of transport and deposition of material, lithification or diagenetic processes. The exploration geologists, and especially sedimentologists, can reliably determine the conditions and the environments of deposition of sediments and the rocks deposited in geologic past through systematic study. The inferences can logically be experimented to demonstrate the depositional environment and process (Fig. 5.5).

FIGURE 5.5 Alternate thin layers of extremely fine-grained calcium carbonate (light and white) and metallic minerals (sphalerite, galena and pyrite—gray, yellow and orange) deposited within week duration out of fine clayey diamond drill cuttings passing through zinc-lead mineralization in limestone host rock. The deposition occurs within the drill-sump of return water. The particles settle in rhythmic layers due to difference of Sp. Gr. between limestone and metallic minerals and dried fast at day temperature of ~50 °C. The recently formed compact sediment resembles laminated limestone. Image is taken at drill site of Lennard Shelf Exploration Camp, Meridian Minerals Limited, Western Australia in midsummer of 2011 by the Author.

5.2.4. Lithification

Lithifaction is a complex set of physical and chemical processes known as *diagenesis*. These are processes by which bulk, soft, water-saturated loose deposits gradually become solid sedimentary rock. The major diagenetic processes of lithifaction include compaction and cementation.

Compaction is the process of mechanical compression or packing of soft, loose, porous and water-soaked sediment with increasing pressure due to the weight of new sediments, i.e. due to increased depth of the overlay. The older compact sediments fall deeper under the increasing amount of new sediments, so that water eliminates with compaction flow and rises into the upper layers causing chemical diagenetic changes—the secretion of new minerals. The compaction results decrease in porosity of the sediments and its gradual solidification with the changes in mineral composition, and excretion of new mineral resulting in cementation of the sediments or rocks (Section 5.5.5).

Cementation is the process of excretion and crystallization of minerals in pores of the deposits. The new or authigenic minerals are called *cement* (Fig. 5.18). It leads to a decrease in porosity and to interconnection of individual grains and components in solid rock. There are other chemical diagenetic processes, other than cementation, that play an important role in solidification of deposits and their gradual transition in solid rock. The most important among them are dissolution, pressure dissolution, authigenesis, recrystallization, silicification and dolomitization (Sections 5.5.5, 5.7.1.4, and 5.7.2.1).

5.3. TEXTURE AND STRUCTURE OF SEDIMENTARY ROCKS

The sedimentary structure refers to all the features caused by their mutual relations, spatial distribution and orientation of individual components, as well as external and internal morphological forms of sedimentary rocks. The texture of rock includes the grain size, relationships, distribution and shapes of the mineral components, i.e. internal microdistribution of its constituent parts.

Primary sedimentary structural shapes (layers and laminations) are formed during deposition or shortly after, and certainly prior to compaction and lithifaction of deposits. The primary textural–structural shapes also add up the forms, the appearance and features that are in the sediment or by simultaneous deformation with deposition or shortly after deposition before covering with new sediments. All other forms that in sediments and rocks are formed after deposition, i.e. during diagenetic processes, are called secondary textural–structural forms.

In general, textural–structural shapes and structural characteristics of sedimentary rocks belong to its most important feature. The primary textural–structural sediments form under the direct result of the conditions that existed in transport and deposition of material and the resultant of all processes in the environment of deposition. The secondary textural–structural features are the result of complex diagenetic processes (recrystallization, pressure dissolution, compaction, and chemical diagenesis).

The grain size is the most important characteristic of clastic and calcareous sediments. It is closely related with physical, chemical and hydrodynamic conditions that existed during formation, transfer and deposition of debris. The investigations of grain size of clastic sediments are essentially significant for determining the conditions of weathering and breaking methods and mechanisms of transfer and deposition of material. It is also necessary for the classification and nomenclature of the clastic sediments based precisely on the grain size. The calcareous sediments originate in different conditions and present various textural–structural features, classification based on the size and shape of grains or skeletal and nonskeletal components (e.g. terminology intrasparite, intrasparrudite, grainstone, rudstone, biocalcrudite, biocalcarenites, etc.; Section 5.7.1.3). The degree of crystallinity and crystal size are the main textural features of chemical sedimentary rocks. The size and crystalline forms are microcrystalline, euhedral macrocrystalline, mosaic structure, etc. (Section 5.3.2).

The knowledge of textural–structural characteristics and layer forms of sedimentary rocks are valuable information and necessary for the reconstruction of the conditions and environment of deposition, such as, depth and water energy, transfer method of material and mechanisms of deposition, flow direction, role of organisms, shape of sedimentary bodies, facies, etc.

5.3.1. Bedding

The main structural characteristic of sediments and sedimentary rocks is bedding with wide variations that include irregular, regular or rhythmical and cyclical, gradation, sloppy, flazer, lenticular and wavy, horizontal, oblique and sinuous. The bedding is one of the first characteristic that is observed in the field as a fundamental feature of sedimentary rocks. This phenomenon is more or less clear separation of individual textural–structural, lithological or grain size distribution of unique members or "layers" in sedimentary rocks.

The layer is the geological body generally of uniform composition and internal layer forms throughout the thickness of sedimentary package. The sediments deposited below and above are separated by some discontinuity, change in particle size or mineral composition. The thickness of layer is not always the same size, but varies in wide limits, depending on the morphology of sediments below, mechanism, method and conditions of deposition and textural–structural shapes. The layer can be considered as lenticular body of different thickness and propagation, and in certain circumstances and environment of deposition and wedge-shaped body, e.g. the case for foreset of deltas and underwater dunes (Sections 7.2.2 and 7.2.3).

The separation of individual layers can be mechanically represented by the open layer or layer plane. In the absence of such mechanical separation, the bedding can be clearly identified by changes of smaller and larger components, the distribution of fossils, organic matter, change of color, porosity and methods of cementing or changes in forms, the types of rock and mineral components, changes in internal features (changing layers with oblique bedding and lamination). The layers can be separated from each other with thinner, usually soft, interlayers, like clay, silt, marl or sand, deposited between solid layers of limestone, sandstone, conglomerate and chert.

The sedimentary rocks sometimes do not include any bedding and are massive or nonbedding rocks, which is often the case in late-diagenetic dolomites and breccias. The massive rocks are often without internal organization or internal forms and not just on the rocks with no open bedding. The distinctness of two main beddings is (1) external or irregular, rhythmic or cyclical bedding, and, (2) internal bedding.

5.3.1.1. External Bedding

External bedding with the presence of separation of individual layer is the most significant structural characteristics of sedimentary rocks. The basic unit of the external beddings is the layer created as a result of sedimentation in uniform physical, chemical and/or biological conditions under constant and continued deposition. If any of these parameters changes, then there is a change of sedimentation, thus forming a new layer. The layers differ with regard to thickness by bedding (>1 cm) and laminae or lamination (<1 cm). The vertical layers are classified into two categories, (1) irregular and (2) regular, considering the grain size sorting with different or same lithological or textural–structural characteristics.

1. Irregular bedding includes a completely uneven and asymmetrical vertical sorting of layers of different types of petrographic or textural–structural forms (e.g. ACBADC), as a result of a completely irregular changes in sediment area.
2. Regular, cyclical or rhythmic bedding with the vertical sequence of alternate petrological and/or textural–structural different layers at uniform vertical periodic repetition of such changes (Fig. 5.6). The group of layers in one cycle of repetition is called a sequence, cycle, parasequence, cyclothem, megacycles or megasequence, according to periodic repetition with respect to the dimensions duration and manner of deposition origin.

5.3.1.2. Internal Bedding

Internal bedding includes textural–structural forms within a single or multiple layers. Most often, and for the interpretation of conditions and environment of deposition, the major types of internal beddings are horizontal bedding, convolute bedding, planar cross-bedding, trough cross-bedding, hummocky cross-bedding, flaser bedding, wavy bedding, lenticular bedding and graded bedding.

Horizontal bedding is a type of internal bedding in which each layer consists of many thin (0.1–10 mm) parallel laminae. Each individual lamina is characterized by unique granulometric and petrographic composition (Fig. 5.7). This

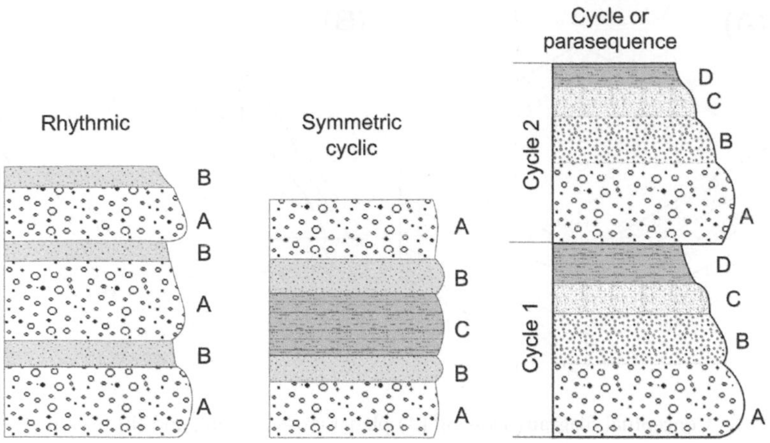

FIGURE 5.6 Regular bedding with vertical sequence of rhythmic, cycles and parasequence style of sedimentary deposition and lithification.

type of bedding typically occurs as a result of faster or slower changes in the deposition of fine-grain detritus, silt and fine sand, or changes in temperature and concentration of water.

Convolute bedding is a special type of bedding inside the layer (thickness >10 mm) or within the lamina (thickness <10 mm). It is characterized by convolute laminae within the layer that is in sequence with the top and bottom layers. The convolute bedding is generally the best and most developed in the fine sand and dust sediments. It occurs as a result of hydroplastic deformation of still unhardened sediment under the influence of strong water currents that flow over such deposits. The bedding is formed by the mutual friction between the current flow and sediment. It can also form in the postphase of sedimentation as a result of hydroplastic deformation due to the sudden displacement of water or the sudden release of gases from the sludge.

Cross-bedding is the most common and most significant layer in the form of sedimentary rocks. It consists of groups of mutually parallel lamina or layers deposited askew in relation to the outer surface layer (Fig. 5.8). The cross-bedding refers to horizontal units that are internally composed

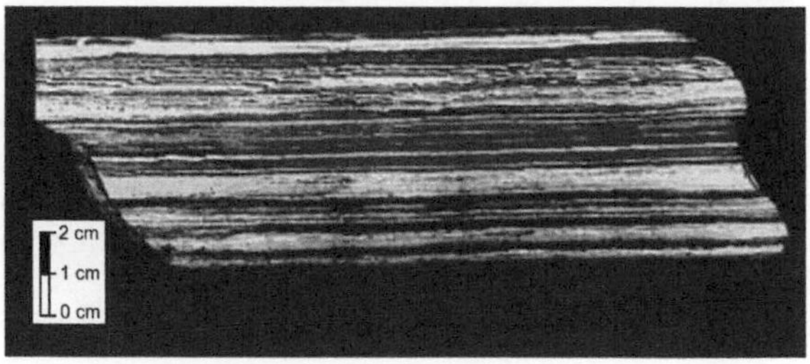

FIGURE 5.7 Horizontal bedding—vertical changes of dark clay-marly laminae and light carbonate laminae.

of inclined layers. The groups of units of the same or similar slope are called *set*. The layers and laminae within the sets that have a slope in the direction of input of materials are known as *foreset*, for example, by precipitation in the delta (Fig. 7.8(B)), large currents or underwater dunes. The cross-bedding can be designated as (1) planar cross-bedding, (2) trough cross-bedding and (3) hummocky cross-bedding according to the shape and characteristics.

Planar cross-bedding is characterized by more or less planar boundaries between sets (Figs 5.8(A) and 5.9) and is characteristic of river sediments, and the most important feature of eolian dunes and foreset underwater dunes.

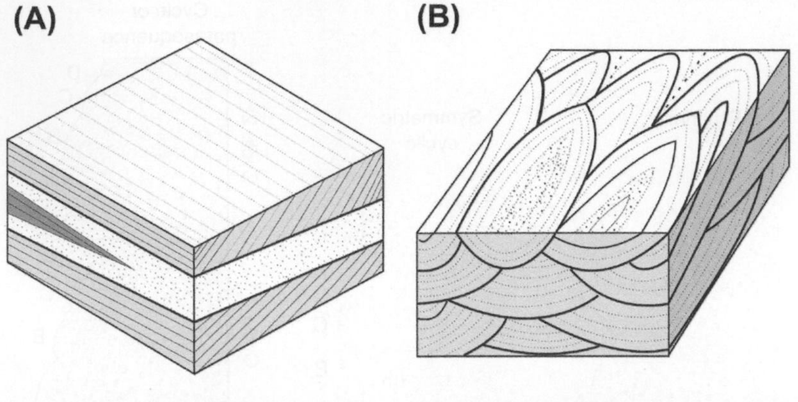

FIGURE 5.8 Conceptual diagram showing (A) planar cross-bedding and (B) trough cross-bedding.

FIGURE 5.9 Field photograph of planner cross-bedding. *Source: Prof. Biplab Bhattacharya.*

FIGURE 5.10 Field photograph of trough cross-bedding. *Source: Prof. Biplab Bhattacharya.*

Trough cross-bedding have surface of sets in shape of trough (Figs 5.8(B) and 5.10). More sets and cosets of similar thickness appear regularly in between layers. This shape can be well observed in cross-section perpendicular to the axis of the trough. It is characteristic of fine-grained sandy river sediments deposited in river beds, and along with planar cross-bedding in sandy sediments of intertwined rivers.

Hummocky cross-bedding is a special form of cross-beddings, characterized by wavy or hilly sets of inclined lamina. The inclined lamina are parallel to the base, and a little wavy, and placed one above the other, alternating convex and concave curved forms, often showing erosion on older lamina. Hummocky cross-bedding is generally found in sediments of granulometric composition to coarse powder and fine sand (0.03–0.25 mm). It occurs in shallow shelf sands below fair weather wave base, which is usually located at 5–20 m depth. It is an essential feature of tempestite or storm sediments, i.e. sediments deposited in stormy waves.

Flaser, lenticular and wavy bedding are of great importance for the reconstruction and interpretation of conditions and hydrodynamic features in the environment of deposition. The gradual transition from flazer through lenticular bedding has a certain interrelationship of deposition of sand and mud (clayey-dusty detritus) and indicates decline in water energy (Fig. 5.11).

FIGURE 5.11 Conceptual diagrams depicting the flaser, lenticular and wavy bedding composed of finely granular sand (light), and clay and mud (dark).

Flaser bedding occurs in fine-sandy and weakly clayey and muddy sediments with wave and current ripple. It is characterized by cross-laminations draped with silt or clay. Flaser beds form in environments where the strengths of current flow fluctuate considerably, thus permitting the transport of sand in ripples, followed by low-energy periods when mud can drape the ripples.

Wavy bedding is genetically similar and associated with flaser bedding. It is characterized by equal amounts of sand and mud in composition. The clay–argillaceous deposits, above sands containing the wave or current ripple, precipitate in the form of continuous layers, unlike flaser beddings. The sandy wavy layers are mutually separated by a layer of clay or argillaceous material (Fig. 5.11). The sand with ripples is deposited in the period of stronger energy of water, and clay and/or pelite during weak energy of water which, in comparison to what in creation of flaser beddings last longer. Wavy bedding in modification of flaser and lenticular bedding usually occurs in the tidal plains and in the tidal environments.

Lenticular bedding is characterized by the appearance of individual, mutually laterally and vertically disconnected lenses of sand deposits within clay–silt residues (Fig. 5.11). It occurs in quiet, shallow water, usually in tidal environment of deposition. The structure is common on the foreheads of the marine delta and small lake delta, where the deposition of mud is predominant with occasional input of sandy accumulation in short periods of excessive supply of water. The sand deposits in the form of isolated unrelated wave or current ripples or lenses, specifically in stronger flow of water.

Graded bedding is characterized by a gradual decrease in grain size from the base to the top layer. The grain size ranges from gravel in the bottom, through the sand to muddy sediments at the top. The deposit usually forms in interval of T_a Boumina turbidity sequences. It follows an initial deposition of large grains and graded progressively upward finer grains with decrease in transport energy as time passes. The differences in speeds of movement control sequence of deposition of coarse and fine-grained debris (Fig. 7.10). The gradation can occur in deposition from the suspension in the final phase of severe flooding and high tides. The gradation can also occur by deposition of volcanoclastic material with volcanic eruptions. The volcano ejects large-size particles at the beginning of the eruption and followed by weakening of the eruption progressively charging smaller material.

5.3.1.3. Upper Bedding Plane Structures

In the upper bedding planes, most important and common forms of structures are desiccation cracks ripple marks, and occasionally even reptile footprints and imprints of rain drops.

Desiccation cracks or *mud-cracks* form when drying mud shrinks. The shape and size depend on their mineral composition, grain size, intensity of drainage, thickness and homogeneity of the layers and deposits. The cracks appear in the upper bedding planes of clay, silt, mud, clayey–sandy sediments, and muddy or micrite formed by early diagenetic limestone and dolomites. The cracks are arranged on the surface layers in more or less regular polygonal shapes (Fig. 5.12), generally V-shape in cross-section and in the lithified sediments. The cracks are

FIGURE 5.12 Desiccation cracks developed in recent sediments at Papuk, the largest mountain range in the Slavonia region in eastern Croatia.

regularly filled with sediment of upper layer rock materials. The formation of cracks are limited to the subaerial conditions in continental environments, especially alluvial flood plain, drained ponds, tidal and supratidal environments, and primarily of vast tidal plain, where water is rapidly lost from the deposits or sediments.

Ripple marks are systems of microridges and valleys, like surface of wavy sea and desert sand dunes, and often observed on the upper bedding planes of sandstone and limestone layers. The ripples (waves) are described by measuring their height or amplitude and wavelengths to ascribe the morphology, such as symmetric, asymmetric, and transverse ripples. These elements, however, provide very important information on the conditions and environment of deposition, particularly on energy and the way of the water flow. Ripples appear in groups and always on large planes (Fig. 5.13). The ripples are created by moving the unbound, mainly sandy sediments with water currents.

The wave and current ripples differ according to the origin:

- Current ripple marks occur in one-way transfer of sand with water flows or currents, i.e. currents which are moving in one direction only for long time, such as in tidal currents. The current ripples are characterized by the orientation of longitudinal axis of ripples transversely to the direction of flow and properly arranged crests and troughs. Little current ripple marks occur in muddy and fine-grain sand and limestone sediments in river environments, on the tidal plains and

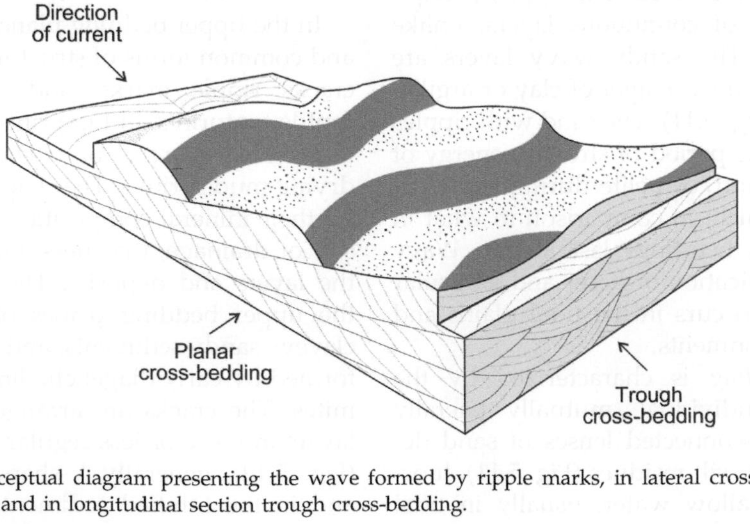

FIGURE 5.13 Conceptual diagram presenting the wave formed by ripple marks, in lateral cross-section showing the planar cross-bedding, and in longitudinal section trough cross-bedding.

sandy beaches. Large current ripple marks, known as mega-ripples, sand-waves underwater (subaqueous) dunes, occur in higher energy of water in river environments, tidal channels and backwaters. These are frequent on the tidal plains, sandy beaches and coastal shallows where the difference in the level of low tide and high tide is >1 m or shallow water exposed to strong tidal currents with waves. The current ripples are excellent indicators of energy and water depth, and if they are asymmetric also can indicate deposits transport directions (Fig. 5.14).

- Wave-formed ripple marks generally forms in weak currents because of relocating sand in oscillation motion of water with waves, i.e. in continuous circular motion of water. These types of ripples are distinguished by long, mutually parallel with arched crests. They are typically symmetrical. The most common ripples are on the tidal plains, beach front or tidal zone and lake beaches and are common in lagoons, sandy beaches below low-tide, supratidal zone and lakes.

Footprints of reptiles, particularly large Jurassic and Cretaceous reptiles, the dinosaurs, are often rare in the upper bedding plane. These marks can be preserved only in special conditions, where the reptile left footprint on the soft sediment in tidal shallows with low water supply. The high-quality and authentic prints are expected in the peritidal sediments formed under conditions of low water energy, especially those containing cyanobacteria (blue-green bacteria and blue-green algae) meadows with mud sticking property. Thereafter, gentle tidal currents bring new sediment that fills and covers the total footprints, and gradually the entire surface layer. The footprints are preserved in such environments excelled by rapid cementation of deposits. These footprints of prehistoric era are particularly significant from the sedimentary and paleontological point (Fig. 5.15).

Raindrop imprints formed during collision of large drops of rain on the soft clay, sand, and rarely on carbonate mud, which is located above the water level. The rain drops can be preserved, only if, new sediment starts to deposit quickly, while there has been no erosion or destruction of the impression. Prints of the rain drops are excellent indicators of environmental conditions and precipitation since they appear only in continental environments or muddy supratidal zone.

FIGURE 5.14 Shallow current ripple bed forms on arenitic sandstone, Kolhan Group, Chaibasa, Jharkhand, India. *Source: Prof. Joydip Mukhopadhyay.*

FIGURE 5.15 Dinosaur footprints at Istra, the largest peninsula in the Adriatic Sea, Croatia.

All these print impressions are authentic indicators of the youngling direction of stratigraphic column.

5.3.1.4. Lower Bedding Plane Structures

There will be many types of inorganic forms on the lower face of the plane surface. These will play significant role in determining the sequence of layers. These forms are found most commonly in turbidite deposits. They are divided into two genetic groups:

1. Traces of erosion resulting from the action of turbidity currents or vortex flow.
2. Traces of erosion resulting from the action of various items that are carried by water currents.

The most common signs of flow, among the traces of erosion, are flute casts, vortex casts and erosional channels, and are described.

Flute casts are triangular or spindle-shaped protrusions on the lower surface of sandstone bed. The flow of input vortex currents is more protruding on a narrow front than on the wider part of the back, which gradually disappears on the flat surface. The flutes are narrow, elongated, straight, parallel ridges generally consisting of till, sand and clay. The length of the flute casts generally ranges between 2 and 10 cm, and sometimes to 1 m. The casts occur in filling of depressions on the muddy bottom at the beginning of movement. The depressions are filled by sandy sediment and lithifaction on the lower surface of sandstone on muddy bottom. It will remain as protrusion which has greater convexity on the input than on the output of the resulting erosion and sand sediment-filled valleys. The flute casts can easily determine the direction of the paleotransport.

Vortex casts are spiral protrusions on the lower surface of sandstone. The casts have shapes similar to the spiral end of snail home. The dimensions and shapes are various. The height of protrusions generally varies from 1 to 3 cm, and their diameter between 6 and 20 cm. The casts are sand-filled depressions formed by erosion of strong vortex currents or turbulent flows. The vortex casts are good forms for determining the paleotransport directions and for reconstruction of hydrodynamic conditions in the environment of deposition.

Erosional channels are erosional forms created with the removal of sediment further into the depositional area from the portion of one or more layers. The channels have width ranging between few decimeters and tens of meters and length between few decimeters and hundreds of meters. The most common occurrences are by erosion of clay or marl bottom.

The most common traces of moving object on bottom are signs of cutting, rolling and pulling.

Groove marks are created when sharp object is dragged across the surface of muddy substrate. The length varies between few decimeters and few meters, width between one-tenth millimeter and several centimeters, and height between several millimeters. The groove marks usually appear in groups with parallel arranged linear prominence. In some places, it gradually transforms into traces of pulling. The marks occur as objects scours out a groove along the top of the bed, which is later filled by coarser sediment. The groove marks are good indicators for determining direction of paleotransport.

Impact casts are short, narrow, asymmetrical shapes embossed on the lower surface. The input end is thin and gradual, and the output end is wide with the sharp end. The lower surface looks like small wedges. The casts appear in groups, often with a different mutual orientation. The impact casts are formed by filling cavities in the muddy bottom, which has left with current carried object into the bottom (Fig. 5.16).

Bounce marks appear on the lower surface as a straight series of small bumps arranged of approximately equal, millimeter to centimeter intervals. The bounce marks occur in filling of depressions or mold, formed on the clay bottom when object is carried by vortex current. It makes contact with clay bottom and then bounces back up in the turbidity flow.

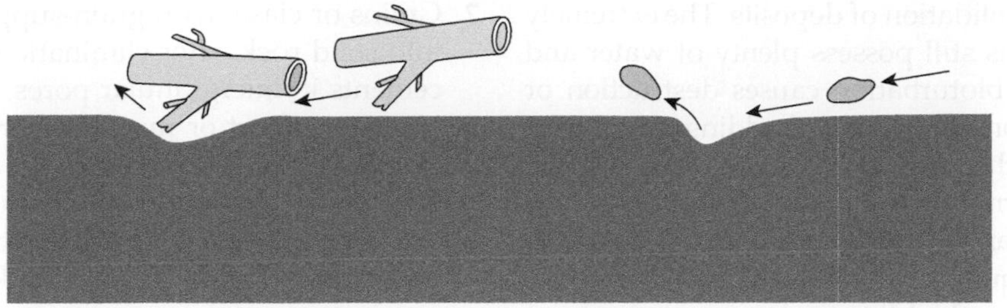

FIGURE 5.16 Conceptual diagram showing the formation of impact casts in soft sediments.

5.3.1.5. *Forms Created by Underwater Slides and with the Destruction of the Layers*

Slump or structure of underwater sliding is created by underwater sliding and the destruction of layers. It occurs in yet unrelated or semiplastic sediments due to the increasing angle of inclination of the bottom. The structure represents the occurrence of more or less deformed layers, often interrupted with continuity of one undeformed layer between mutually straight and parallel (concordant) arranged layers. The mechanism works in sliding of one or more layers of partially lithified or semiplastic sediments on clay substrate. The sliding takes place by gravity at an angle of inclination of the bottom of only 1–3° (Fig. 5.17). The slumps are common in sediments formed by rapid and intense accumulation of deposits, such as deltas and canyons, slopes of carbonate platform as well as upper parts of submarine slopes. The slump scars can be formed by underwater landslides of large dimensions with complex structure. In a way it is created by slip or sliding of massive sediments. The slumps occur along the low-angle inclined slopes in sedimentary rocks. Therefore, the incidence indicates the mechanism either on the synsedimentary tectonics such as subsidence of basin bottom and/or uplift of the coast and the mainland, or deposition in sedimentary body of inclined sides, e.g. the case on the front of deltas fans (Fig. 7.3; Section 7.2.2).

Bioturbation is a common name for all kinds of changes in unrelated sediments and soils, formed by the activities of organisms. It depicts the network of soils and sediments or forms as a result of the life activities (moving, digging, crawling, eating, and making dwellings) left in unbound or poorly consolidated sediments by plans and animals. The organisms itself are not preserved. The largest part of such forms occurs immediately after deposition and in the first

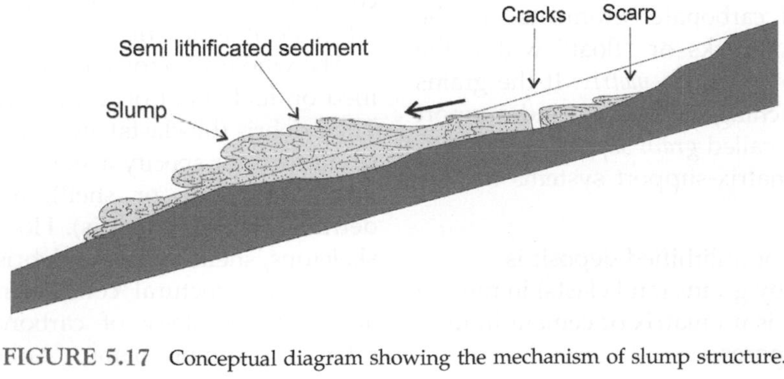

FIGURE 5.17 Conceptual diagram showing the mechanism of slump structure.

phase of consolidation of deposits. The extremely soft organisms still possess plenty of water and oxygen. The bioturbation causes destruction or transformation of primary bedding formed in sediment with inorganic material. Many of the primary internal textural and structural features may disappear by bioturbation process, e.g. horizontal, oblique or sinuous lamination, slope, flaser and wavy bedding.

Bioturbation is very common incidence under slow deposition and helps in the interpretation of conditions and environment of deposition. There is no possibility of inhabiting sediments with organisms, especially mud eaters, in rapid accumulation of deposits. The bioturbation sediments are indicators of slow sedimentation with small amounts of input and sedimentation of detritus in oxidative conditions.

5.3.2. Packing of Grains

The sedimentary rocks, especially clastic and many biochemical and chemical carbonate rocks, have grains, matrix, cement and pores. The grains in clastic sediments are clasts, i.e. solid substances remaining after the physical and chemical weathering of older rocks. The clasts are deposited after the transfer by water, ice or wind, as clastic sediments on land, in freshwater or sea. The grains (particles) in the biochemical and chemical carbonate rocks are the primary structural components formed by deposition of skeletal and nonskeletal carbonate materials within the depositional area. The grains in clastic and biochemical carbonate sediments are the basic structure of rocks or "float" within the dense mass, known as the *matrix*. If the grains or clasts are touching each other and support one another, it is called *grain-support* (Fig. 5.18).

The various matrix-support systems are the following:

1. Clast-support of unlithified deposit is characterized by grains (and clasts) in mutual contact. There is no matrix or cement in the intergranular pores.
2. Grains or clasts with grain-support cemented into solid rock, after elimination of mineral cements in intergranular pores.
3. Matrix-support or mud-support is characterized by the matrix in which the grains of clasts are not touching each other ("swim" in the matrix). Lithification of matrix is resulting in solid rock of matrix-support.
4. A general example of the structural components of a sedimentary rock with grain-support is consisting of grains and matrix.

In the mineralogical and petrological terms, the grain that built sediments and/or sedimentary rocks can be of different composition so that the grains are the following:

1. Individual mineral
2. Fragments of rock
3. The whole skeleton or shell
4. Fragments of skeletons or shells (bioclast)
5. Accumulations of carbonate minerals formed by chemical, biochemical and organic processes of abstraction from the seawater or freshwater.

The mineral grains and rock fragments of silicate composition consisting of quartz and other silicate minerals are usually covered by the common name, *siliciclastic grains* or *siliciclastic detritus*. The rock fragments of carbonate composition, i.e. fragments of older limestone and dolomites, unlike to the siliciclastic detritus, are commonly referred to as *carbonate lithic clasts* or *carbonate lithic detritus*.

The whole skeletons or shells may not be lithified on its habitat or in the location of growth. Their debris (bioclasts) are often matured as fossil detritus, or specify a skeletal detritus (for the whole skeletons or shell), and as bioclastic detritus (for fossil debris). However, the whole skeletons, shells and their debris, as the primary carbonate structural components in petrology and sedimentology of carbonate rocks, have often a common name "carbonate detritus".

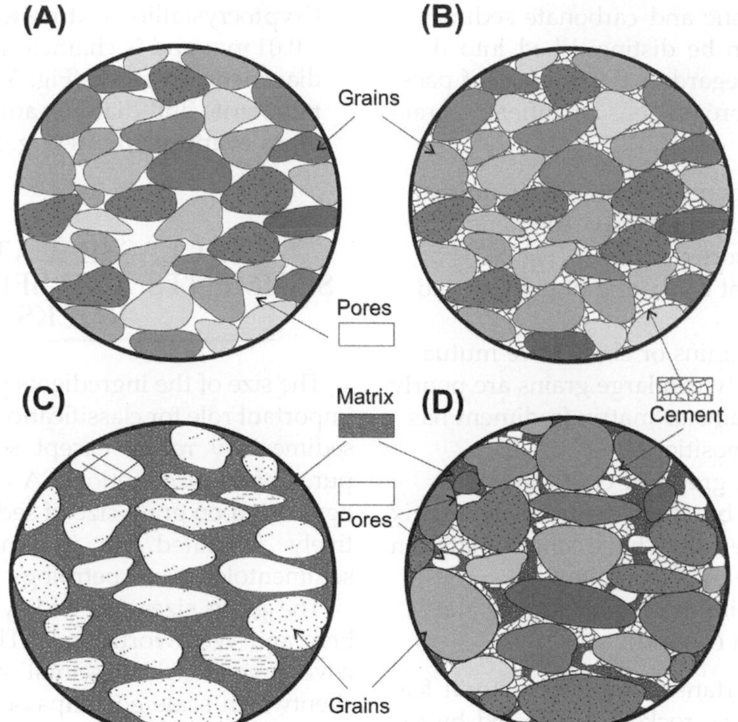

FIGURE 5.18 Conceptual models showing the packing of grains, cementation, matrix formation and lithification of clastic sediment. The various matrix-support systems are: (A) clast-support of unlithified deposit without matrix or cement in the intergranular pores, (B) grains or clasts with grain-support cemented into solid rock, (C) matrix- or mud-support with the grains "swim" in the matrix, and (D) finally, common example of sedimentary rock with grain-support of grains and matrix.

These grains usually undergo transport and deposition under the influence of ocean currents, waves or tides.

Matrix is the fine detritus, transported and deposited together with the grains (Fig. 5.18(C)). The matrix of sandstones is typically silt or clay. The same in conglomerates and breccias are fine sand, silt and clay. In limestone sediments the matrix is lime mud which is limestone lithified in micrite. The detritus with dimensions of <0.030 mm is marked as matrix. The matrix is placed in the sediment or in the interstices of grain or grains "swim" in it. In the interstices grains, the matrix is usually found with grain support. The grains or large clasts often swim in mud or muddy support (Fig. 5.18(C)).

Cement is the mineral substance secreted in pores between grains after their deposition. It is postsedimentation component originated from secretion of mineral substances from pore solutions. The pores are free spaces between the grains in which there is no matrix or cement (Fig. 5.18(A) and (D)). The free spaces are usually filled with gases (carbon dioxide, methane, hydrogen sulfide, and nitrogen dioxide) and/or water or oil.

The distribution and orientation of grains in the sediment depend on the conditions during transportation, deposition and mechanical diagenesis, especially compaction or compaction due to pressure overlays. The effect of compaction is strongest in the clay sediments, and very weak in large clastic and carbonate sediments. The sediments can be distinguished into three main types with regard to the manner of packing, sorting and proportions of different grain sizes:

1. Clast-support: grains or clasts have mutual support, i.e. between the coarse grains or clasts are well sorted by finer (Fig. 5.21) or matrix (sediment consisting of pebbles and sand).
2. Clast-support: grains or clasts have mutual support, and between large grains are poorly sorted small grains or matrix (sediment has polimodal composition).
3. Matrix-support: grains or clasts are not in mutual support but they "swim" in the matrix (Fig. 5.18(C)), i.e. not to have contact matrix in which they "swim" has polimodal composition (with mud, silt and clay and contains a grain of sand).

The most important textural–structural features of sedimentary rocks are reflected by the ways of packaging, and relations between the grains, matrix and cement. The key principles of classification of sedimentary rocks are based on these features. The chemical sedimentary rocks, i.e. recrystallized limestone, dolomite, evaporites and some siliceous sediment, are composed of crystals of chemical origin and not detrite grain. These rocks have crystalline texture according to the size of the crystal and divided into the following three types:

1. Macrocrystalline texture with crystals >0.1 mm and is especially common in late-diagenetic dolomite (Fig. 5.67) and recrystallized limestone, and is a common in some types of anhydrite.
2. Microcrystalline texture with crystal in diameter between 0.01 and 0.1 mm and is common in late-diagenetic dolomite, recrystallized limestone, anhydrite, gypsum and some types of chert (Fig. 5.70), radiolarite and diatomite.
3. Cryptocrystalline texture with crystals <0.01 mm and is characteristic of early diagenetic dolomite (Fig. 5.66), cherts, radiolarite and diatoms and, also certain types evaporite rocks (Fig. 5.69).

5.4. CLASSIFICATION OF SEDIMENTS AND SEDIMENTARY ROCKS

The size of the ingredients plays an extremely important role for classification of sediments and sedimentary rocks, except sediments that are pure chemical secretions. A common terminology, based on grain size of sediments, is qualitatively indicated for sedimentary rocks in sedimentology and petrology (Table 5.1).

The grain sizes are widely used after Atterberg and Wentworth scale. The Atterberg scale covers geometric, decimal and cyclical. The Wentworth scale encompasses logarithmic and geometric based on number 2. The Wentworth and Atterberg scales are shown in Fig. 5.19. The Wentworth scale is used in sedimentology and petrology. The Atterberg scale is usually used in geotechnical, civil engineering, hydrogeology and engineering geology.

There are two major genetic groups of sediments and sedimentary rocks and there are several mixed sediments between these two main groups:

1. Clastic sediments
2. Chemical and biochemical sediments.

TABLE 5.1 Size of Grains in International Language

Greek	Latin	English
Psefit (psephos = gravel)	Rudite (rutus = gravel)	Gravel
Psamit (psamos = sand)	Arenite (arena = sand)	Sand
Alevrit (alevros = silt)	Lutite (lutum = silt)	Silt
Pelite (pelos = clay)		Clay

Wentworth scale

Size (mm)	Name
256–∞	Boulder
64–256	Cobble
4–64	Pebble
2–4	Granule
1–2	Very coarse sand
0.5–1	Course sand
0.25–0.5	Medium sand
125–250 μm	Fine sand
62.5–125 μm	Very fine sand
3.9–62.5 μm	Silt
1/∞–3.9 μm	Clay
1/∞–1 μm	Colloid

Atterberg scale

Size (mm)	Name
>200	Boulder
60–200	Cobble
20–60	Pebble
2–20	Gravel
0.6–2	Coarse sand
0.2–0.6	Medium sand
0.06–0.2	Fine sand
0.002–0.06	Silt
<2 μm	Clay

FIGURE 5.19 Comparison between Wentworth and Atterberg scale: the former is used in sedimentology and petrology, and the later for geochemical, civil engineering, hydrogeology and engineering geology.

5.5. CLASTIC SEDIMENTS AND SEDIMENTARY ROCKS

5.5.1. Genesis and Classification of Clastic Sedimentary Rocks

Clastic (detrite or mechanical) sediments and sedimentary rocks are composed of particles, grains and fragments that resulted from physical and chemical weathering. The physical breakage and destruction of older rocks are exogenous in origin and especially effective. These are solid particles, grains and fragments, i.e. individual particles composed of detrite or mineral grains or fragments of rocks, covered by a group name "clasts". Sediments and sedimentary rocks formed after shorter or longer transfer by deposition on land, in freshwater or sea are called "clastic sediments" or "clastic sedimentary rocks".

The four basic groups of clastic sediments are (Table 5.2) the following:

1. Cataclastic sediments that have been wholly or partly formed by the progressive fracturing and comminution of existing rock by a process known as *cataclasis*.
2. Rinsed residues, divided into coarse (rudite), medium (arenite) and fine (argillaceous) clastic rocks.
3. Residues are the remains of rocks that could not melt at the weathering and usually consist

TABLE 5.2 Two Main Genetic Groups of Sediments and Sedimentary Rocks

Clastic sediments
Cataclastic sediments
Rinsed residues
Residues
Pyroclastic sediments
Chemical and biochemical sediments

of very resistant to chemical weathered minerals (quartz and silicate minerals) or autogenous minerals, mainly clay minerals and aluminum hydroxide.

4. Pyroclastic sediments that are formed by deposition of material of volcanic origin which is ejected by the eruption of the volcano, and that the transfer in air and/or water sediment on land, at sea or in the lake along with the smaller or larger amount of deposition of sedimentary origin, or without material sedimentary origin.

5.5.2. Coarse-Grained Sediments—Rudaceous

The coarse-grained sediments and sedimentary rocks are formed by the accumulation of grain diameter >2 mm. It can be of cataclastic origin and/or belong to the coarse-grained rinsed residue. The coarse-grained clastic sediments have following main types of sediments and sedimentary rocks:

Unbound	Debris	Pebbles	Till and diamictite
Bound	Breccia	Conglomerate	Tillite

Debris are unbound angular clastic sediments, fragments of rock of more than half have a diameter of >2 mm, and dimensions larger than grains of sands. These are the typical accumulation of debris caused by sudden and rapid rockfall under the influence of gravity down the steep slopes and cliffs (stone landslides or avalanches).

FIGURE 5.20 Rockfall on the Dolomite Mountain cliffs with amazing panorama at a vertical height of 2484 m from summit, Sellajoch, Val Gardena, South Tyrol, Italy.

Rockfall or rock-fall or debris is accumulation of freely falling fragmented rocks (blocks) from a cliff face. The rockfalls are detached by sliding, toppling, or falling, that falls along a vertical or subvertical cliff, and proceeds down slope by bouncing and flying by rolling on talus or debris slopes. It ultimately settles on the land or steep rock slopes or at the foot of those slopes and is known as "rock-fall" (Fig. 5.20).

Gravel is an unbound accumulation of rocks, rarely minerals, well-rounded clasts, mostly in diameter >2 mm, and variable amounts of grain sizes of sand, dust, and sometimes clay, dust (mud).

Conglomerate is a firmly linked rock that mainly consists of well-rounded clasts in dimensions of gravel cemented by sand and mud component or rarely without it (Figs 5.21 and 5.22). The boundary between the breccias and conglomerates with certain types of coarse clastics may not be sharp and clear. These two types are of mutual crossings or clasts of particular

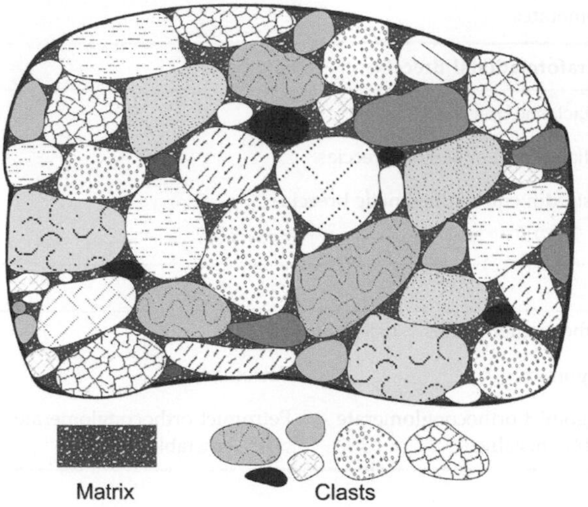

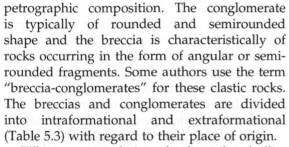

FIGURE 5.21 Conceptual diagram showing type conglomerate consisting of rounded and semirounded clasts cemented in fine-grain matrix.

FIGURE 5.22 Field photographs of conglomerate consisting of assorted grains and pebbles (rounded, semirounded and angular) of quartz (white), jasper (red) within a fine-grained matrix that have become firmly cemented together from Basal conglomerate, Kolhan Group near Jagannathpur, Jharkhand, India. *Source: Prof. Joydip Mukhopadhyay.*

petrographic composition. The conglomerate is typically of rounded and semirounded shape and the breccia is characteristically of rocks occurring in the form of angular or semirounded fragments. Some authors use the term "breccia-conglomerates" for these clastic rocks. The breccias and conglomerates are divided into intraformational and extraformational (Table 5.3) with regard to their place of origin.

Till is an accumulation of unbound and tillite of bound, poorly sorted and nonlayered moraine material dominated by fragments of >2 mm size. Such fragments often show well-preserved stretch marks and wears on one surface, as a consequence of scraping on the bottom and sides of rocks through which the glacier moved. Tillite is characterized by an abundance of fine-grained matrix, and is usually dark gray in color.

Diamictite is an extremely poorly sorted rock. It is composed of blocks of rocks and clay matrix and forms by accumulation and lithifaction of detritus derived from glacial and periglacial processes, as well as mud or detrite flows in subaerial and subaqueous conditions. Most of the diamictite clay matrix consists of different silicate minerals formed during glacial crushing (pulverization), i.e. decay and fragmentation of rocks. The share of clay matrix in relation to the proportion of rock fragments is very low. The blocks present great variability of dimensions and shapes.

5.5.2.1. Intraformational Breccias and Conglomerates

Intraformational breccias and conglomerates are coarse-grained clastic sediments formed by destruction and resedimentation of poorly or incompletely lithified sediment without significant transfer of fragments and clasts within the sedimentary area. Their precipitation occurs immediately after the destruction of the layer and formation of clasts, i.e. in the same stratigraphic unit. Intraformational breccias and conglomerates are usually limited to a narrow sedimentary horizon. It has no significant lateral and vertical distribution. Their origin is strictly limited to certain conditions and environments

TABLE 5.3 Genetic Classification of Breccias and Conglomerates

Intraformational Breccias			**Extraformational Breccias**	
Black-pebble breccias			Cataclastic breccias	
Stormy breccias			Collapse and emersion breccias	
Edgewise breccias			Postsedimentary diagenetic breccias	
Landslide and slump breccia pyroclastic breccias				
CONGLOMERATES				
Paraconglomerates <15% matrix			Orthoconglomerates >15% matrix	
Laminated matrix	Unlaminated matrix		Oligomict	Petromict
Laminated conglomeratic mudrock	Tillite	Tllioid	Oligomict orthoconglomerate <10% unstable	Petromict orthoconglomerate >10% unstable

of deposition (Table 5.3). These rocks are mostly located within the pelitic sediments and marls, often within a very shallow marine limestones and early diagenetic dolomite of carbonate platform. These rocks are excellent indicators for identifying changes in the conditions of deposition, frequent association with short emergence of deposits above the medium tide level or in tidal and supratidal environments exposed to high water energy, where they are usually deposited in stormy waves.[50,51]

Black-pebble breccias (and conglomerates) with black fragments resulting in resedimentation of erosion residues reductive black limestone deposits of coastal wetlands, marshes rich in organic matter and pyrite. It typically is located in the activity of bacteria that reduces sulfates. The rock forms by resediments depressions in tidal channels at peritidal, particularly tidal and shallow subtidal environments. The black marshes and ponds gradually change to brackish or freshwater ponds. It may result in sags and depressions on subtidal zone in areas with humid climates and reductive conditions. The carbonate sediments at the edges and bottom of the ponds and marshes are mostly all-black due to the abundance of organic matters. The storm or high tidal waves erode and break off pieces. The black pebbles wash into depression of tidal channels and deposit along with other carbonate, clay and sometimes, detritus, forming black-pebble breccia. The rocks are well isolated from the oxidation process due to the rapid covering of new sediments, so as to preserve their black color.

Stormy breccias and conglomerates or storm-tide deposits are a special kind of intraformational breccia or conglomerate. The storm breccias form in flooding and accumulation of deposits and fragments of the tidal plain and the shallowest parts of the lagoon edge or flat low coastline at storm waves and storm tides.

Edgewise breccia with flat fragments is a special type of intraformational breccia, whose origin is associated with resedimentation of flat muddy or carbonate-rich sediments. Such fragments occur in the superficial parts of the clayey and muddy sediments in their fractured desiccation cracks as a consequence of sudden drying up of deposits on river banks, flood plains and edges of lakes or ponds during low water levels. This is particularly common in the tidal and peritidal zone, and the muddy and carbonate tidal plains. The desiccation cracks can easily break down into flat fragments in high tides or storm waves that accumulate, usually far from the foundation of the depositional area.

Mudstone intraformational conglomerates are composed of spherical, ellipsoidal mud pebbles of mudstone or muddy limestone. Some pebbles may show clear traces of plastic deformation, generally within squeezing initially stronger pebbles in softer ones or kneading pebbles in the compaction. Muddy pebbles form by the destruction of incomplete lithificated muddy sediments with increasing water energy (strong tidal currents and waves, especially storm tides and storm waves) in the shallowest parts of the depositional area, i.e. the shallowest of the shallow subtidal and lower tidal zones. Their relatively frequent occurrence within the peritidal carbonate sediments is usually in connection with swallowing and/or sea-level fluctuations. A residue within the flood plain is related to the erosion action of the rivers in the rising of the water.

Land-slide and slump breccia results by accumulation of rock material with translational or rotational sliding destroying the mass of larger or smaller fragments in the form of olistostrome and slumps accumulated on land or underwater at the bottom of the slope. The slump breccias occur at the bottom of the submarine slopes of the accumulation of large amounts of sediment. It rotates during sliding down the slope and in its base accumulates in the form of deformed layers. If the deformed sediments have large dimensions, they are called "mega-slumps". Some parts of stronger lithified layers slide down the slope, break and fit into the homogenized plastic deposits making slump breccia.

5.5.2.2. *Extraformational Breccias*

Breccia is a general term for more or less tightly bound clastic rocks composed of angular to semirounded rock debris and cement or matrix (Fig. 5.23).

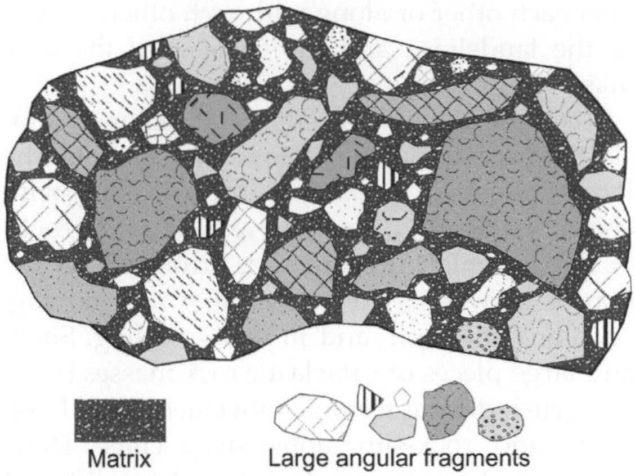

FIGURE 5.23 Matrix support breccia composed of rock debris in the form of large angular fragments of limestone and dolomite in the fine-grained matrix.

Breccias, in geological terminology, are often called by prevailing petrographic type of fragments. They can be dolomite breccia, limestone–dolomite breccia, etc. The sedimentological and petrological classification of breccias is based on the mode of their origin and is divided into the following:

1. Cataclastic breccias
2. Collapse and emersion breccias
3. Postsedimentary diagenetic breccias
4. Pyroclastic breccias.

5.5.2.2.1. EXTRAFORMATIONAL BRECCIAS

Extraformational breccias are coarse-grained clastic sediments containing clasts resulting in destruction and resedimentation of older rocks deposited in some other older geological formation. Thus, clasts originate from weathering of older rocks located outside of formation in which they are deposited. A good example for extraformational breccias is "Jelar breccias" whose clasts are deposited during the Oligocene (~33–23 million years ago) and are formed in destruction of limestones and/or dolomites deposited during the Jurassic (~200–145 million years ago), Cretaceous (~145–65 million years ago) and the Paleocene–Eocene (~65–32 million years ago). Extraformational cataclastic breccias contain clasts whose origin is related to the processes of breaking and crushing rocks with the movement of the rock mass over each other or along with each other, as well as the landslides and the collapse of the rock mass.

Tectonic is the most important factor of cataclastic, because in the tectonic movements, the largest range of rock mass moves along with enormous energy. Breaking and crushing rocks in the tectonic movements is strongest on the border between two masses that are moving, i.e. along the fault, and in the wrinkling. Small and large pieces of cataclastic rock masses break and crush during tectonic movements like landslides and rock-slip along steep cliffs. These broken rock pieces are transferred by different mechanisms, accumulated in the breccias zone and lithified. The clasts constitute special kind of cataclastic breccias.

Rockfall breccias are resulted by cementation of rock debris which pour down the steep slopes and accumulate at the base of such slopes in the form of large rockfall fans. Rockfalls are typically located at the foot of steep cliffs or ravines in between the steep rock and cliffs (Fig. 5.24). These breccias are commonly associated along with subaerial spending and strong erosion of rocks on a steep relief, along with more or less continuous tectonic uplift. The debris may move down the slope and reach in the lake, sea or river environments and/or switch to debrite flow.

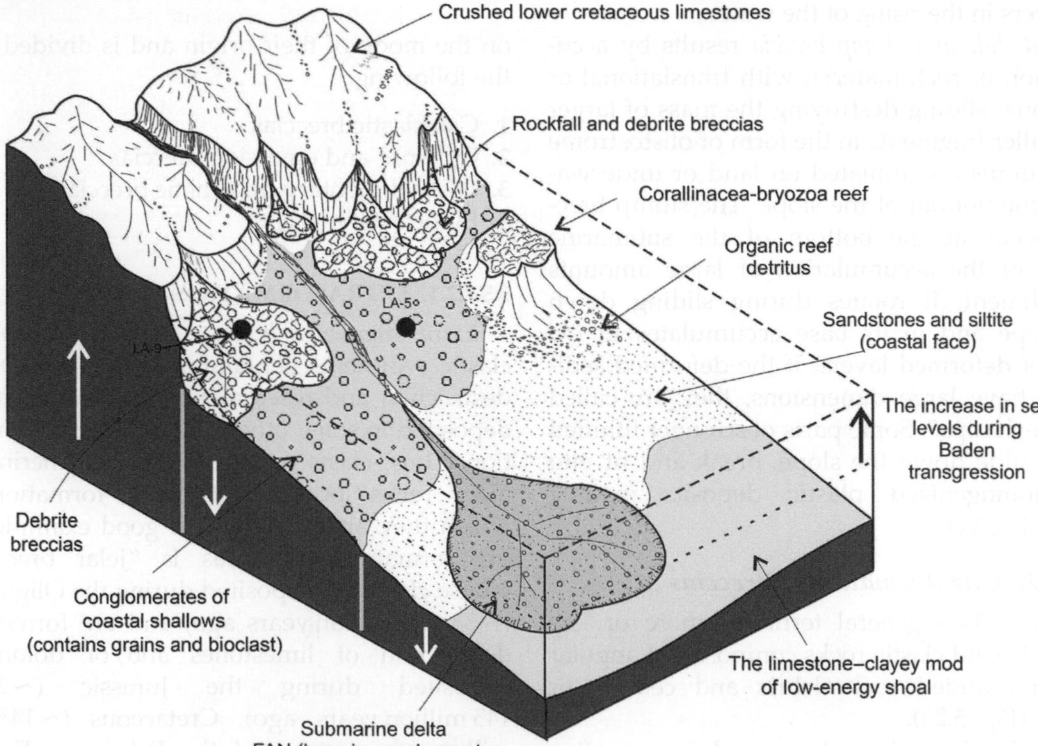

FIGURE 5.24 Idealized model of depositional facies distribution of Miocene sediments in to oil and gas field at Ladislavci, Croatia (not to scale).

Debrite breccias are formed by debrites cementation, i.e. the rock debris that has been transferred down the slope in debrite flows or in flow of the rock debris. The debrite breccias contain more matrix, usually silty material with a bit of sandy detritus, unlike rockfall. These are mainly located at the foot of submarine slopes and submarine canyons (Fig. 5.24).

The debrite breccias often include large clasts of olistolith (Fig. 5.24). An olistolith or olistostrome is a sedimentary deposit composed of a chaotic mass of heterogeneous material, such as blocks and mud. It forms by mutual enclosure relation of clasts and the abundance of finely disintegrated material or matrix of the same lithological composition. It may show similar shapes of bodies and the environment of deposition. The debrite and rockfall breccias are clearly distinguished from tectonic breccias. Rockfall and debrite breccias belong to dolomite and limestone–dolomite breccias, reservoir rocks from many oil and gas reservoirs. With the clasts of smaller or larger dimensions, such breccias occasionally contain larger blocks, even large olistolith (Fig. 5.24). The outline of sedimentary body of such breccias is wedge-shaped, fan or completely irregular, depending on the morphology of the slope or terrain at the bottom of the cliff. If debrite and/or rockfall debris plunges into a river or sea, it will be partly or completely processed into the gravel and in turn change to breccia conglomerate or conglomerate. It contains a different proportion of clasts, already in the river flow or shallow sea, as well as marine debris. The breccia conglomerate, conglomerates of coastal shallows and front beaches are formed by mixing the debrite and/or rockfall material with river sediments, coastal pebbles, shallow-sea carbonate, and fossil detritus. The coastal breccia conglomerates are suitable loci for oil field, e.g. Ladislavci oil and gas-fields, Drava depression in the south of the Pannonian basin, eastern Croatia (Fig. 5.24).

Fault and tectonic breccias are related to the tectonic zone of breaking, faulting, folding, wrinkling and pulling. These natural and mechanical phenomena crush crustal rocks into fragments and debris; dissolve the minerals later by the circulation of pore waters and finally able to be cemented in the solid rock as breccias. The solid lithified breccias is formed due precrystallizations of finely disintegrated calcareous detritus material that has emerged as a product of intense tectonic crushing, cataclastic and mylonitization. Tectonic breccias or clasts are frequently separated by only a system of tectonic cracks and not significantly moved. The tectonic breccias are characterized by clast support, and the matrix support arises in case of stronger crushing or grinding of rocks into "stone dust".

It only happens in the strongest zones of tectonic crushing or shearing during diagenesis, recrystallization and cementation of such small "stone flour".

5.5.2.2.2. EMERSION AND COLLAPSE BRECCIAS

Emersion breccias (karst breccia) are a special genetic type of carbonate breccia form in complex processes of physical and chemical weathering of limestone in subaerial conditions in the Earth's surface or in aerated zone. The emersion breccias are located in present-day karst environment in irregular shape as inserts or inlays, e.g. within the Mesozoic limestones of the Adriatic carbonate platform (AdCP). The AdCP is one of the largest Mesozoic carbonate platforms of the Perimediterranean region. The deposits comprised a major part of the entire carbonate succession of the Croatian Karst Dinarides Mountain chain with thickness up to 8 km and age between Middle Permian and Eocene.

Collapse breccias (breccias of dissolution and collapse) contain cemented clasts created by breaking of layers and rock masses during subsidence of layers after partial dissolution of fractured rock mass in the basement (melt breccia), collapse (collapse breccias) or cracking and crushing of quickly lithified surface part of sediment in mechanical diagenesis (evaporite collapse breccias). The collapse breccias are related to the processes of rock collapse, either by erosion or chemical corrosion processes. This is typically the case in karst caves, and in dissolution of layers of salt or gypsum. This is followed by collapse or subsidence of the roof clastic or carbonates rocks with their tearing into blocks and smaller or larger clasts and collapsing blocks in large corrosion cavity or cave. The collapse may occur on the steep cliffs, composed of two or more different petrographic types of rocks differently resistant to erosion or chemical weathering. It is possible, for example, the collapse of soft marls or mudstone, and firmly cemented sandstone in flysch or in turbidites.

5.5.2.2.3. POSTSEDIMENTARY DIAGENETIC (TECTOGENIC-DIAGENETIC) BRECCIAS

Postsedimentary diagenetic breccias form as a result of strong tectonic crushing of some parts of the rock mass, and followed by intensive diagenetic allochemical processes. The significant features and mechanisms are simultaneous processes including corrosion of the edges and corners of carbonate fragments, dolomitization of limestone fragments, calcitization or dedolomitization and occasional silicification of dolomite fragments. The process further continues in fine matrix disintegration and the multistep cementation of primary and secondary pore with secretion of calcite or ferrocalcite from pore solution saturated in Ca-bicarbonate.

Pyroclastic breccias are composed of coarse clasts, originate from volcanic eruptions and accumulate on the land, in freshwater or marine environments, after a short or longer transportation. These breccias are more fully discussed as part of pyroclastic sediments (Section 5.6).

5.5.2.3. Extraformation Conglomerates

Conglomerate is a solid rock formed in cementation and lithifaction of clasts with angular shape, i.e. pebble sizes of gravel (>2 mm). The mutual relations of pebbles and grains, matrix and cement distinguish between orthoconglomerates and paraconglomerates (Table 5.3).

5.5.2.3.1. ORTHOCONGLOMERATES

Orthoconglomerates are firmly cemented coarse-grained clastics which are characterized by clasts support and mainly composed of pebbles of gravel dimension and not more than 15% fine-grained matrix (argillaceous and clayey detritus), and cemented with the chemically extracted mineral cement (quartz, calcite, opal, etc.).

The oligomict and petromict conglomerates (Table 5.3) are distinguished by the mineralogical and lithological composition of pebbles resistant to wear, i.e. the amount of pebbles and grains of quartz, quartzite and chert. Oligomict conglomerates consist of >90% to wear-resistant pebbles and have simple petrographic and mineralogical composition; consist of pebbles and grains of quartzite and chert. The intergranular pores have relatively high content of chemical secreted cement. Pebbles and grains of oligomict conglomerates are the most stable remains of intensive spending of large amounts of older rocks. The rock represents the most resistant remnants of resedimentation rock debris after several cycles of transportation and deposition. Therefore, it symbolizes high degree of sedimentological maturity. The degree of sedimentological maturity is determined with content of the most resistant components as a result of several cycles of resedimentation. Pebbles of chert may be the remains of wear large masses of limestone containing lenses, nodules and concretions of chert, quartz and quartzite pebbles, remains of granite, gneiss and other metamorphic rocks that are criss-crossed with veins of quartz, zones or lenses of quartzite, or remains of quartzite inserts in phyllite, or chlorite schists.

Orthoquartzose conglomerates generally do not contain pebbles larger in diameter (8–10 cm) and do not appear in thick layers. It is characterized by a well sorting, a high degree of roundness and clasts support. The grain sizes of orthoquartzose conglomerates make a gradual transition in coarse quartz sandstone. It is typically found as thin layers within coarse-quartz sandstone of alluvial deposits or marine beaches deposited by high energy of water.

Petromict conglomerates hold more than 10% of the chemically wear unstable pebbles and grains of various petrographic and mineralogical compositions. The rocks are characterized by clasts support, and in the intergranular pores have excreted chemogenic cement: calcite, quartz, opal, and dolomite. The petromict conglomerates are mixtures of metastable pebbles (clasts) of different types of igneous, sedimentary and metamorphic rocks and grains primarily of quartz, ±feldspar and mica. The rocks are usually dominated by one petrographic type of pebbles, such as limestone (Fig. 5.25) or crystalline schist and quartzite of high degree of metamorphism. This is the most common and widespread type of conglomerate. Petromict conglomerates are characterized by relatively large pebbles, in some cases with a diameter >20 cm, as well as poor level of sorted sand grains in the interspaces between pebbles (Fig. 5.25). It mainly belongs to river sediments (alluvial fan), delta (delta head, the slope of the delta), and coarse-grained, rarely medium granular turbidites.

FIGURE 5.25 Petromict orthoconglomerate predominantly composed of perfectly rounded pebbles of Jurassic and Cretaceous limestone.

5.5.2.3.2. PARACONGLOMERATES

Paraconglomerates are a special type of conglomerates with a muddy or matrix support containing more than 15% clay-dusty (muddy or pelite) matrix whose share is often higher in relationship to the total volume of pebbles sizes of gravel (Table 5.3). These are in reality the sediments, which are not incurred in ordinary conditions of transportation and deposition of clastic material, but mostly a combination of iceberg transport and water floods in rapid melting of glaciers, sudden torrential flows at the foot of the mountains (piedmont zone), debrite flows or alluvial fans in sudden floods. This type of conglomerate has significantly lower distribution with respect to its other counter parts.

5.5.3. Medium Granular Clastic Sediments—Arenaceous Rocks

The arenaceous rocks include all those classic rocks with particle sizes range generally between 2 and 1/16 mm. The most common arenites are graywacke and sandstone followed by calcarenites (carbonates and limestone), oolitic iron ores and glauconite beds.

5.5.3.1. The Composition and Distribution of Sandy Sediments

The clastic medium granular sediments are represented by sands as unbound sediments and sandstones as solid rocks. Sands and sandstones are sediments that are predominantly composed of detrite grains sizes of sand, i.e. a grain in diameter between 0.063 and 2 mm. The rocks are characterized typically by dominance of sand-size grains with minor share of powder-size clay particles and of tiny gravel.

The primary material, grains of sand, are derived from weathering component of any rock. The mineral composition of sands and sandstone can be very different and complex depending on the parent rocks, manner of weathering, transfer and deposition. The clasts that make up sand residue or sandstone include mineral grains and rock fragments of siliciclastic and carbonate composition, as well as fossil remains of the skeleton and shells of organisms, i.e. fossil detritus. Siliciclastic components include all grains of quartz, silicate minerals and rock fragments containing quartz and silicate minerals (all clasts, muddy and clay matrix), and ingredients left after the physical and chemical weathering of silicate minerals and rocks, which are transferred to precipitation area from land (terrigenous components). Carbonate components or carbonate detritus are carbonated grains: mostly fragments of limestone, dolomite and fragments of calcite and dolomite minerals remaining from wear of carbonate rocks and minerals, primarily calcite, dolomite and siderite veins. Carbonate detritus in its origin may be either of the following:

1. Extrabasinal arises from the physical and chemical weathering of older limestone and dolomite on the mainland (terrigenous components).
2. Intrabasinal belongs to ooids, oncoids and pellets formed in the surrounding shallows or even intraclasts that originate from the destruction of carbonate rocks within the depositional area and are nearly as old as the sand in which deposited. It is often the case in Badenian sediments in Pannonian basin, east-central Europe, in which siliciclastic material derived from weathering of older crystalline and lower Miocene rocks on mainland. The intrabasinal carbonate detritus from the destruction of reef Badenian limestone from coastal shallows and underwater reefs. These are calcarenaceous sandstones (Section 5.5.3.4).

Fossil components or fossil detritus include the fossil remains of flora and fauna in the form of whole shells and/or skeletons or their fragments known as *bioclasts*. The fossil detritus in sandy sediments may originate from resedimentation from older rocks or carbonate detritus. It may be intrabasinal belong to planktonic and benthic organisms residing within the depositional area. The redeposited fossil detritus from older Baden corallinacea-bryozoa ridge rocks are often found in Sarmatian and Pannonian sandstones and intrabasinal fossil components (bioclasts of corallinacea, bryozoa, echinoderms, and molluscs) in Baden biocalcarenites sandstones of Pannonian basin.

The essential ingredients of sands and sandstones are quartz, feldspar and rocks fragments ± micas, carbonate and clay minerals, and heavy minerals (density $>2.85\ g/cm^3$). Certain types of sandstone can contain a substantial proportion of muddy matrix, fossil detritus or glauconite.

The salient features of sands and sandstones are the following:

1. Quartz is the most abundant element on the sands and sandstone derives from the wear of acid igneous rocks, crystalline schist and older sandstone.
2. Feldspars are particularly abundant ingredients of some sands and sandstone, especially molasse type, whose detritus derived from severe physical wear and rapid deposition at the foot of mountain massifs built from neutral and acidic igneous rocks and gneisses.
3. Excerpts of quartz and feldspars originate from wear of mafic intrusive and intermediate extrusive (volcanic) igneous rocks, numerous sedimentary rocks (in particular, siltstone, sandstone, chert, limestone and dolomite), as well as many metamorphic rocks (especially quartzite, phyllite, mica schist and gneiss)

are primary ingredients of many sands and sandstone.

4. Micas, especially muscovite, are regular ingredients of nearly all sands and sandstones, usually with a small share.
5. Clay minerals and chlorite in some types of sandstone (graywacke) are present in large amounts and in some types (arenaceous rocks) in minor amounts or even completely absent.

Sandstones are divided in two main groups according to relative content of grain sizes of sand and mud matrix: pure sandstone or arenites and impure sandstones or graywacke (Fig. 5.26).

5.5.3.2. *Arenite Sandstones or Arenaceous Rocks*

Pure sandstones or arenites are classified in to five types according to the proportion of the major components of quartz, feldspar and rock fragments:

1. Quartz arenites containing >95% quartz.
2. Lithic arenites contain <75% quartz and rock fragments has more than feldspar.
3. Arkosic arenites contain <75% quartz and feldspar has more than fragments of rock.
4. Sublithic arenites contain quartz between 75% and 95% and rock fragments have more than feldspar.

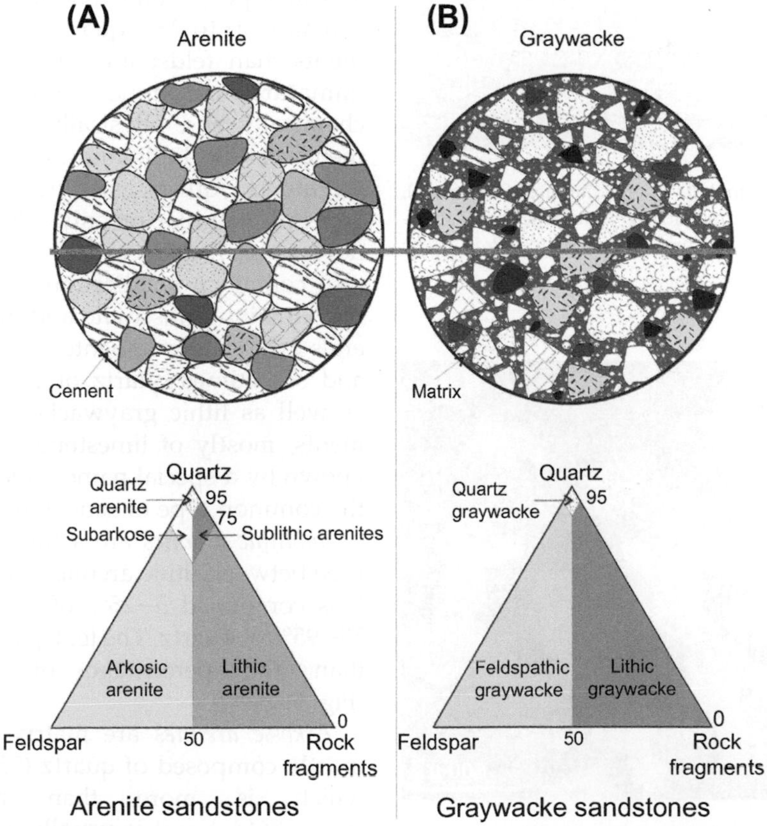

FIGURE 5.26 Conceptual diagrams showing the basic classification of sandy sandstones as (A) arenites, and (B) graywacke with further subdivisions.

5. Subarkoses contain between 75% and 95% quartz and feldspar has more than fragments of rock.

Quartz arenites contain a high proportion (>95%) of well-sorted and rounded detrite quartz grains in association with stable accessory minerals and rock fragments as well as quartz, opal or calcite cement (Figs 5.27 and 5.28). This distinct type of sandstone attains the highest degree of purity and sedimentological maturity, considering the unique mineralogical composition, and the ingredients belong to the most stable grains. Sand grains are remaining after an intense chemical weathering and the long transfer from the source rocks to the place of deposition. The final products go by strong and long-term chemical and physical weathering, abrasion and sorting of debris, often after several cycles of resedimentation. The most resistant detrite grains, mainly of quartz and rarely fragments of quartzite, remain stable even after passing more cycles of wear, transfer and deposition of sediments.

FIGURE 5.27 Laminated quartz arenite (sandstone), composed of +90% detrital quartz from Srisailam Formation, Chitrial, Andhra Pradesh, India. *Source: Prof. Joydip Mukhopadhyay.*

FIGURE 5.28 Photomicrograph of thin section of quartz arenite comprising of +90% quartz (q), chert (Ch) and minor accessory minerals in layered form. *Source: Prof. Sukanta De.*

Lithic arenites are the frequent and most widespread type of sandstone in the lithosphere. It contains up to 75% quartz and more rock fragments than feldspars (Fig. 5.26(A)). These are immature sandstones, which include many chemically and physically unstable rock fragments. In the lithic and sublithic arenites rock fragments, quartz and feldspars grains are generally angular or only a little rounded, and never well-rounded. The constituent minerals also contain a smaller amount of slips detrite mica, mainly muscovite and biotite. These minerals are usually cemented with calcite cement, and sometimes quartz or opal. Lithic arenites as well as lithic graywacke contain rock fragments, mostly of limestone and dolomite, and known by a special name "calclithite". They are the common type of sandstone in the tertiary.

Sublithic arenites are a transitional sandstone type between lithic arenites and quartz arenites. It is comprised 5–25% of rock fragments and 75–95% of quartz. The feldspars fraction is lower than the percentage of rock fragments (Fig. 5.26(A)).

Arkosic arenites are matrix poor sandstones mostly composed of quartz (75%), and feldspar which is more than rock fragments (Fig. 5.26(A)) and is usually cemented by fines of quartz, calcite and feldspars (Fig. 5.29). Feldspars can be completely fresh, and usually

FIGURE 5.29 Arkosic arenite: medium to fine angular to subrounded grains of quartz (q) and feldspar (microcline "m"), cemented by silica and little interstitial sericitic (s) matrix with extensive sericitization of feldspar grains (s) at Proterozoic Kaladgi basin, Karnataka, India. *Source: Prof. Sukanta De.*

belong to potassium-rich alkali feldspar (microcline) and acid plagioclase (albite and oligoclase). The colors of arkosic arenites are reddish, reddish brown, pink or rarely light red. The reddish and pink colors are derived from pink microcline or hematite and limonite. In addition to quartz and feldspar arkosic arenites and subarkoses include detrite mica (muscovite and biotite), which are typically oriented parallel to the layers.

Mineral composition and structure of feldspathic arenites clearly indicate that the parent rocks from which detritus emerge are granites and/or gneisses. It also indicates that the original rocks are extensively consumed in terms of steep terrain and cold or arid climates where chemical wear of feldspars was limited or prevented by rapid transport and deposition.

Subarkoses are feldspathic sandstones with the mutual proportions of quartz and feldspars make the transition from arkosic to quartz arenites. The share of quartz varies between 75% and 95%, and contains more feldspars than rock fragments (Fig. 5.26(A)).

The arenites are comparatively of low cost depending on grain size, color, size of blocks and quality, and widely used for constructional purposes. The common uses are as building material for domestic houses, palaces, temples, cathedrals, mosque, ancient forts, monuments and minarets (Fig. 5.30), ornamental fountains, statues, roof tops, grindstone, blades and other equipments.

5.5.3.3. *Graywacke or Wackes*

Graywacke is a variety of impure sandstones and is generally characterized by its hardness, dark color, and poorly sorted angular grains of quartz, feldspar, and small rock of lithic

FIGURE 5.30 The Qutab Minar, a 72.5 m high, 379 stairs five story victory tower is located at UNESCO World Heritage Site, Delhi, India. It was built in 1193 AD by Qutab-Ud-Din-Aibak, the first Muslim Sultan of Delhi. It is the highest stone tower in India, made from red and buff sandstone and famous for its architectures.

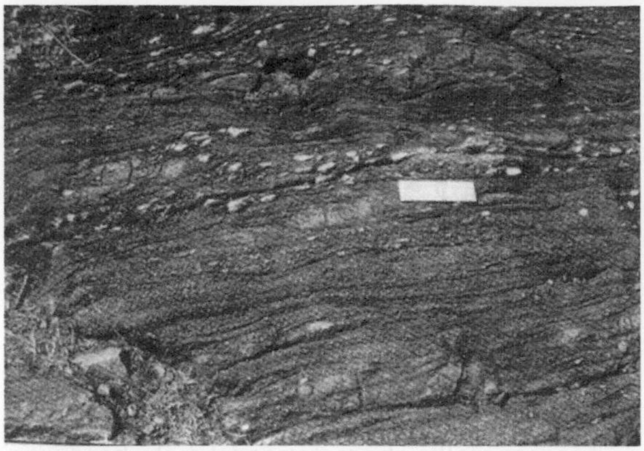

FIGURE 5.31 Dark color graywacke with poorly sorted grains set in a compact fine clay/muddy matrix showing graded bedding structure at Zawar, India.

fragments set in a compact fine clay and muddy matrix (Figs 5.26(B), 5.31 and 5.32). The term *graywacke* (from the German "*graywacke*") in the geological literature was first enacted in the eighteenth century, for the dark gray, solid lithified poorly sorted sandstone in Hartz, Germany, which contain many angular fragments of rocks, grains of quartz and clay–sericite–chlorite matrix that comes from spending of unstable rock fragments. Graywacke in the Earth's rocky crust is a very widespread type of sandstone and

FIGURE 5.32 Photomicrograph of lithic-graywacke composed of angular quartz grains, fragments of quartzite, crystalline schist, and clay matrix.

shares 20–25% of all sandstone. Graywackes are classified in to three groups according to the proportion of main components of quartz, feldspar and rock fragments (Fig. 5.26(B)):

1. Lithic graywacke containing <95% quartz and more rock fragments than feldspar.
2. Feldspathic graywacke containing <95% quartz and more feldspar than rock fragments.
3. Quartz graywacke containing >95% quartz.

Lithic graywacke are matrix-rich sandstones that with the quartz (up to 95%) contain more rock fragments than feldspars (Figs 5.26(B) and 5.31). Lithic graywacke belongs to the group of sandstones of low level of maturity due to large amounts of matrix, particularly clay minerals, illite, and metastable fragments of rock. The rocks are characterized by a poor sorting, and dark gray to dark green color due to clay–chlorite matrix and high content of dark rock fragments. The rocks may be dominated by debris of volcanic rocks (diabase, spilite, keratophyre, dacite and porphyry), followed by fragments of schist of low and intermediate level of metamorphosis (slates, phyllite, quartz–sericite, mica schist and quartzite) and sedimentary rocks (cherts, siltstone, shales and sandstones). Quartz is generally the most abundant element in the sand fraction of detritus and share is generally higher than 50%. Feldspars generally contain only acidic to neutral plagioclase with little of K-feldspars.

Lithic graywacke matrix forms by synsedimentary muddy and/or clayey detritus (protomatrix) which is converted into a dense mixture of chlorite, sericite and quartz (orthomatrix) during the diagenetic processes. This is typical of many Paleozoic and Mesozoic lithic graywacke. However, part of the chlorite–sericite matrix can come from diagenetic changes of unstable rock fragments.

Feldspathic graywacke matrix contains a considerable amount of feldspar and rock fragments in addition to quartz (up to 95%). The proportion of feldspars can vary in wide limits and is always greater than the percentage of rock fragments (Fig. 5.26(B)). In some variations of feldspathic graywacke, feldspars are even more abundant than quartz. Detrite micas (muscovite and biotite) are often present. Matrix of feldspathic graywacke is similar to that of lithic graywacke, generally thick fine-to-microcrystalline mixture of clay, chlorite, sericite, quartz and carbonate minerals (often siderite), and pyrite. Clay minerals, especially kaolinite group, originate from chemical weathering of kaolinitization of feldspar (Section 5.2.1.2). Feldspathic graywacke represent much less common type of sandstone in relation to the lithic graywacke, quartz graywacke are very rare type of sandstone.

5.5.3.4. Mixed or Hybrid Sandstones

Sands and sandstones which, in addition to quartz, feldspar, silicate rock fragments and mica, contain a substantial proportion of detritus chemical and/or biochemical origin or materials of other origin, are not included in the standard classification of sandstone. These rocks of new composition belong to a special group of sandstone. This particular group consists of mixed or hybrid sandstones that includes different calcarenaceous, green and phosphate sandstones.

Calcarenaceous sandstones are the genetic groups of mixed hybrid clastic and chemical and biochemical rocks. It is composed of a mixture of grain (>50% siliciclastic or quartz, feldspar, rock fragments, and mica) and limestone grain of chemical–biochemical in origin (10–50% bioclasts, fossils, intraclasts, pellets, ooids and oncoids). Calclithite that belongs to either lithic graywacke or lithic arenite contains fragments of older limestone and/or dolomite. Calcarenaceous sandstones, with siliciclasts, contain a significant proportion of fossil debris and ooids and/or oncoids and pellets. The mixed rocks gradually change to biocalcarenite limestone with the increase of limestone grains of intrabasinal origin. It will no longer remain as calcarenaceous sandstone if the share of limestone grains of intrabasinal origin exceeds 50%. The new rock is limestone biocalcarenite type.[50,51]

The fossil detritus in calcarenaceous sandstones are mostly shell of benthic foraminifers, echinoderms skeletal debris, corallinacea, bryozoa, molluscs and gastropods. In general, calcarenaceous sandstones are cemented with calcite, and occasionally may also contain fine-grained clayey–calcareous (marly) matrix.

Green sandstones contain a considerable amount, and in some places more than 50% of spherical, oval beads of glauconite accumulations of material or a mixture of glauconite, chlorite, smectite and seladonite, in addition to siliciclasts (quartz, feldspars, rock fragments, and mica). These grains are distinctly green or dark green color, and the sandstones have been named as *green sandstones*. Glauconitization is a very slow process. Green glauconitic grains forms by glauconitization processes in marine environments under low reductive conditions at normal salinity and low speed of deposition over a very long time, say several hundred thousand years. The parent materials for the origin of green glauconitic grain are biotite, pellets, wrapped grains, foraminifers, volcanic glass and volcanic ash that undergo diagenetic changes.

Phosphate sandstones are siliciclastic sandstones that include calcium phosphate (apatite) or contain a substantial proportion of phosphate detritus or phosphate ooids as cementing material.

5.5.4. Fine Granular Clastic Sediments—Pelite

Pelite is clayey fine-grained clastic sediment or sedimentary rock, i.e. mud or a mudstone.

5.5.4.1. Classification of Pelitic Sediments

The fine-grain clastic sediments or pelite mainly consist of silt and clay with a grain size of <0.063 mm (Table 5.1; Fig. 5.19). There are several different types of pelitic sediments

TABLE 5.4 Classification of Pelitic Sediments on the Basis of Mutual Interest of Silt and Clay

Silt		100–66%	66–33%	33–0%
Clay		0–33%	33–66%	66–100%
Unrelated		Silt	Mud	Clay
Related	Homogeneous	Siltstone	Mudstone	Claystone
	Lamination or fissility	Silt shale	Mud shale	Clay shale
Slates		Quartz slate	Slate	

according to the proportion of silt and clay, the degree of lithifaction and the features, as given in Table 5.4. Pelite sediments containing more than two-third silt components are divided according to the degree of lithifaction on the powder or as a loose silt and siltstone as related to sediments. If related or lithified rocks show lamination, therefore are not homogeneous, it is termed as *leafy siltstone*. Similarly, pelite sediments that contain more than two-third clay component, given on the degree of lithifaction, known under the name of the *clay*, as an unbound, and *claystone* as bound rocks. If their lithified version features laminated structure, it is called *clay shale*. Pelite sediments, which contain between one-third and two-third silt and clay components, are divided into the "mud" as the loose sediment and "mudstone" as a solid rock and in the case of lamination is called *mud shale*. *Loess* is a special kind of siltstone of Aeolian origin and the *marls* are mixed or hybrid rocks consisting of clay and carbonate, mainly calcite, component with variable share of powder.

5.5.4.1.1. CLAY AND CLAYSTONE

Clay and claystone generally contain predominantly one of the following three groups of minerals: illite, smectite (montmorillonite) and kaolinite group and a smaller or larger proportion of chlorite and rare glauconite. Chlorites and glauconite in claystones occur during diagenetic processes.

Illite mineral group is typical of the marine clay deposits. Illite, in the claystones, is mainly derived from the diagenesis of kaolinite by chemical weathering of feldspar. Smectite (montmorillonite) group of clay minerals contain up to 20% water and absorb Ca and Mg. Clay and claystone mainly composed of this group of clay minerals and are called *bentonites*, and form as a result of alteration of acidic tuffs and volcanic glass (Section 5.6.3). Kaolinite group of clays is typical for kaolinite rich or pure kaolinite clay known as *kaolin* (Fig. 2.16). Clays containing kaolinite group of clay minerals are characterized by a light or milky white color and in contact with water become remarkably plastic. These minerals are used as a highly valued raw material in ceramic production and with a higher proportion of powder in the manufacture of bricks and tiles. Clay and claystone rich in kaolinite group minerals precipitate in freshwater and not in marine environments because kaolinite quickly transforms into complex clay minerals in seawater. The basic characteristic of the clay with water is to become plastic, can knead and shape, after drying and firing to retain shape. This makes them perfect for pottery, porcelain, ceramic products, sculptures, tile and brick.

5.5.4.1.2. SILT AND SILTSTONE

Silt is loose pelite sediment and *siltstone* is pelite rocks, containing >66% silt grain size (particle size 0.004–0.063 mm). Siltstones are rocks, according to granulometric measurements, chemical composition and textural–structural features, that make the transition from fine-grained sandstone in the mud and clay rocks. The dominant component of siltstone is angular grains of quartz, significantly associated with the tiny grains of feldspar and mica flakes, and up to 33% clay. Some types of siltstone containing a substantial amount of carbonate, mainly calcite cement or fine-grained carbonate detritus (carbonate mud deposited along with grains of silt size), and such a rock is called *calcareous siltstone*. The calcite cement can be paved with authigenic quartz, opal or chalcedony, or sometimes mineral binder which originated from the diagenetic processes of clay minerals, i.e. sericite, chlorite and illite (Section 5.5.5.2). Siltstones are generally massive, thickly layered, strongly lithified, homogeneous, sometimes horizontally or obliquely laminated rocks. Siltstones are often represented and deposited together with the sludges, mainly in lacustrine and marine basins.

5.5.4.1.3. SHALE AND MUDSTONE

Shales are thinly laminated fine-grained pelite clastic rock composed predominantly of siliciclastic materials by granulometric composition of mixtures of clays and particles size of powder, or silt. Shales can be grouped as clay and mud shale based on the mutual shares of particles (clay and particles of powder). Shale laminations are not always just a consequence of the way of deposition. The deposition is not exclusively related to synsedimentary processes. The thin laminations of most of the shales originated during earlier geological periods are the result of mechanical diagenetic compaction processes occurring due to high pressure at greater depths of covering, which leads to the destruction of loose packed structure of particles "house of cards" in plan-parallel order (Section 5.5.5.2, Fig. 5.38).

Shales are the most common sedimentary rocks in the Earth's crust. It occurs in lithifaction and complex diagenesis of water-rich mud and powder-clay sediments (Fig. 5.38). Mineral composition of shale is diverse and variable. Shales form by combination of the composition of detritus particles and the chemical diagenetic processes. The essential ingredients are clay minerals and illite, quartz, significant amount of feldspars, chlorite and sometimes carbonates. The share of clay minerals, quartz, feldspars, chlorite, muscovite and carbonate is an important factor in the degree of lithifaction and shale laminations. The young (Tertiary) shales prevail illite, kaolinite and smectite (montmorillonite). The older shales typically contain 20–30% quartz, 5–30% detrite feldspars and 15–35% minerals of a complex group of illite–smectite–muscovite, kaolinite, chlorites, carbonates, oxides, hydroxides of iron, organic matter and sulfides. In shales of the Paleozoic age or in those at depths >3–4 km, the proportion of typical clay minerals (kaolinite and montmorillonite group) is insignificant because of their transition processes in the diagenetic chlorates and sericite or muscovite (Section 5.5.5.2).

The shales can have different colors due to the content of organic matter and oxides of some metals Black shales usually contain organic matter (carbon) and/or pyrite and are formed in reducing conditions. Red color of shale is the result of high content of ferric oxide, mainly hematite, and refers to the oxidative conditions during wear and sedimentation, which predominates in continental depositional environments. Green shales contain glauconites and chlorine and come from moderate reducing environmental conditions.

Oil shales (Fig. 5.33) contain a high proportion of kerogen (mixture of organic chemical compounds) and other naphthenes, and are carriers of potential reservoirs of raw materials or are petroleum source rocks. The term *oil shales* is used for all laminated pelite sediments, and also for laminated marls and dolomitic limestone, from which oil can be extracted by heating. Oil shales are dark gray to black color due to the high content of naphthenes and other kerogen.

Oil shales generally belong to the lake and marine sediments and occur in protected anaerobic low-energy lake, river, delta and marine

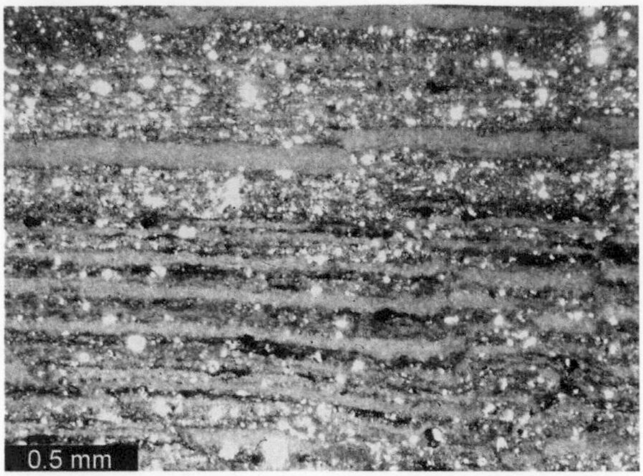

FIGURE 5.33 Oil shale composed of clayey and silty laminae saturated with kerogen and naphthenes (black), Slavonia crude oil and gas field, Croatia.

environments. Mudstones are, unlike the shale, homogeneous, solid lithified rocks that contain mixture of particles clays and powder (from one-third to two-third clay and powder, Table 5.4). The oil shales show homogeneous texture and granulometric and mineral composition is almost identical to geologic young muddy shales.

5.5.4.1.4. LOESS

Loess is a homogeneous, nonlaminated to thickly layered, poorly lithified, well sorted and extremely porous pelite-clastic sediment. The granulometric composition of loess is characterized with a high content of grain sizes of medium and coarse powder (silt). The diameter of grains is predominantly 0.015–0.05 mm. Loess usually contains small grains of powder sizes (0.004–0.015 mm) and from 10% to 20% particles sizes of clay, and sometimes even smaller share of the fine sand grain size (0.063–1 mm).

The predominant mineralogical composition of loess is detrite grains of quartz over to detrite feldspars, usually in the ratio of 4:1. The share of calcite, mainly of authigenic origin, varies in the range between 10% and 30%, and mica and clay minerals between 10% and 20%. An important feature of loess is its extremely high porosity, typically being 40–60%. The pores of loess retain water due to capillary forces, enrich with Ca hydrogen carbonate in the periods of drought and secrete calcite which cements grains of dust and clay particles. The enriched solutions typically circulate only along easily permeable parts of loess, and calcite secretes from the pores. The pores water cannot uniformly rise by capillary forces or just secreted around some of the carbonate grains. All these limitations strongly restrict loess in homogeneous cemented throughout the area. Therefore, loess undergoes irregular concretions due to uneven cementing areas and greater wear and erosion of uncemented parts resulting morphological formations known as *loess dwarfs*.

Loess forms by deposition of Aeolian powder material transferred by wind from large distances. The powder originates from the sludge left over after the flooding of vast valleys and drying of this sludge after the withdrawal of water in river beds. Wind and air currents rise and spread dry powders over long distances and deposit on land or in water. The largest amounts of loess deposited in the Quaternary, especially in the Pleistocene, in the ice ages, when the climate was dry and windy. A huge amount of sludge was deposited by melting of ice and flooding of river valleys during the interglacial periods.

5.5.4.2. *Marlstone*

Marls are mixed carbonate-clay rock and are composed of cryptocrystalline or microcrystalline calcite and siliciclastic detritus of pelitic dimension, primarily clay, with larger or smaller portions of powder (Fig. 5.34). Part of calcite can be of chemogenic in origin, arise from the secretion of the sea or lake water, while some may be the tiniest carbonate detritus of lime sludge. Marl is usually considered a rock that contains between 20% and 80% clay and 80% and 20% calcite.

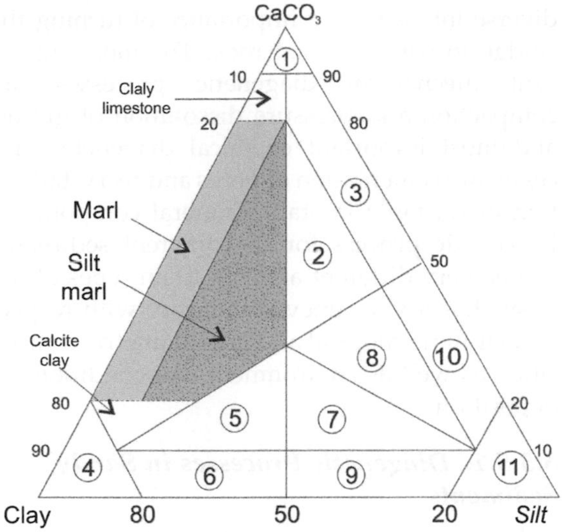

1 - Limestone
2 - Claly–silty limestone
3 - Slity limestone
4 - Clay
5 - Calcite–silt clay
6 - Silt clay
7 - Calcite claly siltstone
8 - Claly calcite siltstone
9 - Claly siltstone
10 - Calcite siltstone
11 - Siltstone

FIGURE 5.34 Detailed classification and nomenclature of the limestone–clayey–silt sediments.

Rock, originally comprised calcite and clay with mutual relationship and chemically equivalent to marl, undergoes diagenetic process in the greater depth transforming typical clay minerals in the illite, chlorite, sericite, muscovite, and designated as marl or marlite. A real "clean" marls, composed only of clay and calcite, are rare in the nature. Much more common are dusty marls containing between 10% and 33% of siliciclastic grain sizes of powder, with calcite and clay between 20% and 80% (Fig. 5.34).

Marls are the most common insulator rocks in the deposits of oil and gas and are the primary raw materials for manufacturing of cement.

5.5.4.3. Organic Matter in the Argillaceous Sediments

Organic matter is present in small amounts in almost all the sediments, and a significant proportion is virtually located mainly in argillaceous rocks, especially mudstones, shales and marlstone. Such rocks can therefore be a source of crude oil and are named *oil source rocks*. The old name for more diagenetically tough rock is *oil shales*. *Sapropelic* is often used to name for the mudstone rich in organic matter. Organic matters in sediments are located in four basic forms: kerogen, asphalt (bitumen), crude oil and natural gas, which consist of a wide range of complex hydrocarbons.

Kerogen is a solid dark gray or black organic substance that contains hydrocarbons insoluble in the common organic solvents such as ether, acetone, benzene and chloroform. It has complex organic composition and is believed to originate mainly by wind inflicted spores and pollen of plants and very small aquatic plants (algae) that are deposited along with winds issued powder. The kerogen is necessary for their fossilization in anaerobic conditions under an anaerobic environment. Three different types of kerogen with regard to the origin of organic matter:

1. Algal kerogens (Fig. 5.33) that generate oil (characterized by high values of the ratio of hydrogen/carbon between 1.0 and 2.2 and a low ratio of oxygen/carbon, <0.1).
2. Mixed kerogens that generate oil or gas (characterized by average values ratio of hydrogen/carbon between 1.0 and 1.7 and the average values of relations oxygen/carbon between 0.0 and 0.2).
3. Humic kerogens that generate gas (a low ratio of hydrogen/carbon between 0.5 and 1.0 and a high ratio of oxygen/carbon between 0.07 and 0.25).

Asphalt or bitumen is sticky, black and highly viscous liquid or semisolid form. It is similar to kerogen in composition, but is soluble in the common organic solvents. It contains 80–85% carbon, 9–10% hydrogen, 2–8% sulfur, and negligible amount of oxygen and nitrogen. In the sediments is typically found in the pores, tectonic cracks and crushed zones.

Crude oil or *fossil fuel* is the name of the hydrocarbons that are flammable liquid at the normal pressure and temperature. It occurs in the sediments and rocks as fills in primary and secondary pores. It contains 82–87% carbon, 12–15% hydrogen and traces of sulfur, nitrogen and oxygen in the form of four types of very complex molecules of each variable: paraffin, aromatic hydrocarbons, naphthenes and asphalt. It is recovered from the parent rocks at the temperatures between 60 and 120 °C.

Natural gas is the name of naturally occurring hydrocarbon gas mixture consisting primarily of methane with other hydrocarbons, carbon dioxide, nitrogen and hydrogen sulfide. The gaseous hydrocarbons contained in pores of sediments and sedimentary rocks. It is recovered from rocks at the temperatures of about 120–220 °C because at these temperatures, kerogen is not as inert with respect to the generation of carbon.

5.5.5. Diagenesis of Clastic Sediments

Diagenetic processes convert loose, unbound, water-saturated packages of sediments to the firmly lithified sedimentary rocks by either of the system:

1. Early diagenetic processes that occur in a completely unrelated, pore-water-saturated sediments at shallow depths overlap, i.e. a small thickness of overlays.
2. Late-diagenetic processes at greater depths overlap, i.e. below the thick layers of overlays in already partially lithified rock.

In both cases, the sediment is subjected to mechanical and chemical processes arising from the depth of the overlay, composition of deposits and pore-water. This is also influenced by other physical–chemical and geological conditions of diverse intensity and importance of turning the sludge to solid lithified rock. The most significant mechanical diagenetic processes are compaction and pressure dissolution of grains, and most important chemical diagenetic processes are cementation of pores and recrystallization of unstable in stable mineral components. Diagenetic process for the different sediments can be very different and in certain types of deposits has a very uneven intensity with respect to their mineralogical and granulometric composition as well as environment and conditions of deposition.

5.5.5.1. *Diagenetic Processes in Sandy Sediments*

Early diagenetic processes in sandstones include all reactions between the mineral grains of sand and pore-water contained in the sand from the time of deposition to the moderate depth of the overlay. There are other reactions related to the life activity of bacteria. Early diagenetic processes in the sands are significant for the further course of diagenesis because porosity of sediment may change due to early diagenetic cementation (porosity reduction) and/or dissolution of certain mineral grains (increasing porosity). Such processes also affect significantly the late-diagenetic processes that occur when sand sediment reaches a greater depth of overlay.

The processes of compaction of sand begin almost immediately after deposition, and ending at deep covering after pressure dissolution of grain and almost complete cementation (Figs 5.36 and 5.37). The compaction processes or mechanical diagenesis of clean sandstone have significantly minor role than the chemical diagenetic process, unlike pelitic sediments, and clayey sandstone.

Comparatively loose and loosely packed sediment emerges with high porosity during the deposition of sand in the water or air. Well-sorted sand grains with high degree of sphericity have intergranular porosity of about 40% after

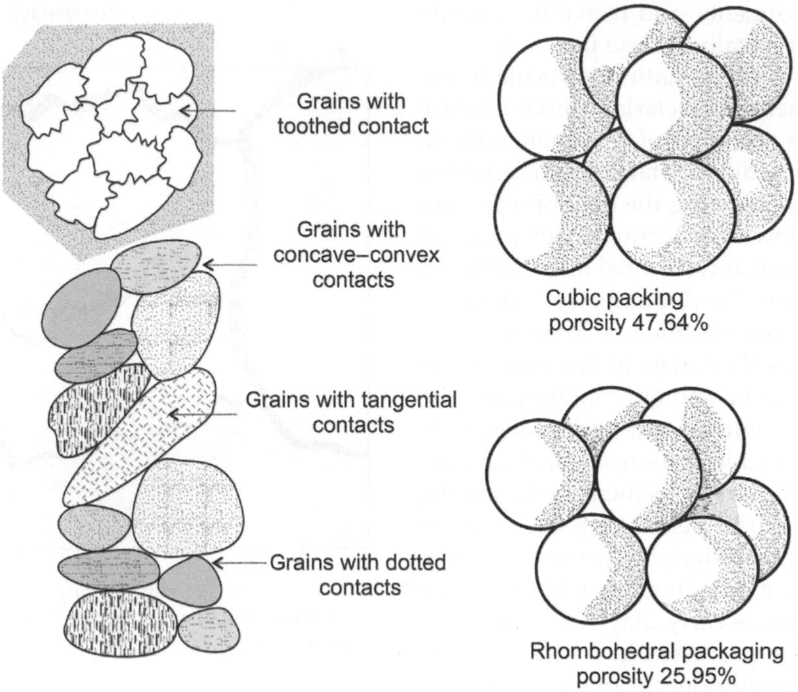

FIGURE 5.35 The main types of grain contacts in the sediment and relative packaging porosity due to compaction.

deposition. The same sands without cementing in the deeper parts of sediments possess about 15% porosity. Therefore, increasing the depth of the overlay due to compaction reduces the porosity of the sands.

Reduction of porosity due to compaction of the sand at the very beginning is substantially different for the fine-grained and coarse sand (Fig. 5.35). However, after the sands have been deposited for 1000–1500 m of thickness of new sediment, the porosity of the coarse-grained sands reduces in relation to fine-grained sands in the absence of significant amounts of clay or carbonate. Specifically, well-sorted coarse grains slide more easily during pressure, and deploy loose cubic form in denser rhombohedral packing. Compaction of fine-grain sands containing clay matrix is higher.

Chemical diagenetic processes can also start immediately after the deposition of sand, with reactions of pore water and sand grains. These are reactions of dissolution of mineral grains as well as reactions that cause the secretion of new authigenic minerals, in the form of cement and pushing one mineral with other such as feldspar with kaolinite. Chemistry of the initial pore water in sand is similar to that in water where sand is deposited. Marine pore water can circulate a few inches below the layer of sand and there in the pores of the sand cause early diagenetic cementing with carbonate or phosphate excretion. The aragonite or Mg-calcite cement (known as *beach rock cements* type) is usually secreted during early diagenesis in sea sands (siliciclastic and carbonate).

Pore water associated with deeper currents extrudes by compaction of clay deposits and moves into younger sediments at higher level. It can cause the release of mineral cement near the border of the sediment and water. In this way, carbonate cements, and hematite, limonite and Mn-oxides typically secrete in sands.

Oxidation of organic matter, particularly the life activity of aerobic bacteria, causes increase of CO_2 resulting secretion of carbonate cement. The bacteria that reduce sulfate by production of H_2S contribute to lowering the limit of redox potentials. This allows the formation of pyrite in the presence of iron and the total iron excretes in the form of sulfides. The dissolution of $CaCO_3$ excretes siderite below this zone, or in freshwater.

The sands, deposited in evaporite and sabkha environments (coastal saline), contain evaporite pore water with high content of dissolved substances that can exude carbonate calcite, aragonite, dolomite, Mg-calcite cements, and sulfate, anhydrite, and barite cements. Sandstones in the zone above the underlying water (arid environment) excrete calcite, hematite, limonite and manganese oxides as early diagenetic cements.

Late-diagenetic processes in the sandy sediments occur at greater depths overlay with two important factors.

1. General increase in pressure and direct pressure on the grain contacts, which causes severe mechanical compaction and pressure dissolution of sand grains at their points of contact to which it transmits pressure.
2. Temperature increase due to increased solubility of many mineral grains and mineral ingredients that contain constitutively water, lose water and transform it into new stable minerals under such conditions.
   Constitutional water is extracted from the minerals when heated, and it is in atomic state, mainly as OH-groups.

General increase in pressure at greater depth of the overlay occurs due to the weight of deposited sediment. This hydrostatic pressure causes denser packing by increasing the surface area of "grain on grain" contacts, reducing the pressure between the grains, reducing the thickness and increasing the level of lithification in sandstone (Figs 5.36 and 5.37).

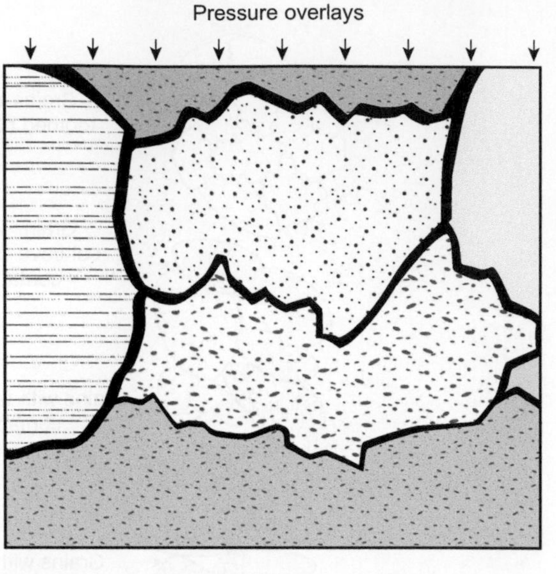

FIGURE 5.36 Pressure melting of quartz grains in contact with surrounding grains or "grain on grain" due to pressure overlays.

Pressure dissolution is partial dissolution of the sand grains, usually quartz, on the grain contacts through which pressure overlays is transmitted. This extends the melting area of grain contacts in the form of a toothed grain encroachment into the other grains (Fig. 5.36). This reduces the intergranular porosity and thickness while increasing the level of lithifaction in sandstone (Figs 5.35 and 5.37(B)). The melting points of grain to grain shape changes due to their reduction, thinning and mutual interference in one another, or grains become flatter. Pressure dissolution at greater depths covering 1000–1500 m is the most significant factor in compaction of sandstones.

Besides these effects of compaction, pressure dissolution has another important role in diagenesis as the dissolution releases silicon dioxide related to silica acid (H_4SiO_4). This mobile acid reextracts the same or neighboring sand layer in the form of quartz cement around quartz grains or "regeneration edge" or secondary growth of quartz.

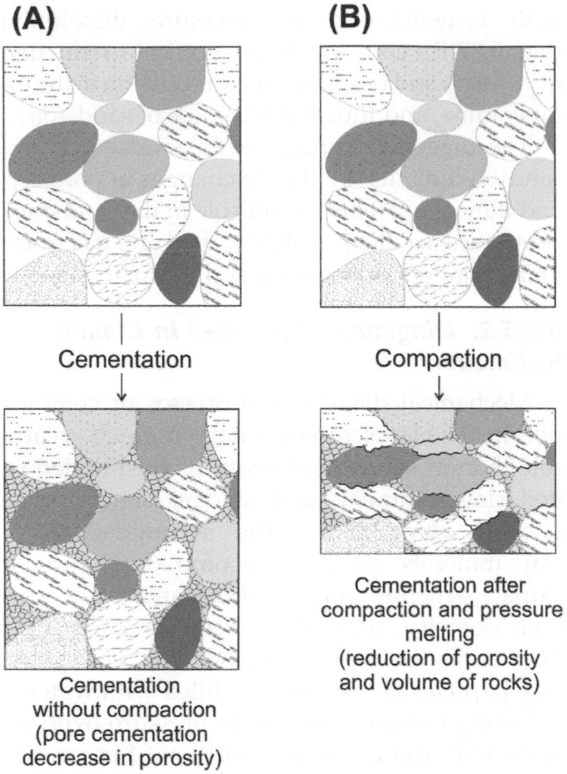

FIGURE 5.37 Conceptual diagrams showing the relationship of porosity of sandy sediments in their cementing without or stronger compaction at shallow depths covering. (A) Excretion of mineral cement in intergranular pores involving large quantity and flow of pore water for sufficient mineral cement and filling the pore spaces between sand grains. The volume of sand before and after cementing in fact remains constant with decrease only in porosity. (B) Compaction with cementation and pressure cause significant decrease in porosity and to considerable reduction in the total volume, and thus the thickness of sandstone sequence.

The rise in temperature has great influence on the chemical diagenetic process in the sandy sediments at greater depth. The effects of higher temperatures by increasing the depth of the overlay manifest the following:

1. Changes in the solubility of minerals as a function of temperature:

The solubility of mineral components of sandstone increases with increasing temperature. The pore water is enriched by ions in compaction currents that can excrete new authigenic minerals, particularly quartz cement. Cementation of sands with quartz cement is extremely slow and time-consuming diagenetic process due to relatively low concentrations of silicon in these solutions.

2. Facilitating the incorporation of highly hydrated "cations" in the lattice of carbonates:

Highly hydrated cations that at low temperatures prevalent on the surface of the Earth are in a melted state, such as Mg^{++} and Fe^{++}, cannot be in the presence of marine pore water incorporated into the carbonate lattice. However, with increasing temperature they become less hydrated and already at a temperature of 60–100 °C (the depth of coverage of 2–3 km) are excreted as Mg and Fe-carbonate ferrocalcite, and siderite cements. There are numerous examples of such rocks, such as, in deep wells of gas fields at Molve, Kalinovac, Croatia.

3. Squeezing the OH-group (constitutional water) from clay minerals and their transformation into new stable minerals (illite, muscovite, and chlorite):

A rise in temperature and pressure causes the formation of higher density minerals that contain water or constitutively contain very little. It is in sandstones with clay matrix (graywacke) in diagenetic processes at greater depths overlay manifest with transformation of clay minerals from the smectite/montmorillonite group and kaolinite in stable minerals from the group illite and chlorite, as well as muscovite, i.e. sericite matrix.

The research to establish changes in composition and stability of clay minerals with increasing temperature and pressure at increasing depth of covering shows that smectite (montmorillonite) and mixed layer of clay minerals become unstable at temperatures between 60 and 100 °C, which corresponds to the depth of the overlay of 2–3 km, and transforms to illite and chlorite. Similarly, kaolinite becomes unstable at the temperatures between 120 and 150 °C, which correspond to the depth of the overlay between 3 and 4 km, and it is transformed to illite.

Illite gradually transforms to muscovite if the pore solution containing enough K and Al. The kaolinite and illite are common ingredients of graywacke sandstone. This process typically causes sericitization of matrix, i.e. the conversion of clay minerals in fine-grained cluster of illite and muscovite, and is commonly called *sericite matrix*. Sericite is the name for the small mica flakes that are not specifically identifiable microscopically.

Cementation is the most important diagenetic process by which loose, scattered sand converts into tightly bound rock sandstone. This process occurs during the early and late diagenesis under conditions of greater depth overlay. Cement can be the authigenic mineral that has caused the reduction of intergranular porosity, therefore the mineral that separates the solution from the pores between the grains (intergranular pores) or in the pores inside the grains (intragranular pores). Sands can be cemented and transformed into sandstone in two fundamentally different ways:

1. Only by the secretion of cement in the intergranular pores of sand: this is going on with bringing of cations and anions in the melted state by circulation of pore water or diffusion of ions (Fig. 5.37(A)).
2. Pressure melting of mineral grains in the pressure points and reexcretion of minerals, usually quartz, in the form of cement.

In the first case, i.e. in excretion of mineral cement in intergranular pores of sands from the solution requires a large amount and flow of pore water to allow the extraction of sufficient quantities of mineral cement for filling the pore spaces between sand grains. The volume of sand before and after cementing in fact remains constant with decrease only in porosity (Fig. 5.37(A)). In the second case, i.e. compaction with cementation and pressure dissolution (Fig. 5.37(B)) comes with a significant reduction in porosity and to significantly reduction in the total volume, and thus the thickness of sandstone.

The diagenetic processes in sandstones and sandy sediments are due to changes in porosity, and they play an important role in the sandstone properties, i.e. the possibility of oil and gas reservoirs, aquifers of drinking and thermal water.

5.5.5.2. *Diagenetic Processes in Clayey Sediments*

Mechanical diagenetic processes or compaction in the clay sediments have a much greater role than in sandy sediments, as freshly deposited clay sediments and sludges signify loose packing components. The accumulations of clay minerals form honeycomb or "house of cards" structure (Fig. 5.38(B)) and have very high porosity, typicallybetween 70% and 85%. The pores between honeycombs aggregated clay particles are completely filled with water.

Strong compaction due to pressure overlays starts with the gradual deposition of increasing amounts of new sediment. The loose packing of particles in the honeycomb, or "house of cards" is not stable and particles began to restructure in parallel with each other schedule and significantly reducing the porosity. Primary porosity of clayey sediment and sludge is appreciably higher than the porosity of newly deposited sand. A honeycomb packed clusters of clay particles restructure or "crash" in parallel position (between 100 and 200 m, Fig. 5.38(B)). Clusters are in such a destruction of the structure oriented perpendicular to the largest surface bearing pressure, causing laminated sediment (mudstone becomes mud shale). Simultaneously with the restructuring of particles, the other important diagenetic process takes place by displacement of pore water or another fluid (e.g. oil) that filled pores in the mud. The first process causes compaction of sediment and reducing porosity. The second process causes a strong flow of water or pore fluids.

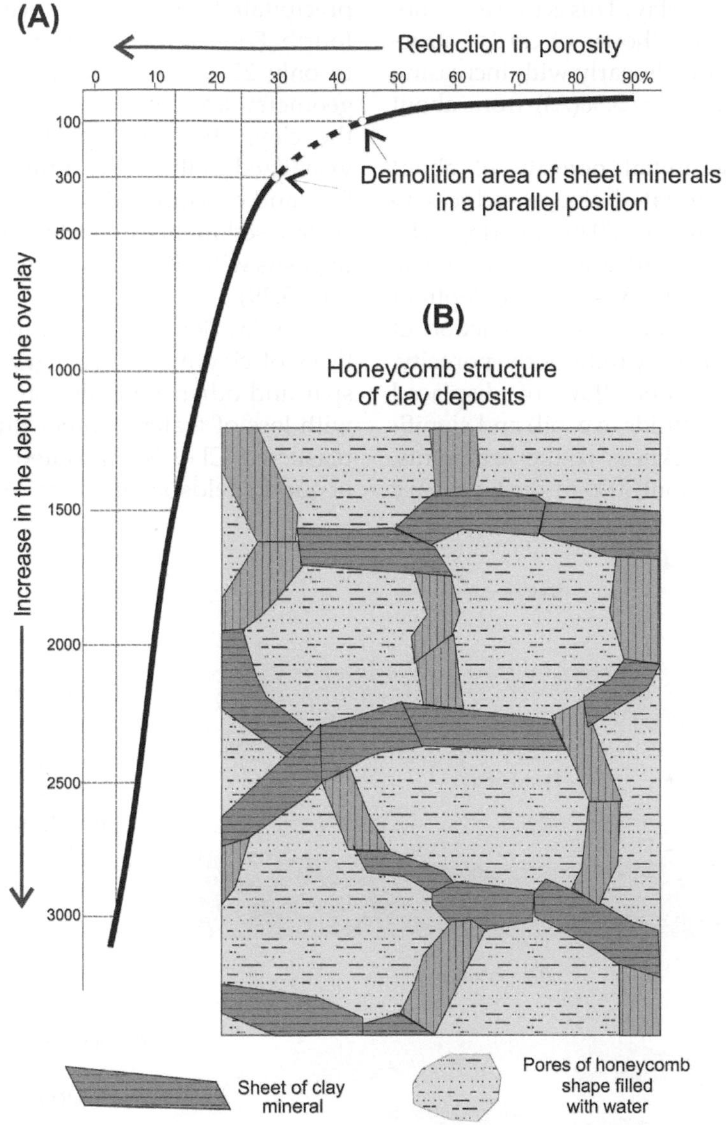

FIGURE 5.38 Schematic diagram showing honeycomb structure of the clay deposits related to increasing depth of deposition and reduction in porosity: (A) changes in porosity of clayey sediment depending on the increasing depth of the overlay as a consequence of compaction and restructuring of particles and (B) honeycomb structure.

Experimental studies show the clay compaction process to be highly compacted clay at a pressure of 50 MN/cm^2 (MN = MegaNewton) as in nature, covering equivalent of about 250 m. The curve of the general reduction in porosity of clayey sediment with increasing depth of the overlay (Fig. 5.38(A)) shows that in the beginning of the clay sediments overlay with new sediments, the porosity decreases very rapidly to a depth of 100–200 m at a small increase in depth of overlay. This is due to demolition of honeycomb or "house of cards" structure. Porosity decreases linearly with increasing depth of the overlay (Fig. 5.38(A)) from about 300 to 3000 m.

In this way, the initial porosity of about 80–85% of just precipitated sludge reduces to about 43% at a depth of 100 m covering. The porosity continues to reduce to 30% at 300 m, 13% at 1600 m, and finally 3–4% at the depth of 3000 m (Fig. 5.38(A)). Thus, the compaction of clay sediments significantly reduces the porosity, causes severe compaction flow of displaced pore water (and other fluids, e.g. oil) and significantly reduces the thickness of the sediments. In-depth coverage of about 3000-m-thick primary precipitated sludge is reduced by about three-fourth. For example, thick mud of 100 m changes to only 25–30-m-thick layer of clay shale. The geometry and shape of sedimentary bodies in the clay, shale and pelitic sediments can be visualized in this way. If turbidity or submarine fan sand or sand body occurs within such deep-seated sediments or sedimentary package, the deposits will assume a convex or lenticular shape (Fig. 5.39).

The clay deposits are compacted until the particles of clay mixed with grains of quartz, feldspar and other minerals come to closest contact with loss of water or other fluids lead to loss of plasticity. Clay is imprinted in the interspaces of quartz, feldspar and other mineral grains sizes

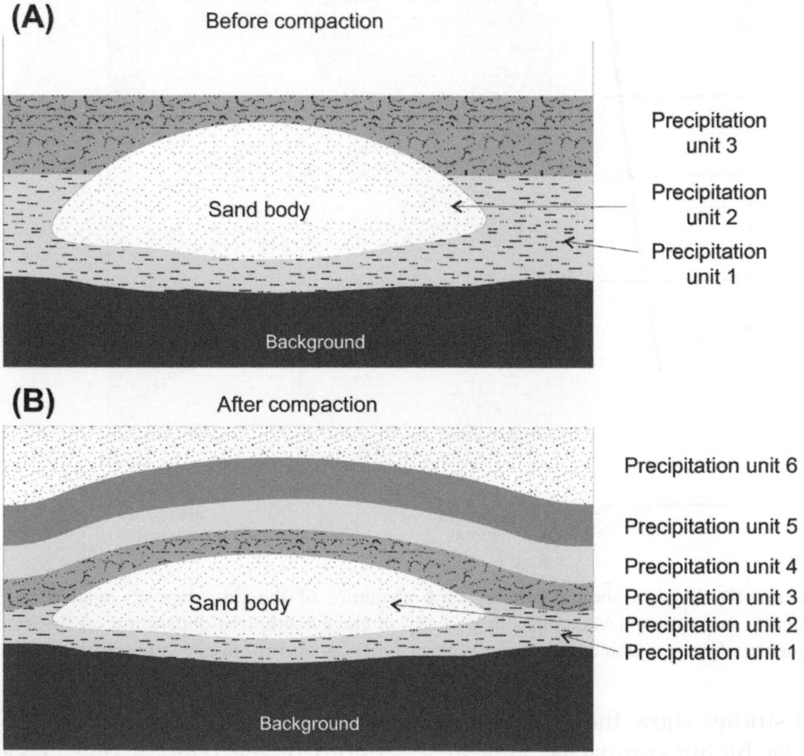

FIGURE 5.39 Experimental diagram showing the shape and thickness of (A) primary precipitated sediment and sand body and (B) changes due to the different effects of compaction on the clay and sand deposits with increasing depth of the overlay.

of silt and fine sand at high pressure. The grains are embossed on the clay and the sediment hardens mechanically. If the process of compaction by the high-pressure overlays continues, there would be deformation of certain components, pressure dissolution, and at even higher pressures resulting cracking of mineral grains.

Mechanical compaction of clay sediments has large role in their diagenesis. It is not the only important diagenetic process as it is regularly followed by chemical diagenesis. These processes are caused by changes in the chemistry of pore water when sediment comes under compaction currents. In the early stage of diagenesis, the clay sediments have high porosity with possible compaction flow of pore solution, and intense ion-exchange. The sediment is increasingly dominated by reducing conditions, and negative Eh-potential with increasing the depth of the overlay, oxygen deficiency. In the late stages of diagenesis, the porosity of clay sediments reduces and pore flow intensity. The pressure and temperature grow, and thus the speed of chemical reactions. In the final reduction of the porosity of only 0.5% at the depth of 6000–9000 m and at temperature of 220 °C, diagenetic process gradually disappears and metamorphic processes begin.

Chemical diagenetic processes in clayey sediments, due to the instability of clay minerals at high temperature and pressure, kaolinite that occurs in large quantities during the weathering process and immediately after deposition at greater depths overlay is no longer stable. The clayey sediments at greater depth covering typically do not contain kaolinite. The kaolinite completely disappears and is transformed into chlorite and illite at depths >3000–5000 m (Fig. 5.40).

Geologically old clay sediments (Paleozoic and Mesozoic) that have undergone intense diagenetic changes (mudstone and shale) usually have simpler mineral composition such as illite, muscovite and chlorite. Smectite, kaolinite and muscovite transform into more stable illite, muscovite and chlorite at higher temperatures (Fig. 5.40).

A good example of changes in clay minerals with increasing depth of the overlay is Miocene marls and marlite from deep wells and oil fields in eastern Slavonia, Croatia. Marlites (altered marl), illite and chlorite are reported as clay minerals by derivation of diagenetic processes at depth of 1300–1500 m. Chlorite occurs most intensely in the late stage of diagenesis at greater depths covering with transformation of kaolinite, montmorillonite and clay minerals. Part of the chlorite may occur in the marine spending in the early stage of diagenesis. Early diagenetic chlorites are usually rich in magnesium, and late diagenetic in iron. Chlorite is easily transformed into vermiculite and smectite, and in clay sediments often occur in disordered interstratified mixed layer minerals from a group of chlorite–vermiculite and chlorite–smectite.

Shale, formed from mud at greater depths overlay, contains with the stable quartz, the new stable minerals illite, muscovite and chlorite (Fig. 5.40).

5.5.6. Residual Sediments: Laterite, Kaolin, Bauxite and Terra Rossa

Chemical weathering of some rocks (as is more fully explained in Section 5.2.1.2) creates three groups of products of weathering:

1. Ions in the dissolved state: mainly released from rocks and hydrated alkali and alkaline earth elements (Na, K, Li, Ca, Mg, and Sr) and silicon in the form of silicic acid (H_4SiO_4).
2. Authigenic minerals: particularly clay minerals (kaolinite and seladonite or montmorillonite) and aluminum hydroxides.
3. Residues or waste of the rocks that in the spending did not dissolve (usually those containing quartz and resistant silicate minerals, especially mica).

The second and third groups are residual sediments or residues.

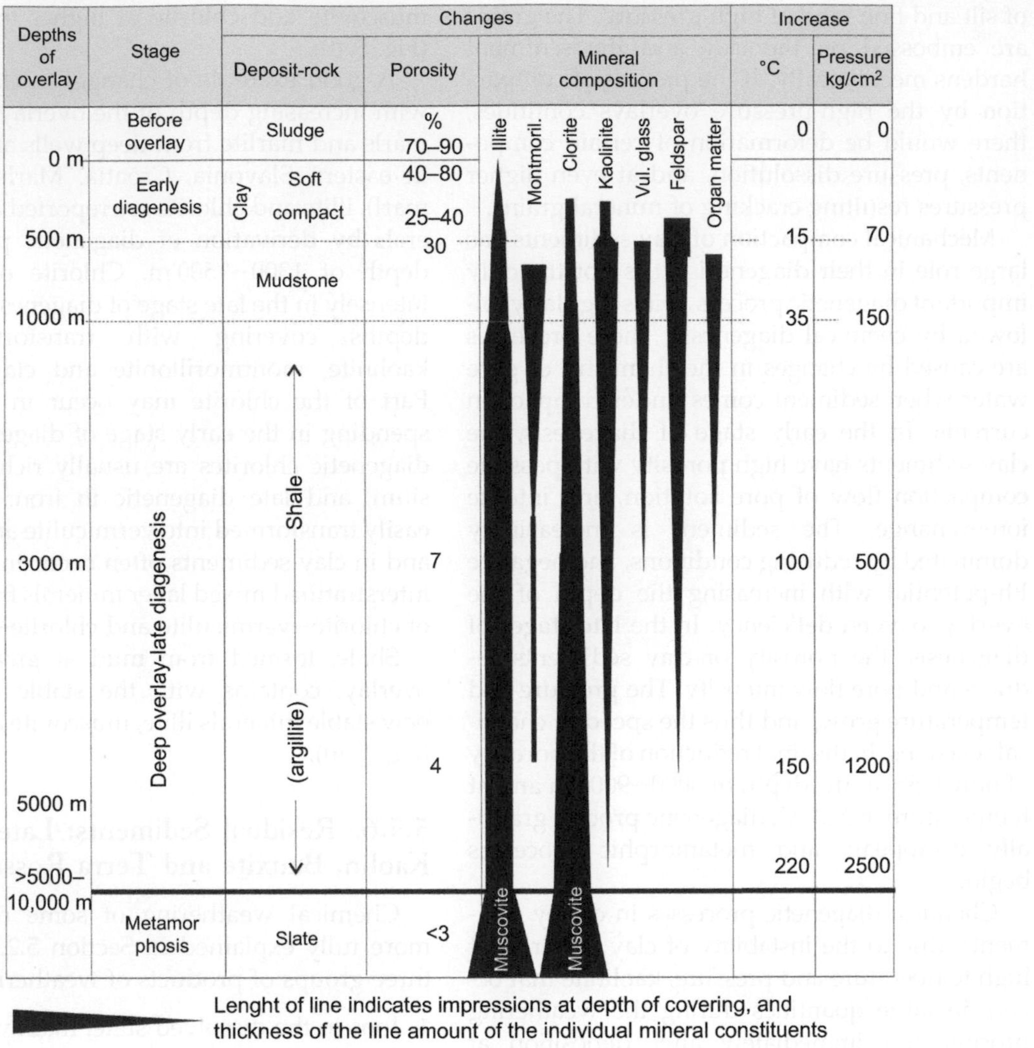

FIGURE 5.40 Diagenetic changes in mineral composition, porosity and rock types, depending on the depth of cover of clayey sediments. *Source: Revised by Füchtbauer and Müller.*[18]

In the initial stage of chemical weathering of mafic, neutral and ultramafic igneous rocks rich in olivine, pyroxene and amphibole, form authigenic minerals from group of chlorite and clay rich in iron and magnesium. Kaolinite, smectite and illite-clay are weathered products of acid igneous rocks and feldspar-rich granite-gneisses. In the advance stages of weathering, the clay is partially washed out in the form of colloidal particles, and also remains in the form of residual deposits. All magnesium and calcium minerals are leached out if the process continues uninterrupted. Quartz is the only left over or the final product or residue from primary mineral composition of rocks and newly formed authigenic minerals of the kaolinite group, boehmite,

FIGURE 5.41 Laterite soil (deep red brown color), chemical residual product from layered ultramafic complex, contains rich nickel resource from upper levels of Sukinda chromite deposits/mines, Orissa, India. *Source: Ref. 25.*

gibbsite, limonite and hematite. Strong chemical weathering, hot and humid air, and little or practically no erosion or removal of products of wear is necessary for the origin of such residue. This procedure generates residual sediments of laterite type, which are often economically significant mineral resources. The most important residual sediments petrologically include laterite, residual clay or kaolin, terra rossa (or "red soil", weathering of limestone) and bauxite.

Laterite soils are reddish-brown color (Fig. 5.41), which are products of strong chemical weathering of mafic and ultramafic rocks rich in olivine, pyroxene and hornblende. Laterite is rich in iron hydroxides, nickel, copper, chromium, platinum–palladium and aluminum, to which also contains small amounts of humus, quartz, calcite, clay and other minerals. Laterites are widely distributed and best developed on large plains made of basalt and basic intrusive ultramafic rocks in areas with humid tropical climate and low relief (India, Africa, and South America) with weak erosion. Laterite can be a potential source of high-value metallic minerals.

Kaolin or *China clay* is residual sediment consisting of pure kaolinite (Fig. 2.16), i.e. clay minerals from a group of hydrated aluminosilicates $Al_2Si_2O_5(OH)_4$, as a residual product of chemical weathering of feldspar, mainly from granite rocks. Kaolin deposits are usually formed by deposition of kaolinite after its shorter transfer by water from the point of spending or granite gneiss, mostly in lacustrine environments. Kaolin layers are typically located along the lake sands, sludges and coal or peat. Kaolin is a valuable mineral raw material for the manufacture of ceramics, especially porcelain, and the raw material in paper production.

Red Mediterranean soil, also known as terra rossa (Italian for "red soil"), is a soil classification that has been formally superseded by the formal classifications of systems such as the FAO soil classification, but that is still in common use. FAO stands for "Food and Agriculture Organization" of the United Nations.

Red soil (terra rossa) is, in geological terms, fine-grain sediment reddish brown and yellowish-red color, which is as clay-dusty cultivable soil located on calcareous, karst terrain of the Mediterranean area. Terra rossa is the chemical weathering product of limestone under oxidizing conditions excelled by Mediterranean climate.

The distinctiveness of red soil is its red color due to soil processes peptization of amorphous iron hydroxide and the formation of tiny crystals of hematite and goethite in tiny, dense ground mass of soil. With respect to granulation of red soils belongs to the fine-grained pelite sediments, because they consist of particle size <63 mm and very small, often insignificant share of the fine sand. Mineral composition of terra rossa is usually as follows: dominant are mica minerals (mica and illite), quartz and clay minerals (kaolinite and disordered kaolinite), and the much smaller proportion of hematite and goethite as well as amorphous substances, plagioclase and K-feldspar.

Red polygenetic soils mainly derived from powder materials that cover on the limestone and dolomite surface applied by wind, and by precipitation in cavities during heavy rains. It is often mixed with a small amount of indigenous soil generated from weathering of carbonate substrates. It also results in prolonged and repeated process of resedimentation. The origin (pedogenesis or soil evolution) is essential Mediterranean climate, good permeability carbonate base for a strong drainage, pH around 7 (roughly neutral) of pore solution, strong carbonate leaching, long-term (>10,000 years) suitable conditions for the formation of hematite and goethite as well as long (>10,000 years) suitable conditions for the genesis of kaolinite and generally accumulation of clay minerals.

Bauxites are rocks that contain minerals mostly from the group of aluminum hydroxide, mainly gibbsite ($Al(OH)_3$) or aluminum oxide hydrate boehmite (AlOOH) and rarely amorphous gel ($Al(OH)_3$). In addition to aluminum and bauxite minerals regularly contain variable amounts of kaolinite and halloysite, quartz, aluminum chlorite, hematite and goethite, and as the minor ingredients rutile and anatase. Bauxites are used for obtaining aluminum ore and also as refractory bricks. Bauxites arise in two mutually substantially different geological conditions:

1. the intensive chemical consumed silicate rocks of igneous and metamorphic origin to transform so-called laterite bauxites or silicate bauxites, and
2. karst on carbonate rocks, and are known as *karst bauxites* or *carbonate bauxites*.

Laterite bauxites are typical of tropical regions of South America, West Africa, India and Australia, and the massifs of Arkansas (USA). Karst bauxites are very abundant in the Mediterranean region, the Urals, West-Indian islands and in East Asia.

The process of formation of aluminum hydroxide in bauxite is associated with the hydrolysis of clay minerals, mainly kaolinite. For such a process requires underlying material, mainly clay minerals and large amounts of water to remove the silicon in the form of silicic acid, which requires a long geological time. Previously, it was thought that the karst bauxites occur by hydration process of clay material that is exclusively insoluble left over of karst and exposed to emersion limestone and dolomite. In recent times, there is more evidence that the parent material for the origin of karst bauxite may largely derive from small material Aeolian origin, therefore, of fine-grained materials or powders were either terrigenic or volcanic origin issued by wind, and only part of the insoluble residue of limestone and dolomite.

5.6. VOLCANICLASTIC ROCK

5.6.1. Definition and Origin of Volcaniclastic Sediments and Rocks

Volcaniclastics contain more than 25% of the ingredients of volcanic origin (fragments of volcanic rock, volcanic glass, and volcanic ash); a material was ejected by volcanic eruptions, and transported by air, water or pyroclastic flows to the place where it was deposited (on land or at sea). In some places, such material can be mixed and resedimented along with greater or lesser amounts of sedimentary material detrite or biochemical in origin. In volcaniclastic ingredients, pyroclastic in origin are the following:

1. Lithoclasts, i.e. fragments of volcanic rocks ejected during volcanic eruptions.
2. Crystal clasts or crystals that are crystallized in the lava before eruption. In the pyroclastic sediments came in more or less intact, or perished condition. Most often these are sharp-edged, angular fragments of quartz crystals, feldspar, amphibole, biotite, pyroxene and olivine.
3. Vitroclasts or fragments of volcanic glass, which are generally smaller than lithoclasts and crystal clast, usually sized between 0.1 and 0.4 mm. They are angular, irregular or

angular wedge-plate sections of acid, neutral and basic volcanic glass.

Tephra is a synonym for pyroclastic materials and pyroclastic sediments and in general for reservoirs of pyroclastic material regardless of the size of the fragments and particles. Volcaniclastic sediments contain fragments and particles of volcanic origin (volcaniclasts) that termed as pyroclasts or hydroclasts considering the place and mode of origin. The pyroclasts are products of volcanic eruptions on land and hydroclast fragments and particles occur in volcanic explosions on the contact of lava and water (submarine volcanism). The rapid cooling and mechanical granulation of lava occur in the contact with water.

Scoria is the name for a dark gray and black pyroclastic accumulation takes place at eruptions of neutral and basic lava.

Pumice stones are extremely porous, vesicular and light volcaniclastic material of bright color that floats on water (Fig. 5.42). It is composed of pyroclasts of different sizes and shapes, and arises from the stronger viscous acid, silica-rich, and neutral lavas.

Volcaniclastic sediments or tephra are broadly divided into three genetic groups with

FIGURE 5.42 Pumice stone, highly porous, vesicular and light volcaniclastic pyroclasts with low density (<1), and float on the water.

respect to the origin and the primary mode of transportation and deposition of pyroclastic materials:

1. Volcaniclastic sediments originated from pyroclastic flows.
2. Volcaniclastic sediments formed by deposition of pyroclastic material from the air.
3. Volcaniclastic sediments resulting from the turbulent flow of low density and high speed.

Volcaniclastic sediments originate from pyroclastic flows resulting from volcanoes hot, gas-rich pyroclastic flows and ash fragments flowing or rolling and crashing down the slope of volcanic eruption or by a similar mechanism of gravity flows. The main components of these flows are volcanic gases and primary volcanic material predominantly acidic composition. The dimensions vary from small grains to large blocks. Such flows occur from subaerial or submarine environments (Fig. 5.43). The sediments of volcaniclastic material deposited by mechanisms of one, several or more pyroclastic flows or ash flows and pumice-rich are called *ignimbrites*.

Volcaniclastic sediments formed by deposition of material falling from the air are the result of accumulation of pyroclastic material ejected by volcanic eruption high into the atmosphere. It is a fine-grain volcanic ash which after the eruption makes cloud of lapilli and volcanic ash into the atmosphere and is transferred over long distances of several hundred to several thousand kilometers away from the eruption site. The farthest reaching are the tiny particles of ash from which arise fine-grain tuff in the vicinity of the eruption precipitated by lapilli, or lapilli tuffs (Table 5.5). In this way, fine-grain tuffs usually form thin bands within the land, lake and marine sediments in connection with each eruption. The tuff settles on very large areas as the mark layers, i.e. layers formed by deposition from the same stage of volcanic eruptions, and

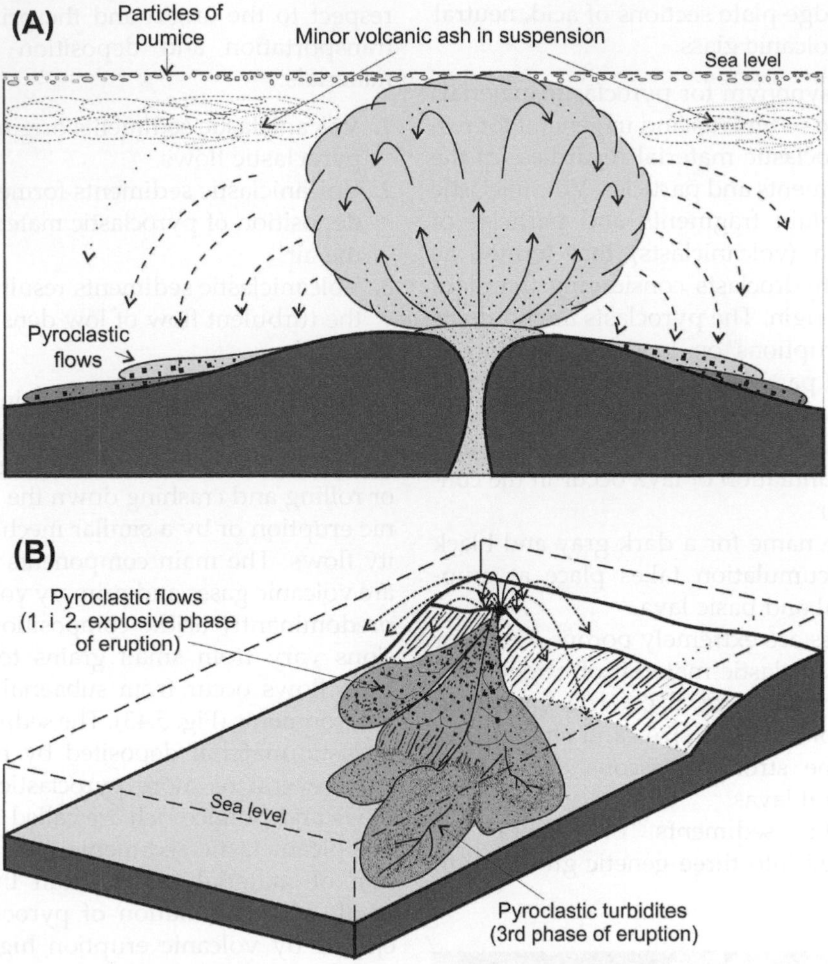

FIGURE 5.43 Deposition of volcaniclastic sediments in undersea volcanic eruptions: (A) ejection and suspension of volcaniclastic material with seawater, and (B) pyroclastic flows occurred in the first and second phases of the eruption and volcaniclastic turbidites occur in the third—low explosive—the phase of the eruption.

TABLE 5.5 Classification of Volcanoclastic Granulometry Sediment (Tephra) and Volcanoclastic Sedimentary Rocks

Particle Size	Type of Clast	Name of Sediment	Consolidated Rock
>64 mm	Volcanic bombs blocks	Agglomerate	Agglomerate
2–64 mm	Lapilli	Lapilli tephra	Lapillistone
0.063–2 mm	Large volcanic ash	Coarse-grained volcanic ash	Coarse-grained tuff
<0.063 mm	Fine volcanic ash	Fine-grained volcanic ash	Fine-grained tuff

Source: Ref. 30.

have defined the exact time of deposition of layers in which they are located.

Volcaniclastic sediments, resulting from the turbulent flow of low density and high speed, are characterized by thin and irregular layers. The sediments are precipitated from the turbulent flows generated by different mechanisms, primarily the strong interactions of submarine eruptions and the surrounding water. It mainly consists of poorly sorted sand and fine gravel (0.063–4 mm), with different composition and origin, with greater or lesser amount of pyroclasts from the last eruption, and the prevailing amount of clasts derived from older volcaniclastic and effusive from previous eruptions. The complete sedimentary cycle of volcaniclastic sediments deposited by submarine volcanic eruptions can be assumed in three different phases (Fig. 5.44).

Phase	Depositional Activities
I	Volcaniclastic precipitated from pyroclastic flow in the most intense phase of the eruption.
II	Deposition lapilli and volcanic ash from seawater during each new eruption of pyroclastic flows and no sedimentary material from turbidity flows.
III	Deposition of pelagic sediments, with brief interruptions of deposition of fine volcanic ash and/or pumice.

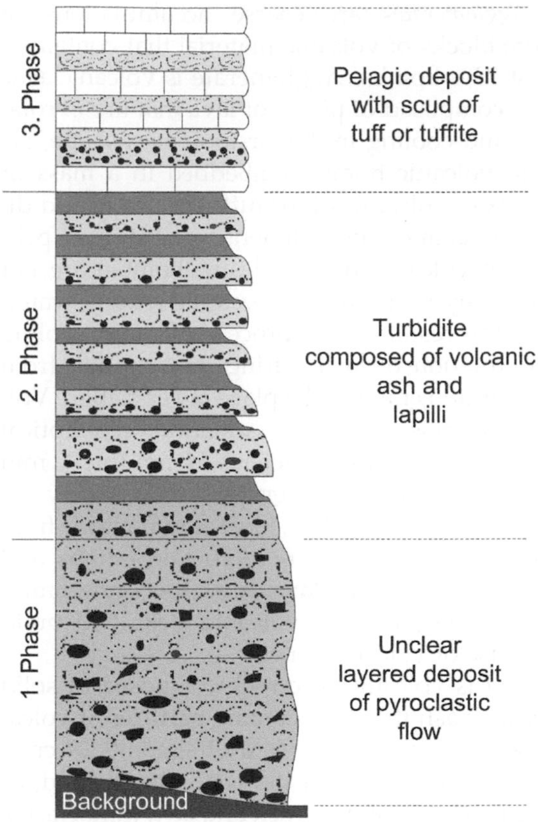

FIGURE 5.44 The complete sedimentary cycle of volcaniclastic sediments deposited in submarine volcanic eruptions. *Source: Modified and supplemented by Schmidt*[39]; *Einsele.*[13]

5.6.2. Composition of Volcaniclastic Sediments and Rocks

Volcaniclastic material or tephra is divided into volcanic bombs, lapilli and coarse and fine volcanic ash based on the grain size. Their precipitation and lithifaction make various pyroclastic rocks such as agglomerates, volcaniclastic breccias, lapilli tuffs, coarse-grained and fine-grain (pelite) tuff (Table 5.5). A mixture of clastic, biochemical and chemical material is called *tuffite*. The common types of volcaniclastic rocks are volcanic breccias, agglomerates, tuffs and tuffite.

Volcanic breccia consists of angular and semi-angular fragments and volcanic ash, i.e. the ejected material in eruption. Individual blocks and fragments of volcanic breccia are typically embedded in the matrix, i.e. lithified volcanic ash with tiny fragments of volcanic glass, which sometimes can be mixed with material of nonvolcanic origin, such as clay, silt and marl. In certain types of breccia, matrix may have vesicular (porous) structure or the structure of pumice.

Agglomerates are coarse accumulations of large blocks of volcanic material that contain at least 75% bombs. Agglomerate is volcaniclastic rock composed of pieces of lava that are in rotation and cooling in the air took on a shape, i.e. from volcanic bombs, embedded in a mass or matrix of volcanic ash or tuff. The shape and dimensions of clastic sediments consisting of pebbles of volcanic rock (volcanic bombs) are not caused by rounding and wear activity of water. It is the result of the process of rapid cooling and rotation of lava during the eruption from the volcano crater to the place of deposition. Volcanic bombs are nearly spherical or elliptical piece of lava with diameter >64 mm (or, 32 mm by some) which is erupted completely or partially in molten state, like the pyroclastic fragments. Agglomerates may contain bombs and fragments of older lava from the same crater and/or fragments of volcanic rocks that build the base of volcanic cones.

Tuff is volcaniclastic rock composed of solid volcanic ash that may contain particles of volcanic glass (vitro-clasts), small fragments of crystals formed in lava (crystal clasts) and/or fragments of volcanic rock and lava (lithoclasts). The various tuffs will be designated as rhyolite, dacite, andesite, trachyte and basaltic based on the composition of the mother volcanic eruption consisting of acid, neutral or basic lava (rhyolite, dacite, andesite, trachyte and basalt). Tuffs, which contain mostly of crystal clasts, are called *crystal tuffs*. The one predominantly composed of particles of volcanic glass (vitro-clasts) will be called *glassy* or *vitro-clastic tuffs*. The one predominantly contains lithoclasts are called *lithoclastic* or *lithic tuffs*. There would be mutual transitions members such as crystal-lithic tuff and crystal-vitroclastic tuff.

Sillar-tuffs are glassy tuffs in which lithifaction is mainly the result of crystallization in pneumatolitic activities. They consist of aggregates of angular, cuneiform, often elongated and curved shards of volcanic glass and are rich in pumice fragments in all stages of breaking. In them are also numerous small fragments of oligoclase and small amounts of biotite flakes.

Merged or *welded-tuffs* occur in lithifaction of hot ash which was hot at the time of deposition. The particles of pumice and small fragments of glass languished in soft ash in the lower parts of the mass because of its weight. Matrix of welded tuffs is porous in its top layers and easily crushed. Matrix will be less porous and harder at the bottom. Most abundant and important ingredients of welded tuffs are fragments of volcanic glass, followed by crystal clasts of quartz, sanidine, biotite and oligoclase.

Tuffite material is a mixture of volcanic and sedimentary origin, or rock that contains ingredients between 25% and 75% of the volcaniclastic origin and 75–25% ingredients of sedimentary origin. Sediments containing 10–25% material from volcaniclastic origin are called as *tuffite* or *tuffite marls* and *tuffite sandstones*.

5.6.3. Alteration of Tuff

Of all volcaniclastics, tuffs and tuffites are the least resistant to chemical weathering. The processes of chemical modification of tuffs are a direct consequence of their composition, structure and physicochemical conditions and environment of their origin and geological age. Alterations of tuffs, especially with volcanic glass, are the result of chemical reaction of glass with water so that the alternating process of dissolution of volcanic glass with the process of excretion of authigenic minerals in places where glass is melted. The most common products of such changes in tuffs are the zeolite group of minerals (Table 2.12). If temperature further increases with the depth of the overlay, they cross in to chlorates, quartz and albite.

Neutral and acidic volcanic glass gives different products of changes in relation to basic volcanic glass. These differences can be observed in the early stages of change.

Alterations of acid glassy tuffs primarily depend on the pH of pore water, seawater or freshwater. The glassy acidic tuff of ~10,000 years old changes to cluster of alkali-rich zeolites (phillipsite and clinoptilolite) in the presence of basic pore solution/water with pH > 9.5. This is common in case in many lakes of arid climatic regions. Authigenic zeolites with initial high salinity are transformed to "analcime" ± quartz± or in K-feldspars ± quartz.

Acidic volcanic glass, under conditions of sea/freshwater with typical low pH, primarily changes to smectites (mainly montmorillonite group) ± opal, cristobalite and zeolite. The acid tuffs alter to bentonite due to the effects of seawater or freshwater. Bentonite (or smectite deposits) is a clay similar to montmorillonite, zeolite, cristobalite, chalcedony and opal. The process of alteration of acidic glassy tuff to bentonite takes several million years at pH 7–8.

Alteration of basic tuffs, especially basic volcanic glass of subaquatic foundation, predominantly composed of basaltic glass, occurs much faster than the changes of acid tuffs. It changes quickly to palagonite. Palagonite is a name for the brown, yellow or orange-gray resinous mixture of different minerals from the group montmorillonite, zeolites, mixed-layer clay minerals, chlorite, limonite and goethite. Palagonitization is the process of alteration of basalt glass and glassy tuffs into palagonite. This is hydration process that occurs with the addition of water and removal of alkali and alkaline earth ions, silicon and sometimes aluminum and oxidation of iron with excretion of zeolite, calcite and minerals of montmorillonite groups.

5.7. CHEMICAL AND BIOCHEMICAL SEDIMENTARY ROCKS

Chemical and biochemical sedimentary rocks belong to the endogenous sediments, i.e. sediments that occur predominantly inorganic chemical or biochemical processes. The rocks are divided into carbonate, silicon and evaporite sediment (Table 5.2) based on the chemistry of essential petrogenic minerals, organogenic components, and the system of secretion or precipitate.

1. Carbonate sedimentary rocks

Carbonate sedimentary rocks include limestone, dolomite limestone and dolomite (Table 5.2), i.e. rocks composed predominantly (>50%) of calcium carbonate minerals, calcite, Mg-calcite and aragonite, or dolomite minerals. It may also include variable proportion of siliciclastic material dimension silt, sand and clay, and authigenic noncarbonate minerals.

5.7.1. Limestone

Limestones are carbonate rocks predominantly organic, to a lesser extent, inorganic origin, in which the dominant component is the mineral calcite. They originated in lithifaction of aragonite, calcite and/or magnesium-calcite sediment. Limestones with calcite may also contain magnesium calcite, rarely aragonite and dolomite. Dolomite-limestone is composed predominantly of calcite.

5.7.1.1. Mineral Composition, Physical, Chemical and Biological Conditions for Foundation of Limestone

The limestone deposits are composed of calcite, aragonite and magnesium calcite, or only one or two of these carbonate minerals, lithified calcareous sediments (limestone) mostly contain only calcite. The other two minerals, aragonite and magnesium calcite, transform easily into stable calcite during diagenetic processes. The dolomite limestone composed of calcite and dolomite forms by late-diagenetic dolomitization. The calcite, aragonite and magnesium calcite are mainly excreted by the sea- or freshwater containing Ca-hydrogen by biochemical or organic, to a lesser extent and inorganic processes.

The secretion of calcite or aragonite depends primarily on the molar ratios of Mg/Ca. The secretion of aragonite is possible in all warm shallow seas with high molar ratio of Mg/Ca compared to the normal ratio (the world's oceans ratio is 5.26). Calcite and low-magnesium calcite secrete at a temperature of about 20 °C and at molar ratio of Mg/Ca <1, as is the case in freshwater lakes and rivers. It also excretes from the seawater with lower molar ratio of Mg/Ca at lower temperature (~10 °C) in deeper water with lower pH (but not <7.8) in the presence of SO_4^- anions. Excretion of high-magnesium calcite from seawater is mainly regulated through a tendency of organisms to build their skeletons from magnesium calcite. The percentage of isomorphic blended $MgCO_3$ in magnesium calcite depends on the temperature of seawater. The warmer water may contain high-magnesium calcite up to 28 mol.% $MgCO_3$. It can be concluded as follows:

1. Aragonite is excreted in the warm and shallow sea with at high ratio of Mg/Ca.
2. Calcite and low-magnesium calcite are excreted in cold or deep sea, where temperatures are lower, as also in lakes and rivers.

Excretion of Ca-carbonate from a solution saturated in calcium hydrogen carbonate takes place according to the following chemical reaction:

$$\underset{\text{calcium hydrogen-carbonate}}{Ca(HCO_3)_2} \longrightarrow \underset{\text{aragonite or calcite}}{CaCO_3} + CO_2 + H_2O$$

It is evident from this reaction that the secretions of calcite or aragonite in water containing dissolved calcium hydrogen carbonate take place, if from the hydrogen-carbonate somehow CO_2 or water is removed.

Removing CO_2 from the sea- or freshwater in nature can be caused by the following:

1. Bacteriological and photosynthetic processes of plants and cyanobacteria (blue-green bacteria and blue-green algae).
2. Heating of water.
3. Reduction of atmospheric pressure.
4. Spraying water into droplets in the waves or waterfalls.
5. Evaporation.

A good example for the extraction of carbonate by mosses and water plants which in photosynthetic processes contribute in the formation of calcareous matter on the waterfalls of rivers and lakes (e.g. Krka and Plitvice Lakes, Croatia).

The biogenic origin for most of the marine and some freshwater calcium carbonate is clearly established. The inorganic origin of many marine and surface limestone precipitations is difficult to prove. The majority of calcite creation gathers from meteor pore water under the surface of the Earth formed by inorganic processes.

5.7.1.1.1. SECRETION OF CARBONATE IN SHALLOW SEA

More than 90% of recent carbonate sediments are the result of biological or biochemical processes in marine, mostly shallow-sea environments. Their occurrence and distribution within the world's seas are directly determined by the growth and development of organisms whose life processes, especially photosynthesis and building skeletons and shells, related to the Ca-carbonate. Growth and development of such organisms are conditioned with temperature, climate, concentration and salinity of seawater. The existing seawater organisms and their preference for the construction of the skeleton or shell play an important role in the formation of the primary mineral composition of limestone deposits, especially those mainly composed of finely crushed skeletons. The favored minerals are aragonite, calcite or high-magnesium calcite. Many plant and animal species are directly or indirectly involved in the formation of carbonate sediments, or limestone. These are organisms that build their skeletons of aragonite, calcite, and thus lithified at the site of the growth form of limestone reefs. The greatest amount of diagenetic limestone deposits are caused by deposition of shell and skeleton, or their bioclasts excelled by the activity of waves, currents and bioerosion in small sections of crushed parts.

Many organisms, especially algae, cyanobacteria, mosses and grasses, to a large extent are indirectly involved in the genesis of carbonate sediments. The most significant photosynthetic processes of plants extract CO_2 and thus induce the secretion of $CaCO_3$ from sea- or freshwater containing calcium hydrogen carbonate. The photosynthetic process can source the release of 2800 g carbonate from today's tropical, warm, shallow seas, sea grass with an area of 1 m^2 in 1 year. One can get the picture of the importance of plants in limestone formations when this figure is deduce to a total area of shallow marine and counted the time of thousands and even millions of years.

5.7.1.1.2. SECRETION OF CARBONATE IN DEEPER WATER

The carbonate production is much smaller in deeper water because it depends directly on the degree of saturation of water in calcium hydrogen carbonate, which significantly reduces as the depth increase. The shallow sea is saturated, and the deeper parts of the seas and oceans in the World are poorly saturated with calcium hydrogen carbonate. Therefore, it is difficult to excrete Ca-carbonate in deeper sea.

The water depth at which the solubility of carbonates is equal to their excretion, it is called the *calcite compensation depth* (CCD). The water contains an excess of dissolved calcium hydrogen carbonate and excretion of Ca-carbonate is possible and stable above that depth (separation of calcite). Ca-carbonates are unstable below the CCD and dissolve because the water is supersaturated with calcium hydrogen carbonate, and cannot excrete. The solubility of calcite follow almost linear trend with increasing depth of the sea until just above the CCD border. Thereafter solubility of calcite sharply increases with a small increase in depth. The seawater will have no more calcite when CCD border line solubility reaches the absolute maximum (Fig. 5.45).

CCD boundary line varies depending on latitude, temperature and salinity of oceans and seas of the present day World (Fig. 5.45). In the equatorial belt of the Pacific Ocean, CCD is located at depths between 4500 and 5000 m. CCD is located between 4400 and 4900 m in the Atlantic Ocean between 40° north and 40° south latitude. CCD in the equatorial zone is at depth slightly more than 5000 m, and at diminished depth at higher latitudes up to 2000 m, and latitudes over 60° and much shallower than 1000 m.

Aragonite compensation depth (ACD) is significantly shallower than the CCD: the Atlantic

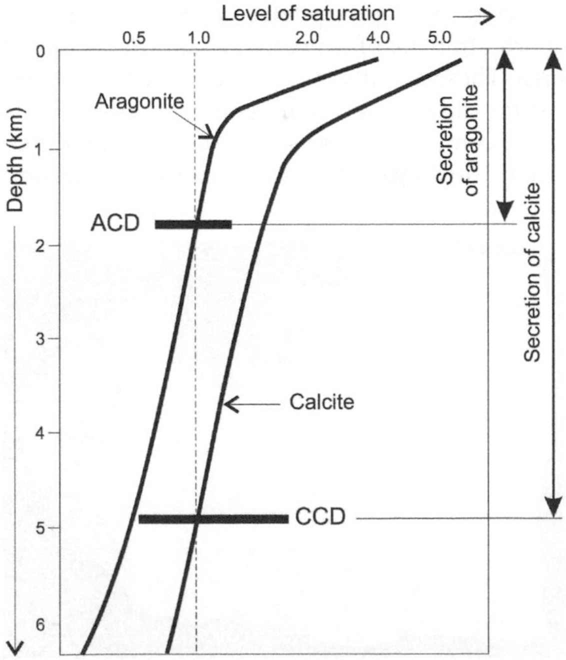

FIGURE 5.45 Position of calcium compensation depth (CCD) and aragonite compensation depth (ACD) boundary lines in deeper water.

Ocean in temperate latitudes is located at depths of 1700–1800 m (Fig. 5.45).

The positions of CCD and ACD boundary lines in the seas varied throughout Earth's geological history in rather wide limits. In the geologic past, the position of CCD has varied between 3000 and 5000 m. CCD and ACD boundary lines determine the stability fields of calcite or aragonite, in the sea from which it is clear that at deep sea (below CCD border line), there is no carbonate sedimentation (Fig. 5.45).

5.7.1.2. The Structural Components of Limestone

Limestones are composed of carbonate grains, limestone mud, and of subsequently extracted authigenic carbonate minerals. Carbonate grains or particles and very fine lime mud, or matrix, generally in the limestone is called "micrite" (Fig. 5.46), belonging to genetic group of primary carbonate structural components. Micrites are incurred and precipitated after a longer or shorter transfer of water in the same depositional area. These are all aragonite, calcite, magnesium calcite grains. The authigenic carbonate limestone components are subsequently, after deposition, during diagenesis extracted calcite and aragonite cements, which are commonly named as *sparite*.

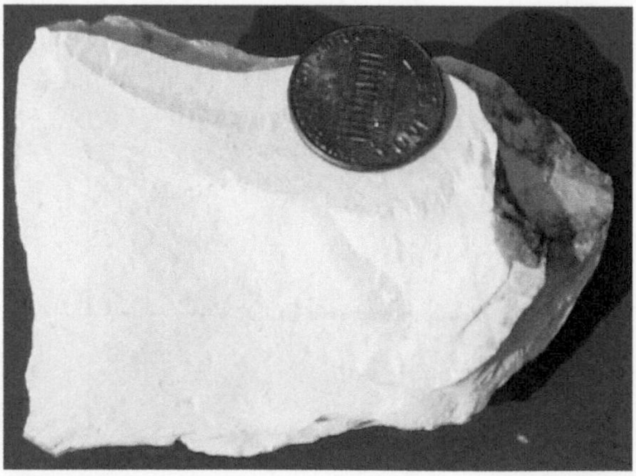

FIGURE 5.46 Micrite is a compact fine-grained limestone constituent consist of calcareous particles ranging in diameter up to 4 μm formed by the recrystallization of lime mud.

Micrite is nowadays generally understood as a very small matrix limestone or lithified lime mud, which consists of carbonate crystals or particles of diameter <30 μm (Figs 5.46 and 5.56). It is dense, in the transient light of microscope slightly transparent, calcite mass composed of allotriomorphic to hipidiomorphic calcite crystals with each other straight or bent contacts. Before lithifaction, it was fine-grain lime sludge—a mixture of tiny particles of aragonite and/or magnesium calcite or calcite. After lithifaction in limestone, micrite contains only cryptocrystalline or microcrystalline calcite and low-magnesium calcite as the primary unstable aragonite and high-magnesium calcite during diagenesis transformed into the stable calcite or low-magnesium calcite.

The origin of primary structural components of limestone and Ca-minerals can be organic (biogenic), inorganic and mixed inorganic–organic. Complete sharp division of the organic and inorganic compounds is not possible because of tight intertwining ways of their foundation so that they cannot always be distinguished. Carbonate mud, which is, for example, formed by excretion of carbonates in inorganic processes, and carbonate mud originated biomineralization, i.e. secretion of aragonite and calcite in the photosynthetic processes of algae and seaweed, together are impossible to differentiate. Moreover, none of these two sludges can be distinguished with petrographic microscope, even the sludge occurs by bioerosion. Therefore, the primary carbonate structural components of limestone are divided usually into skeletal and nonskeletal.

Nonskeletal components of limestone are the primary structural components, i.e. those are of inorganic origin, often laminated (Fig. 5.47) and that clearly do not originate from skeletal material of microorganisms, animals or shells of calcareous skeletons of plants. The group of nonskeletal components that are in the form

FIGURE 5.47 Inorganic well laminated crystalline limestone composed of white calcite ($CaCO_3$) and reddish ankerite [$Ca(Fe,Mg,Mn)(CO_3)_2$].

of beads or particles belong intraclasts, pellets, peloids, grapestone grain and coated beads (ooids, pisoids, and oncoids).

The skeletal-limestone ingredients are those which consist of one or more of carbonate skeletal debris or small shells or skeletons (Fig. 5.48). The skeletal components are common named as fossil, fossil debris, biodetritus and bioclast. A special type of primary structure is the stromatolites, and the primary structural component can be both skeletal and nonskeletal origin in the matrix or micrite.

FIGURE 5.48 Skeletal-limestone, apparently laminated, with debris of small shells and skeleton, represents Devonian reef complex along the northern margin of the intracratonic Canning Basin at Lennard Shelf hosting rich zinc–lead mineralization, Australia.

Genesis of carbonate mud, or matrix, or in the solid limestone micrite is very different. Micrite may arise by mechanical fragmentation of the skeleton, direct biogenic accumulation of small fragments of skeletal calcareous algae and coccoliths, chemical secretion of aragonite in the warm seas, excretion of small calcium carbonate crystals in the photosynthetic processes of plants, secretion of Ca-carbonate in the activity of bacteria, the accumulation of very fine detritus formed in processes of bioerosion of limestones caused by fungi, sponges, algae and other organisms that drill and destroy the foundation on which grow (bioerosion). Coccoliths are composed of thin calcite rings and discs of diameter 20–20 μm, gathered in a cluster of organisms Cocolithophridae, which are of great importance as components of many of the sea (pelagic) limestones. Carbonate mud also occurs in abrasion and mechanical crushing of limestones.

Intraclasts are carbonate grains formed within the depositional area by resedimentation lithified fragments of carbonate sediment, which occurred immediately after the destruction of sediment deposition. It can be very different in size, shape and internal structure depending on the composition, structure and texture of carbonate sediment destroyed, transferred and resedimented and deposited (Fig. 5.49). The composition, structure and lithofacies type of intraclasts are typically corresponds to the layer with activity of waves and ocean currents.

Pellets and peloids are spherical, ellipsoidal, and cylindrical or spindle-carbonate grain diameter mainly from 0.1 to 0.5 mm, rarely up to 2 mm, which is characterized by micrite internal structure. It consists of densely packed cryptocrystalline to microcrystalline carbonate containing an increased proportion of organic matter. Pellets are important and frequent primary structural carbonate components of shallow marine limestones and early diagenetic dolomites and recent carbonate sediments.

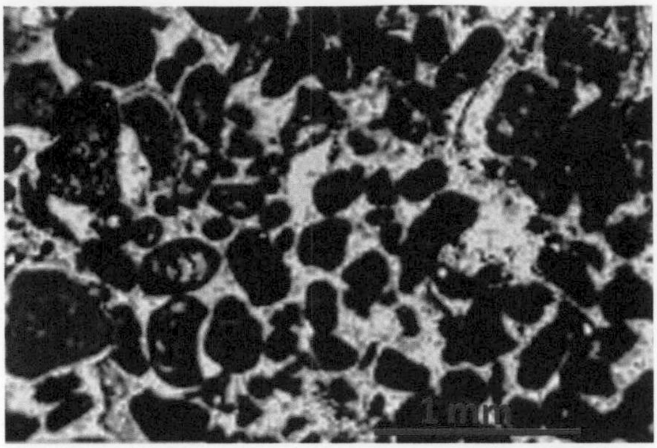

FIGURE 5.49 Intraclastic greystone: poorly sorted, rounded intraclasts with micrite structure (dark grains).

Fecal pellets are incrusted, fossilized feces of organisms that fed with sludges. They have spherical, ellipsoidal and well-rounded shapes. The shape, dimensions and internal structure are uniform in the same rock (Fig. 5.50). These are incrusted, fossilized feces and undigested remains of carbonate mud fed on sludge. They occur in all environments of deposition, in the shallow and deeper water, lived in large quantities of organisms, and preserved only under certain conditions. They are important indicators of environment and conditions of precipitation. The fossil preservation is usually possible only in the lower tidal region in the shallowest part of subtidal zone with low water energy, rapid lithifaction and cementation of deposits. The fossil preservation will be difficult in deeper and shallow water with increased strong water energy because the sediments will disintegrate/crush into loose carbonate mud before compaction, cementation and lithification. Specifically, in order to preserve fecal pellets, each pellet must move quickly from soft to the solid grain or must be cemented fast immediately after it was expelled from the organism for which optimal conditions exist in shallow water with low water energy supersaturated with calcium bicarbonate. Pellet is called only those of fecal origin and peloid are all other similar grains formed in some other way.

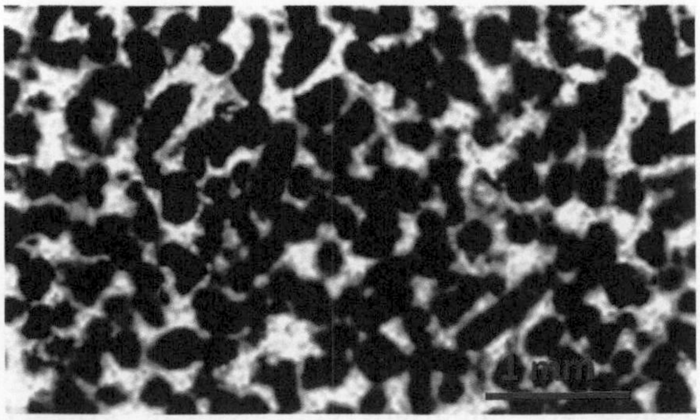

FIGURE 5.50 Pellet greystone: fecal pellets of uniform shape, dimensions and internal structure cemented in mosaic calcite cement.

Peloids are spherical and hemispherical micrite carbonate buildup in diameter usually 0.05–2 mm resulting in incrustations of blue-green algae. Unlike the fecal pellets, peloids are characterized by irregular shapes and different sizes, therefore, are not uniform in size, shape and internal structure.

Coated grains are specific type of carbonate grains of different origin and consist of clear membranes around some core. The coated grains include ooids, pisoids, oncoids or coated bioclasts. The rocks consisting mainly of them are called oolite, oncolite, pisolites. So, just rock, and not single grain, a continuation of "lite" (from the Greek "lithos" = rock).

Ooids are properly shaped, generally oval to spherical grains that consist of a core and multiple concentric membranes or laminae of different thickness (Fig. 5.51). Individual laminae can be thinner in places where the core is irregularly convex. Membranes those are located directly around the core outline the contours and shape of the nucleus, while those farther away from the core tend forming grain as close to a sphere (Fig. 5.51). The core of ooids usually is a pellet, a piece of the skeleton, foraminifera shell or some other skeleton, and grains of sand (quartz, rock fragment, and

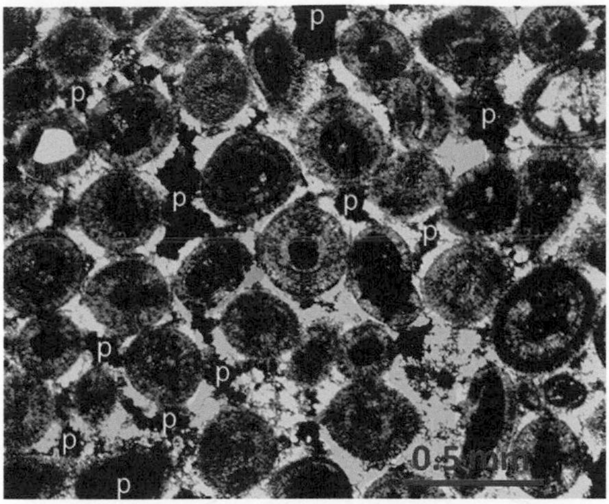

FIGURE 5.51 Ooid limestone composed of spherical ooids cemented with mosaic calcite cement (white). Part of the pore is not cemented (p).

feldspar). Ooids occur in warm, shallow seas with an average annual temperature above 18–20 °C and depths of <2 m, with a low calcium hydrogen carbonate in seawater without significantly elevated salinity. In these marine environments, it is necessary for running water with occasional changes in the intensity of its energy. The presence of granules (pellets, skeletons, skeletal fragments, and quartz grains) serves as core of ooids. The presence of organisms (bacteria and algae) removes CO_2 from the water causing the secretion of calcite or aragonite. It is believed that the growth of marine ooids need between 100 and 1000 years. The growth of ooids as shown by their concentric structure is not continuous, but in the growth of the membranes or concentric laminae, there is a relatively long time lag—no growth phase—in which the surfaces were exposed to weaker abrasion or bioerosion operation of endolith organisms.

Oolites, a sedimentary rock formed from ooids (Figs 5.51 and 5.57), play an important role in the reconstruction of the conditions and environment of deposition, particularly with regard to depth, salinity and water energy. It also serves as a significant source reservoir rocks for oil and gas due to their often extremely high primary intergranular porosity.

Pisolites (from the Greek "Pisos"—peas) are covered grains very similar to ooids, which, unlike ooids, are not primary marine structural components of limestone. Pisolite incurs during the diagenetic processes in caves, and "vadose zone" (or unsaturated water zone) under the influence and effect of freshwater on land or in marginal zones of marine, terrestrial and lake environment, as in the vadose zone around the hyper saline and in the zone of capillary lift the underlying water. These are characterized by a clearly visible regular concentric lamina material around a nucleus. The core around which there is one or more fragments of limestone. Pisolites form in caves and incur in geysers, or moving hot water, have regular spherical shape. Pisolites incur in the vadose zone and in quiet immobile water, have an irregular shape of core that resembles to the lamina, i.e. the outer shape of pisolites.

Oncoids (from the Greek "Onchos"—lumps) are grain covered with irregular shapes with carbonate jackets of micrite lamina. Lamina partially lays one over the other, usually without a clear concentric structure (Fig. 5.52), may contain remnants of organic structures as it forms with biogenic processes of algae and cyanobacteria. Many oncoids wrap beads formed by incrustation of larger number of such organisms, most of them but not all, contain a clearly discernible nucleus around which created such a small accumulation of biogenic carbonate material.

Oncoids can have very different forms of regular concentric spherical arrangement as their shape depends on whether they have nucleus or not, what is the shape of the nucleus, and the fact in which the direction fibers are faster growing. If oncoid have no nucleus or it is small, their shape is usually spherical, and construction is nearly concentric. If peloids have a large nucleus, flat or plate shape, which is often the case when the nucleus is bioclast of shell, then

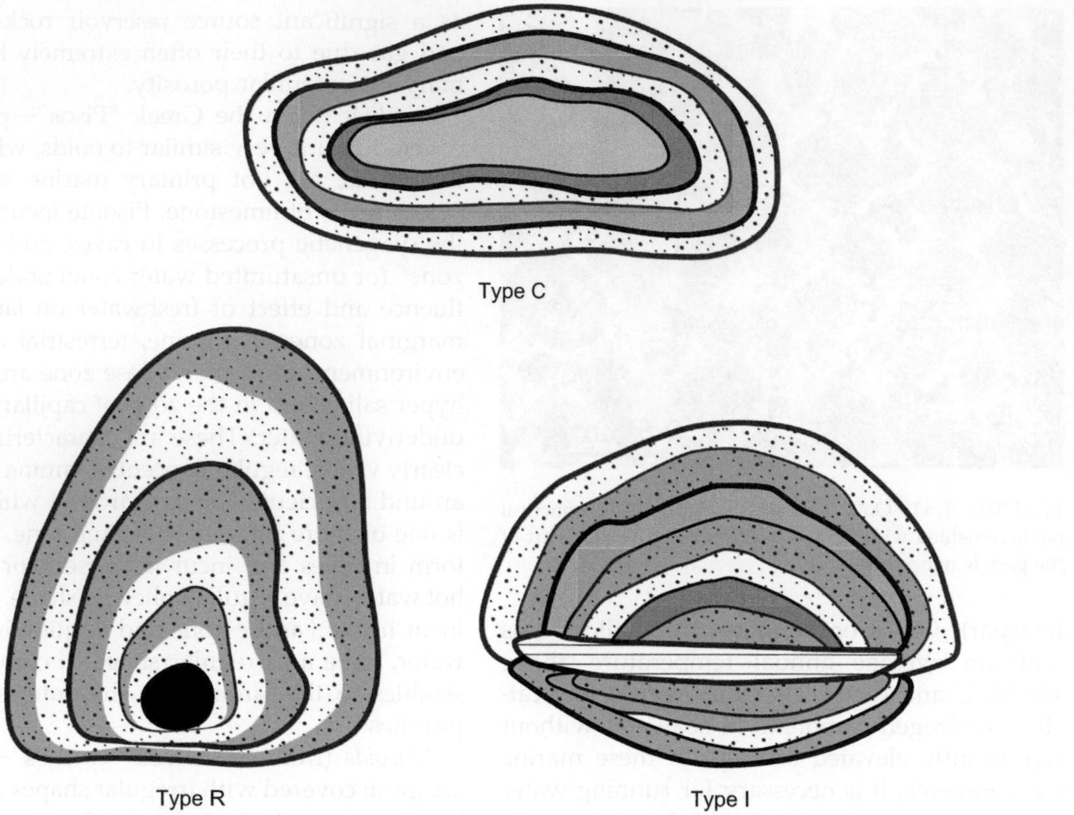

FIGURE 5.52 Structure and internal structure of oncoid type C, R and I.

oncoid is elongated (type C in Fig. 5.52). If oncoids have more irregular shape with a semicircular laminar structure, the environment was calm with only occasional tumbling of oncoid with stronger currents or storm waves (type R and I in Fig. 5.52).

Similar to ooids, oncoids are good indicators of environmental conditions and precipitation, and as a rule they occur in very shallow water, mainly in the lagoons with a low supply of water and sediment accumulation at low speed. Specifically, at the rapid sedimentation, backfilling occurs before finishing the growth of oncoids, and rapid sedimentation prevents the growth of organisms involved in the accumulation of carbonate and growth of oncoids.

Coated bioclasts are grains composed of fragments of skeleton, i.e. bioclast, and thinner or thicker micrite membrane at its surface. Membrane occurs in processes of micritization by activity of cyanobacteria, fungi that drill ground where grows. The life activity of organisms that inhabit the surface of bioclast, resulting small bore in diameter 2–30 μm of tubular shape. The holes are filled with dense fine micrite, probably the product of the secretion of carbonate through the mediation of bacteria after death of organisms. The newly formed bioclasts occur in calm, protected shallows and lagoons with depth not exceeding 15–20 m. The algae, cyanobacteria and fungi that drill the surface cannot settle on grains due to constant wear, abrasion and grind against each other in water with high energy. The coated bioclasts are often found in limestone (greystone and rudstone) deposited on tidal sandbanks and shallow water with high energy and constant activity of the waves. They arrive after the flooding and throwing with severe tidal currents and storm waves from nearby protected shallows or the lagoons environment.

Stromatolites are organic sedimentary structures (Fig. 8.17) formed by trapping, binding and/or secretion of sediment by activity of microorganisms, primarily cyanobacteria. The firmly lithified stromatolite fossils are laminated wavy, thick laminated or dome thick laminated carbonate rocks formed by binding and trapping of carbonate mud and other tiny carbonate deposits on the cyanobacterial mats. Recent stromatolites are composed of organic and inorganic laminae frequently exchanging with each other vertically. Organic laminae contain numerous genera of cyanobacteria. The inorganic laminae include carbonate mud, pellets, tiny skeletons or skeletal fragments of green algae, gastropods, ostracodes, benthic foraminifera of sediments flooded to "cyanobacterial mat" where the fibers are caught on the mucus of cyanobacteria.

Flooding of cyanobacterial mats during high tide and drying during low tide emerge organic (algal and inorganic) limestone lamina (Fig. 5.53). The carbonate sediment (mud, pellets, skeletons and fragments of skeletons) accumulates on the mat during floods. There will be no sedimentation during ebb. The moist muddy soil enables exuberant growth of blue-green algae, cyanobacteria overgrowing all the tide passed sediment. If the tide deposits large sediment on the mat (thick layer), cyanobacteria cannot outgrow, wipe out and generate fenestrations (pores of special forms) creating high fenestral porosity in limestone.

In the tidal and supratidal zone, stromatolites can easily be preserved, since the lithifaction improves with dehydration of residues and

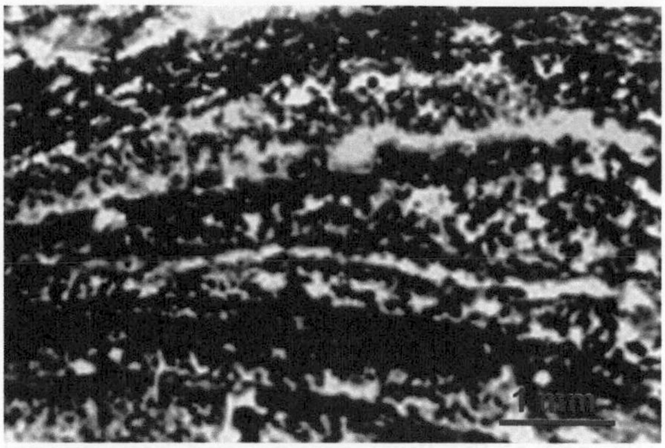

FIGURE 5.53 Stromatolites with changes of dark micrite lamina and light sparite lamina. Light laminae presents fenestrations emerged in decay of "cyanobacterial mats" that are later filled with cement and sparite represents individual fecal pellets.

secretion of calcite, aragonite or magnesium-calcite cement in the pores by evaporation of seawater. Stromatolites also preserve in subtidal zone due to the excretion of such mineral cement from hot calcium hydrogen carbonate supersaturated seawater. The secretion of carbonate has essential role in the process of assimilation of cyanobacteria. The "cyanobacterial mats" are not preserved in the fossil stromatolites due to quick decay after being covered with sediment. The decay forms cavities or fenestrations and subsequently filled with calcite cement. Stromatolites with high fenestral porosity are characterized by high content of cement, various fenestra and modification of laminae composed of carbonate deposits (micrite, pellets, and fossils) with laminae mostly made up only of cement. Stromatolites are extremely important indicators of environment of deposition.

Fossils, or skeletons and shells of organisms or their larger and smaller fragments in most of the limestones are important primary structural components. They are located as follows:

1. Skeleton or shells lithified on their habitat in a position of growth.

2. The whole skeleton and shell before sedimentation transferred by water currents, tides and waves.
3. Bioclasts, i.e. larger or smaller fragments of skeletons and shells.

The limestones of strict textural/structural features are direct result of ecological, sedimentological and hydrodynamic conditions of deposition environment with each of these above modes of occurrence of fossils forms. Therefore, limestones are predominantly composed of fossils, one of the three groups, and have special sedimentological–petrological names (such as rudist limestone and foraminiferal limestones). Sedimentological and petrographic labeling of limestone is based on their textural/structural and genetic features, depending on whether consists of a skeleton and of organisms lithified at the place and position of growth, of entire transported skeleton, or their fragments (bioclasts).

The well-preserved skeletons or shells of organisms lithified in position and place of growth form organogenic ridge (Fig. 7.14), mostly built of limestone (Fig. 5.55). Such sedimentary bodies are different according to morphological features:

1. Biostrome, which takes the form of layers or a large lens with more or less concordant relationship with the rocks (Fig. 5.54(A)).
2. Bioherma, irregular bulging sedimentary body shape caused by lithifaction organisms in the position and location of growth (Fig. 5.54(B)).

Reef-building organisms grow following one type of generation over another form of organogenic reefs. These carbonate sedimentary bodies can be significant reservoir rocks for oil and gas (Section 7.3.1.4), often due to high porosity and permeability over large sizes.

Limestones, which contain mostly bioclasts, are generally called bioclasts-limestone. The limestones, that contain the whole of transported skeletons, are called skeletal-limestone (Fig. 5.57(B)). The limestone, which largely contains well-sorted bioclasts and/or skeletons in diameter between 0.063 and 2 mm, is called "biocalcarenites", where "biocal" indicates biogenic calcareous components, and "arenite" for the sand size. Accordingly, in connection with the dimensions of bioclasts and the skeleton, limestones which mostly contain fossil remains of dimension >2 mm are called "biocalcrudite", and <0.063 mm as "biocalclutite" (Table 5.1).

Siliciclastic terrigenic components of limestone are detrite grains that are transferred to the depositional area by water or air. The material includes mainly quartz, clay minerals, rock fragments, particles of volcanic material and heavy minerals. Most limestones contain little terrigenic detritus, and fine size clay, silt and volcanic ash. The mineralogical and petrographic properties of these fine detritus grains can be investigated by X-ray only in the insoluble remains of limestone after dissolution in acetic, monochloracetic or diluted hydrochloric acid.

Noncarbonate authigenic minerals in the limestones include anhydrite, gypsum, quartz, chalcedony, opal, pyrite, glauconite, tourmaline, albite, K-feldspar, muscovite and zircon. Authigenic pyrite in the form of small grains or aggregates is usually a product of life activity of bacteria by sulfates reduction. Quartz, chalcedony and opal typically occur in processes of silicification or suppression of limestone mud or already hardened limestones. This is due to circulatory pore solutions containing silica acid, dissolve carbonate and secrete opal, chalcedony and quartz. Anhydrite and gypsum in the limestones may occur in the early diagenetic stage in the sabkha conditions. This happens by the secretion of calcium sulfate from highly concentrated solutions in the evaporite conditions, or during late diagenesis by suppressing of carbonates with sulfates with participation of pore solutions that contain sulfate ions.

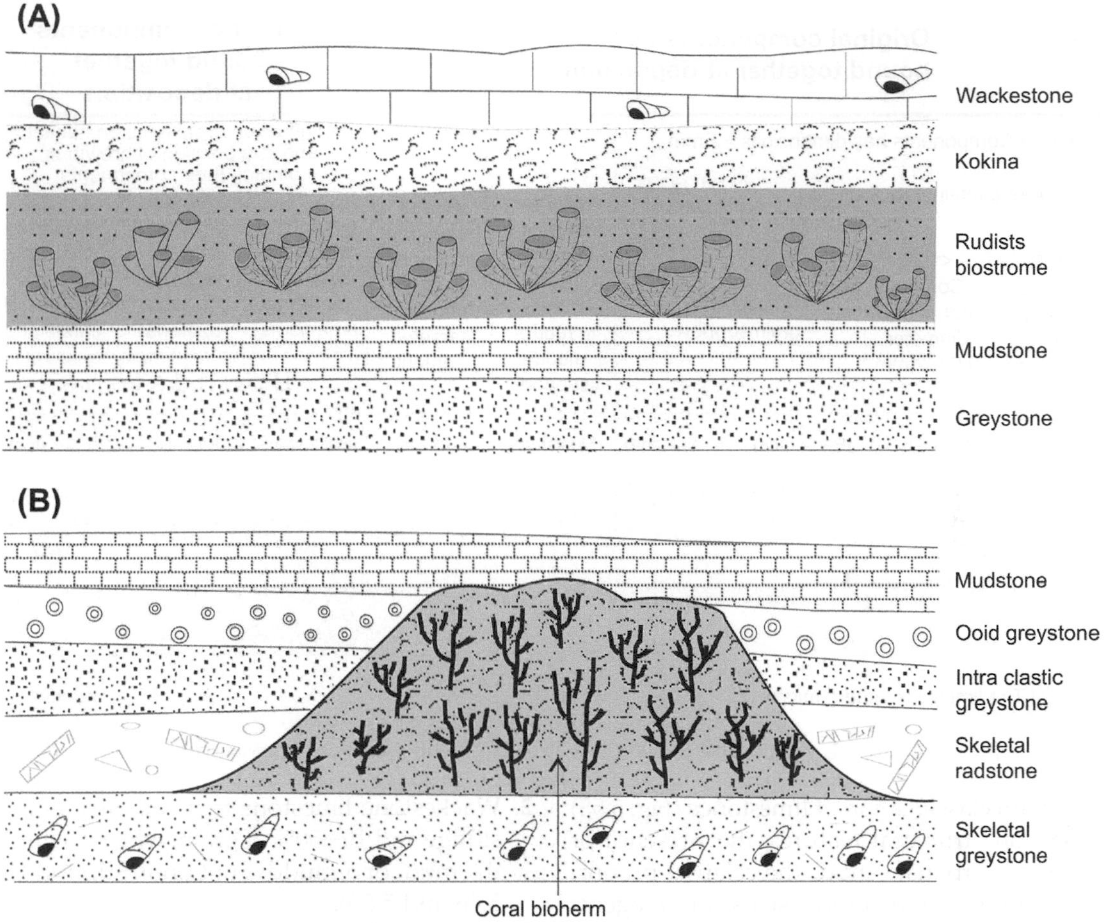

FIGURE 5.54 Morphological deposition of sediments with skeletal/shell organisms (A) regular growth layer Biostrome and (B) irregular growth discordant relation Bioherm.

5.7.1.3. Limestone Classification

Limestones are classified into three main types namely: (1) marine, (2) freshwater and (3) terrestrial with respect to the origin.

5.7.1.3.1. MARINE LIMESTONE

Marine limestones are the most common type of carbonate rocks originated from the sea. Several classifications of marine limestone in the world today exist with the broadest application by Dunham[12] and Embry and Klovan.[14] This is based on texture/structural features of limestone, the relations of primary structural components: grains (intraclasts, pellets, pelloid, wrapped grains, bioclasts, and skeleton), carbonate mud and calcite cement. Dunham classification is more applicable in describing the field and determining the limestone. Other classifications have wider application in the microscopic study of limestone, often used for early diagenetic dolomite. The classification systems are often used in field studies of limestone, as well as for the geology of oil.

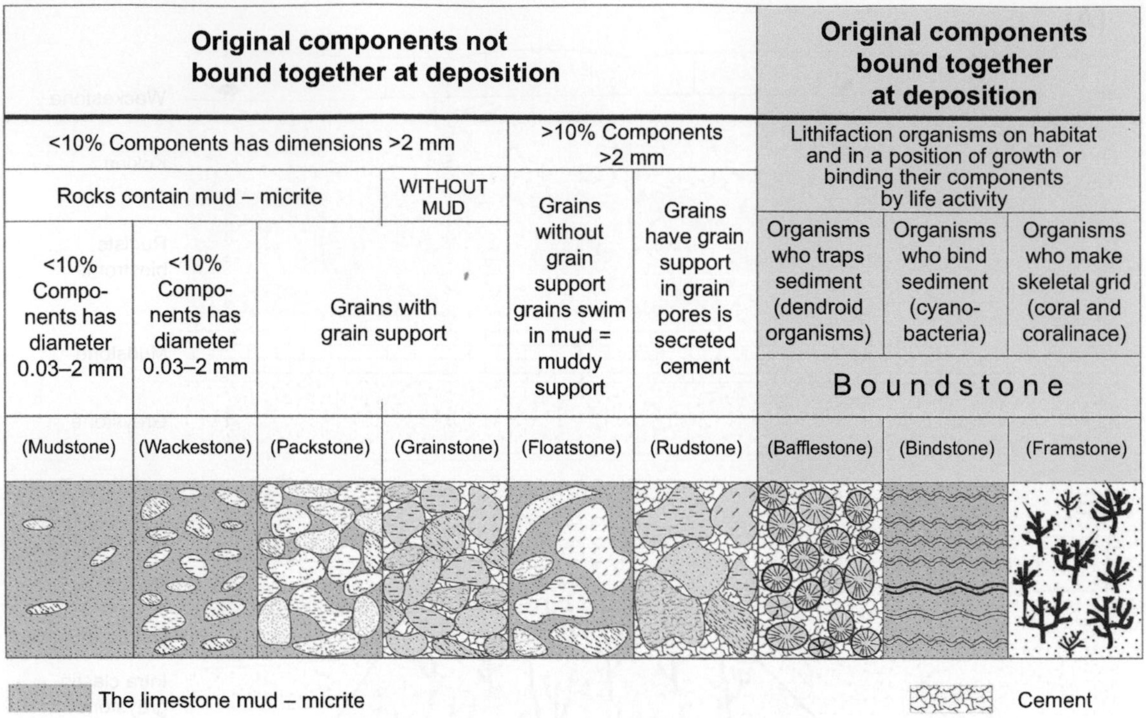

Original components not bound together at deposition
Original components bound together at deposition
<10% Components has dimensions >2 mm
>10% Components >2 mm
Lithifaction organisms on habitat and in a position of growth or binding their components by life activity
Rocks contain mud – micrite
WITHOUT MUD
Grains without grain support grains swim in mud = muddy support
Grains have grain support in grain pores is secreted cement
Organisms who traps sediment (dendroid organisms)
Organisms who bind sediment (cyano-bacteria)
Organisms who make skeletal grid (coral and coralinace)
<10% Compo-nents has diameter 0.03–2 mm
<10% Compo-nents has diameter 0.03–2 mm
Grains with grain support
Boundstone
(Mudstone)
(Wackestone)
(Packstone)
(Grainstone)
(Floatstone)
(Rudstone)
(Bafflestone)
(Bindstone)
(Framstone)
The limestone mud – micrite
Cement

FIGURE 5.55 Limestone classification after Dunham[12] with updates of Embry and Klovan.[14]

Dunham classification of limestones (Fig. 5.55) is based on structural features, the presence or absence of carbonate mud, the relative proportion of grains and mud, signs organogenic bonding skeletal over their development, lithification on the place, and the position of growth. The system is simple and easy to apply, in the field description using limestone magnifier.

Limestone in which the primary structural components or nonskeletal and the skeletal grains/carbonate mud are recrystallized, changed and converted into calcite crystalline mass is called the *crystalline limestone*.

Besides the already mentioned crystalline, Dunham[12] distinguishes five more basic types of limestone:

1. Mudstone limestone which contain carbonate mud and <10% of the grain diameters between 0.03 and 2 mm (Fig. 5.55).
2. Wackestone limestone that contains lime sludge and 10–50% of grain, which "swim" in the mud, or a muddy, not grain support (Figs 5.46 and 5.56).
3. Packstone limestone containing grains, which have granular support, touching each other and support, and lime mud in intergranular pores (Fig. 5.55).
4. Grainstone does not contain lime sludge, but only grains that have the mutual support, and calcite cement secretes in intergranular pores (Figs 5.49–5.51 and 5.57(A)).
5. Boundstone limestone that contains primary skeletal components (fossils) tied together with sedimentation, lithificated on its habitat in the position of growth or the individual components related to organisms, with the sedimentation and formation of biostrome, bioherm (Figs 5.54 and 5.55) or stromatolite (Fig. 5.53).

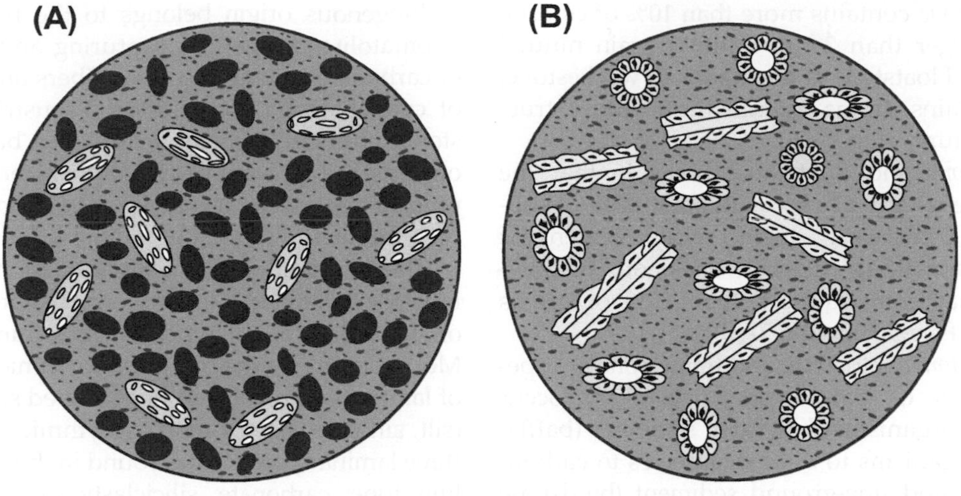

FIGURE 5.56 Micrite limestone wackestone type: (A) Pellet-wackestone is characterized by silty or matrix support of pellets, and (B) skeletal-wackestone containing lime sludge—micrite—in which there are individual skeletons and bioclasts of green algae.

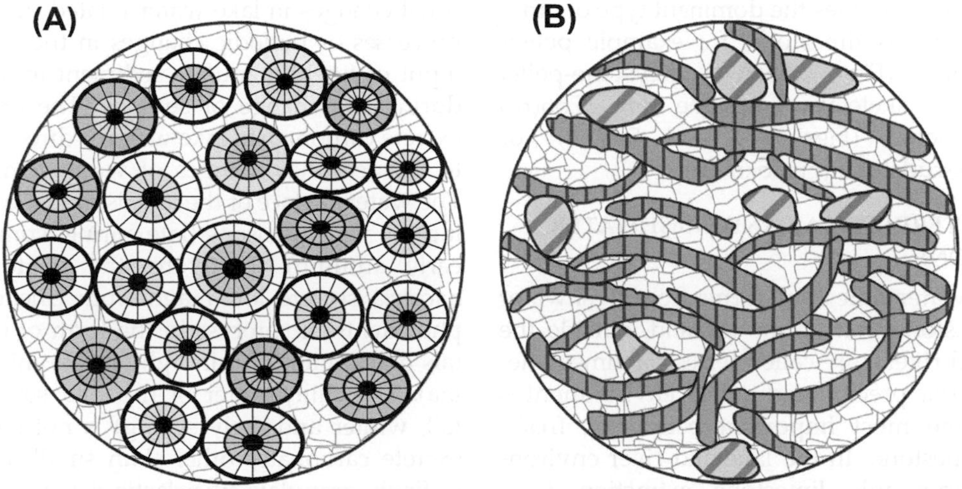

FIGURE 5.57 Greystone types of limestones are characterized by grain support and calcite cement in intergranular pores: (A) greystone-ooid composed of ooids and calcite cement excreted in the intergranular pores, and (B) Bioclastics rudstone contains fragments (bioclasts) >2 mm and cemented with calcite cement.

Dunham classification was updated by the Klovan and Embry (1972) by introducing two new types of rocks: floatstone and rudstone, containing more than 10% of grain diameter >2 mm. Baundstones are divided into three new types: bafflestone, bindstone and framestone, depending on the structure and the manner in which organisms are involved in the formation of these rocks (Fig. 5.55).

Floatstone contains more than 10% of components larger than 2 mm without grain mutual support. Floatstone is analogous to wackestone, but contains grains >2 mm by textural–structural features (Fig. 5.55).

Rudstone and floatstone differs in that the components are >2 mm with grain mutual support and between them calcite cement is extracted making rudstone analogous to greystone that contain more than 10% of grains >2 mm (Figs 5.55 and 5.57(B)).

Bafflestone, bindstone and framestone are special types of baundstone (Fig. 5.55) occur through organisms that catch sediment (bafflestone) organisms to bind themselves to carbonate mud and fine-ground sediment (bindstone or stromatolites), or organisms whose skeletons form the skeletal lattice, such as coral reefs (framestone), as shown in Figs 5.54 and 5.55.

Dunham classification is necessary to use the adjective that defines the dominant type of grain which contains limestone, for example pellet-wackestone (Fig. 5.56(A)), greystone-pellet (Fig. 5.50), skeletal-wackestone (Fig. 5.56(B)) greystone-ooid (Figs 5.51 and 5.57(A)) or bioclastic-rudstone (Fig. 5.57(B)).

5.7.1.3.2. TERRESTRIAL AND FRESHWATER LIMESTONE

Freshwater limestone forms from freshwater and limestone that occurs on land, outside the lakes and rivers, are called terrestrial limestone.

Lacustrine precipitated in lake environment is among the most important petrogenic freshwater limestone. In the lake and river environments, inorganic limestone extraction is a consequence of changes in pressure and/or temperature, and removal of CO_2 from the water due to assimilation processes of plants and/or phytoplankton (such as in Plitvice Lakes, Croatia), evaporation in arid climate areas, or mixing water with different pH, common in rivers and lakes. General textural–structural features of lake limestone are thin lamination and wrapped grain. Large amount of lake limestone of biogenous origin belongs to the freshwater stromatolites formed by capturing and binding of carbonate sediments on the fibers and mucus of cyanobacteria and mosses. Lacustrine limestone often enriched by oncoid and belongs to oncoids group arising by cyanobacteria wrapping of bioclastics or shells of gastropods and rock fragments.

Thin lamination or microlamination of lake limestone manifests as frequent changes of two or three lithological types of very thin lamina. Most common is the two-type rhythmic changes of lamina: carbonate and fine-grained siliciclastic (silt, silt clay, clay or marl). Rhythmic change of three lamina types is often found in the lacustrine limestone: carbonate, siliciclastic and diatomaceous. Such lamination can be identified analyzing the process of formation of "varve" (annual layer of sedimentation/sedimentary rocks). The rhythmic layers are the result of seasonal changes in lake water related depositional processes or periodic changes in the amount of input of finely granulate sediment into the lake during the change of seasons. The amount of excrete carbonate in lake water is directly related to the seasonal changes of water temperature. The water surface of the lake is more heated during the warm seasons (late spring, summer and early autumn). Therefore, the phytoplankton bacteria secrete low-magnesium calcite by the photosynthetic process in the form of tiny crystals (deposition of thin light calcite micrite lamina). The chilly water of the cold seasons (late fall, winter and early spring) is not suitable to excrete carbonates other than small deposition of finely granulate siliciclastic detritus (dust, silt and clay). The lake water provides more luxuriant development of phytoplankton with the increase in the temperature. This causes increased consumption of CO_2 and thus enhances the secretion of $CaCO_3$. The rapid development of diatomaceous flourishing algae and rich diatoms during the spring and early summer can cause increased amounts of deposition of opal skeleton and formation of thin diatomite lamina.

A good example would be that thinly laminated freshwater lake limestone deposits are located in the Sarmatian Pannonian Basin leading to well drilling in many oil fields of eastern Slavonia, Croatia. These rocks record changes in carbonate and clayey lamina bands of thickness between 0.2 and 1.5 mm. Among the grains of biogenous foundation in lake limestone are the most important oncoids incurred through cyanobacteria and green algae in the shallow waves and weak lake water.

Terrestrial limestone includes limestone cover (sinter), travertine, crust limestone and cave limestone or speleothems. Limestone sinters are highly porous, typically soft and form on the waterfalls of rivers and lake by secretion of calcite on moss, cyanobacteria and aquatic plants (Fig. 5.58). This process is particularly intense in the splash of waterfalls. The extraction of $CaCO_3$ due to release of CO_2 from water containing Ca-hydrogen as photosynthetic processes of plants and due to changes in pressure/temperature conditions in the spraying of water or its warming. The famous travertine barriers on waterfalls were formed in this way. The other example is the case of the Plitvice lakes and waterfalls and river Krka and Una, Croatia. These barriers consist of calcite or low-magnesium calcite in the form of irregular masses of limestone mud, micrite cover or shell, secreted on the remains of aquatic plants (moss, grass, fibers of cyanobacteria and branches of trees), as well as in form of sparite crystals that fill the pores of different origin. It also contains small amount of detrite material, mostly quartz grains in size of powder, fine sand, muscovite and clay minerals. It is very common events that thin films or fine carbonate mud (micrite) quickly envelop water plants growing in the river or lake, trees in the water or its branches, fragments of limestone and dolomite, and fragments of destroyed travertine barriers. Micrite shell and irregular masses make irregularly built skeleton or grid barriers, and rapid incrustation occur more or less solid mass from which "grow" travertine barriers.

FIGURE 5.58 The limestone travertine: calcite clusters with numerous small and large holes (molds) of now rotten plant remains, Plitvice Lakes, Croatia.

Large porosity of travertine is partly a result of rotting organic tissues of water plants and/or cyanobacteria, and partly they are dissolution cavity or cavities in which there was CO_2 and/or other gases formed by oxidation of organic matter. It is known that limestone barriers on the waterfalls of rivers and lakes occur mainly at temperatures between 10 and 30 °C with the annual accumulation of carbonates by 1–3 cm "growth" of travertine barriers, thus increasing the level of the lake. However, most of the secretion occurs in hot water during the summer months.

Travertine is lithified firm, spongy, cell built, irregularly laminated or layered limestone formed mainly by inorganic calcite secretion from the warm water around thermal springs, geysers, mineral water rich in carbonate and CO_2 or from hot sulfate springs. With high porosity and relatively high hardness, travertine is characterized by different textural–structural features: a thin and irregular lamination, and cell-like material (Fig. 5.59). Travertine deposition is a consequence of secretion of $CaCO_3$ due to inorganic and organic release of CO_2 at raised temperature in relation to the environment at the time of its outbreak to the surface of the Earth in hot springs. Travertine may be of origin in participation of both inorganic also organic processes related to the activity of bacteria that live in a "meadow" and/or bacterial coating on surfaces repeatedly or continuously covered by hot water. Temperatures of thermal water sources from which travertine arises are ranging between 20 and 95 °C. The annual accumulation (portion of "growth" of travertine in hot spring) is quite uneven and varies from a few millimeters to 20 cm. In many cases there is a connection between the deposition of travertine from hot water and volcanic activity, as testified by the frequent association of travertine and volcanites.

Carbonate components occur by inorganic processes are consist of large rhombohedral calcite crystals in the form of sparite cement or

FIGURE 5.59 Firmly lithified spongy, cell-like built travertine originated by calcite secretion from the warm waters around the thermal springs.

calcite crust with thickness up to several centimeters. It also forms pisoids, partially formed biogenic cryptocrystalline lumps and bacterial pisoids composed of calcite crystals in diameter of 0.5–20 μm.

Crusty limestone includes terrestrial limestone rocks known as "caliche" or "calcrete" which occur in semiarid and arid areas with dry climate and annual rainfall between 200 and 600 mm. The evaporation of water from soil is greater than the total annual rainfall. The pore water is saturated in Ca hydrogen, rises to the surface due to strong evaporation and forms calcite in the form of secreted crust clusters, caliche or calcrete. Crusty limestones usually have small thickness of the crust. This is an important indicator of paleoclimatic conditions, interpretation of environment and carbonate deposition and indicator of fine marine sedimentation, or vadose diagenesis or subaerial spending of limestone.

Cave limestone or dripstone is "stalactites" formed around water dripping, saturated in Ca-bicarbonate in the limestone caves and cavern. Cave limestones that grow from the floor upward are called *stalagmites*. Stalactites often join stalagmites to create "stalagmate" (Fig. 5.60). Stalactites grow due to the secretion of

FIGURE 5.60 Stalactite–stalagmite. *Source: University of Missouri.*

calcite from the evaporation of water droplets hanging from the ceiling of a thin film of water that drenched the rock, and the sudden release of CO_2 from the water saturated in Ca-bicarbonate in the moment of impact of water droplets that cap in a cave.

5.7.1.4. Limestone Diagenesis

5.7.1.4.1. DIAGENETIC ZONES AND PROCESSES OF CEMENTIZATION

The solid limestone rock of present day occur by early and late-diagenetic processes/conversion of water saturated primary hard and soft bulk of limestone sludge and grains under specific zones (location) and environmental condition such as the following:

1. Diagenetic processes in marine zone (1 in Fig. 5.61)
2. Diagenetic processes evaporation zone (2 in Fig. 5.61),
3. Diagenetic processes in condition of mixed zone with meteoric and seawater (3 in Fig. 5.61)
4. Diagenetic processes in condition of meteoric and vadose zone (4a and 4b in Fig. 5.61)
5. Diagenetic processes in greater depth overlay (5 in Fig. 5.61).

Diagenetic processes in marine zone occur in sediments that are soaked with seawater at the bottom of the shallow or deeper sea, the tidal flats and shores. In open marine environments or in the marine area, diagenetic processes strongly depend on water depth and, geographical location, and on tidal flats, shores, the most important factor is climate. The temperature and pressure of seawater has great role in the diagenetic process of limestone sediments due to lowering of the warm water from the surface to deeper/cooler parts of the sea/ocean depth. This further modified by photosynthetic processes with the role of marine plants and animals in the content of CO_2 dissolved in water in the form of hydrogen carbonate. The large differences in physical and chemical conditions prevailing in marine diagenetic zone in shallower areas and at depths usually distinguish between two main marine diagenetic zones: shallow-sea diagenetic zone and deep sea diagenetic zone.

Shallow-sea diagenetic zone is the area significant for diagenetic processes of cementization or secretion of fibrous aragonite and Mg-calcite cement in the pores of carbonate deposits.

Deep-sea diagenetic zone is characterized by pore water unsaturated with aragonite. Only the mosaic (blocky) calcite cement secretes with

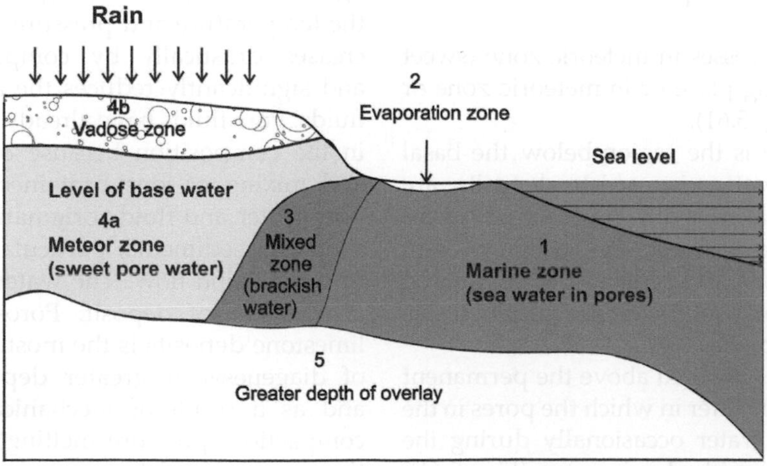

FIGURE 5.61 Schematic diagram of diagenetic zones of carbonate sediments.

the dissolution of aragonite above CCD boundary line. There is no secretion or calcite cement below this line with the dissolution of calcite ingredients. Kohout convection is the circulation of saline groundwater deep within carbonate platforms. In the deep-sea diagenetic zone, very slow process of dolomitization occurs due to thermal convection and saturation of seawater with respect to dolomite (Section 5.7.2.3).

Diagenetic processes occur in evaporation zone when strong evaporation of seawater from the saline or sabkha, and the pores of limestone deposits around in areas with dry, arid climate. In addition to the secretion of aragonite, there is early diagenetic dolomitization (Section 5.7.2.2) and secretions of evaporites minerals say gypsum, anhydrite and halite (Section 5.8.1).

Diagenetic processes in condition of mixed meteoric and marine (brackish) water occur in underground mixed zone where mixing meteoric (rain) sweet and salty seawater. This zone is very variable in shape, spread and geometry in depth and laterally. Depending on the oscillation of sea level, underground mixed zone moves farther from the sea towards the land (relative upper sea level), or from the mainland, extending from the coast towards the sea (relatively lower sea level), which is especially important for dolomitization processes (Section 5.7.2.2 and Fig. 5.67).

Diagenetic processes in meteoric zone (sweet water pore) taking place or in meteoric zone or vadose zone (Fig. 5.61).

Meteoric zone is the region below the basal levels of freshwater, in which deposits are continuously soaked with fresh basic water (Fig. 5.61). There is sweet pore water poor in Mg^{2+} and Na^{+} ions, and generally oversaturated with Ca^{2+} ions, and so is excreted calcite mosaic and "blocky" cements.

Vadose zone is the area above the permanent level of basic freshwater in which the pores in the rock filled with water occasionally during the rainfall and the residual rainwater (Fig. 5.61). Most of the time sediments are in subair conditions, i.e. the pores are either filled with air in the summer and with freshwater in the rainy period. All meteoric (rain) water flow to the sea through the vadose region play key role in the diagenetic process in met-stable carbonate sediments. In areas with heavy rainfall and porous deposit, meteoric water is quickly moving through the surface of the sediment and thereby powerfully dissolves limestone sediments. This process leads to increased concentrations of Ca-hydrogen carbonate, so that evaporation of rainwater in dry periods during the porous carbonate deposits, leads to secretion of calcite in the form of microstalactite calcite, or mosaic cement as well as secretion of vadose ooids.

5.7.1.4.2. DIAGENETIC PROCESSES AT GREATER DEPTHS OF COVERING

Diagenetic processes in deposit located at greater depth are covered with new thicker sediments (Fig. 5.61) and no longer operate diagenetic processes appropriate for surface and subsurface conditions. The depth of the overlay is not yet enough to act metamorphic processes (depth >5500–6600 m). This diagenetic area is interrupted by free exchange of fluid and chemically active atmospheric gases, particularly oxygen and CO_2, and progressively increasing the temperature and pressure. The porosity decreases drastically by compaction processes and significantly reduces the ability to change fluids. Pore fluids have already suffered changes in the composition because of the interaction and mixing of ions contained in the original pore water and fluids originating from the surrounding sediments, particularly those related to compaction flow. The water squeezes in the compaction of deposit. Porosity reducing of limestone deposits is the most significant result of diagenesis at greater depths of covering, and as a result of mechanical and chemical compaction (pressure melting) and cementization as a result of drainage sparite Druze calcite cement.

Compaction includes processes of mechanical compaction, squeezing water and chemical compaction related to pressure dissolution and formation of "stilolite" (serrated surface). Mechanical compaction and squeezing of water have a significant impact only in the mud, deep water carbonate deposit, especially pelagic mud and chalky deposit (Fig. 5.62).

Pelagic mud and chalky deposit i.e. deposits which contain scaffolding of the sea or planktonic organisms, have high primary total porosity (~80%) of which intergranular porosity accounts for ~35% after deposition. The porosity is reduced to ~65% and less, due to the rapid redeployment of squeezing water, reduction of isometric keletal grain size, mechanical reorganization of grains and change of the structure of deposits, at a depth 50 m of covering. At depths between 50 and 200 m began breaking the grain, and at depths >300 m pressure melting processes. The porosity of pelagic mud gradually further reduces (Fig. 5.62).

Shelf carbonate mud (Fig. 5.62) has significantly different compaction in pelagic mud. It mainly consists of elongated particles and needle-like, not isometric, shape, whereby the needles of aragonite act bipolar bind water molecules. The structure of carbonate mud looks like "honeycomb" or "structure of the house of cards" clay deposits. The effects of squeezing the water reduce considerably higher than those at pelagic mud. The primary total porosity of ~70% reduces to ~40% after restructuring of needle particles in the horizontal position at a depth of 100 m and then do not change until the beginning of the process of pressure melting at ~300 m depth (Fig. 5.62).

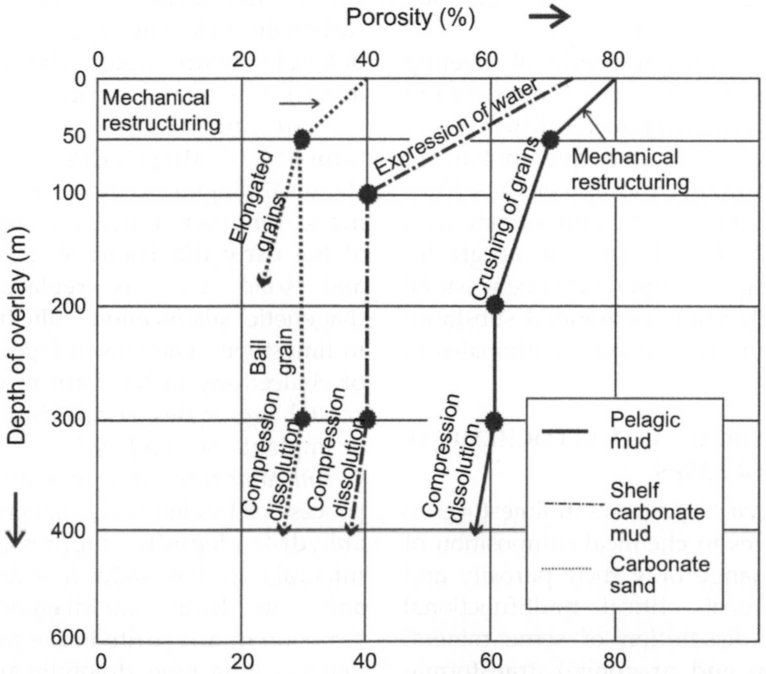

FIGURE 5.62 Diagrams of changes in porosity with increasing depth of the overlay with schematic representation of the main processes that lead to changes in the pelagic chalky mud and deposit, shelf carbonate mud and ooid, peloid and bioclastics limestone sands. *Source: Modified after Scholle[40]; Enoch and Sawatsky.[16]*

Carbonate sands have grain support. After cementization greystone types of limestone, characterized by significantly different effects of mechanical compaction and squeezing of water from pelagic mud and shelf carbonate mud (Fig. 5.62). Primary intergranular porosity of ooid or peloid sands is less than previously stated mud and chalky sludge due to sphere grain shape. There will be stronger decrease of porosity in the first 50 m depth of the overlay. The original 40% will reduce to 30% (Fig. 5.62) due to redeployment of spherical grains of unstable rhombohedral, more porous in stable, less porous cubic grain packing. The pressure melting starts at the grain contacts to about 300 m depth (Fig. 5.62).

Chemical compaction includes processes of pressure melting on contact with the formation of stilolite at depths of covering of few hundreds to a thousand or more meters. Stilolitization significantly reduce the porosity and the total thickness of limestone deposits.

Cementization at greater depths of covering in the limestone sediments are mainly down to the secretion of coarse crystal calcite cement and/or iron dolomite (baroque dolomite). Baroque iron dolomite typically contains 15 or more molar percent $FeCO_3$, and occurs with the participation of hydrocarbons at greater depths of covering at temperatures between 60 and 150 °C or fills voids or mineral substance that suppresses the surrounding carbonates in the rock.

5.7.1.4.3. ISOCHEMICAL AND ALLOCHEMICAL DIAGENETIC PROCESSES

Isochemical diagenetic processes in limestone do not lead to changes in chemical composition of limestone, but change only their porosity and structure. The most critical multifunctional mechanisms are dissolution of some mineral substances (halite and aragonite), transformation of aragonite and magnesium calcite in calcite, secretion of aragonite or calcite cement in the pores (cementization summarized in Sections 5.7.1.4.1 and 5.7.1.4.2), bioerosion, micritization of carbonate grains and activity organisms (endolithic) that drill foundation on which they grow and recrystallization of limestone sludge (micrite) and microcrystalline calcite in microcrystalline and/or macrocrystalline calcite or conversion of limestone type mudstone, wackestone and floatstone in crystalline limestone (Section 5.7.1.3; Fig. 5.55).

Allochemical diagenetic processes in limestone lead to changes in chemical composition of limestone and limestone deposits by circulation of pores solutions that bring into some other chemical compounds and anions (such as Mg^{2+} and Si^{4+} cations and SO_4^{2-} anion, and borrowing Ca^{2+} and CO_3^{2-}). Most significant allochemical diagenetic processes in limestone are silicification, anhydratization and dolomitization.

Silicification is an allochemical diagenetic process in which solution enriched with Si-ions, usually in the form of silicic acid (H_4SiO_4) in carbonate rocks suppress calcite, aragonite and dolomite with opal, chalcedony or low-temperature quartz, meaning, silicon hydroxides or oxides, and dissolved carbonates taking in the form of Ca-hydrogen carbonate. Like other allochemical diagenetic processes, silicification may induce another untied and unlithified deposit at the early diagenetic stage, or already rigid and solid rock is replaced during late-diagenetic silicification. Silicification processes in limestone generates autogenous quartz, opal or chalcedony in the form of single crystals or crystal aggregates, and nodule, lump, lenticular or implants of hornfels.

Anhydratization in carbonate rocks may be the process or forcing early diagenetic carbonate or anhydrite deposits, suppression of carbonate minerals in the solid limestone or dolomite anhydrite during late-diagenetic process, and secretion of anhydrite in the pores, cavities and veins of limestone, dolomite and other rocks in the circulation of pores solution containing sulfate. Sulfate ions-rich solutions usually originate from tidal saline (sabkha), or salt ponds or lakes left in the recesses by tidal environment after the withdrawal of the sea. Early diagenetic anhydratization in such environments requires strong evaporation with increasing temperature and a permanent increase in salinity ("sabkha-conditions"). Late-diagenetic anhydratization occurs during circulation of pores water containing SO_4^{2-}-ions through the rigid carbonate rocks, with the suppression of carbon compounds anhydrite. Such suppression will be of unequal intensity, irregular, and often in the rock selectively anhydratizationed only some components, while others remain largely fully preserved. In late-diagenetic anhydratization with large anhydrite crystals in limestone are usually found numerous tectonic venation, cavities and pores filled with anhydrite crystal aggregates as a result of secretion of anhydrite from solution that are circulated in these rocks.

Dolomitization is the most important and common allochemical diagenetic process that engages the limestone deposits and limestone, and turning them into dolomitic limestone or dolostone or simply dolomite. Dolomitization is a process by which dolomite is formed when magnesium ions replace calcium ions in calcite. It involves substantial amount of recrystallization.

Limestone enjoys major shares of sedimentary rocks in the Earth's crust. It is easily identifiable due to softness, color, texture and instant effervescences with hydrochloric acid (HCl). The natural landscape of limestone gives an excellent panorama in the country sides (Fig. 5.63).

Limestone is readily available, relatively easy to cut into blocks/carving and long lasting. Limestone is very common in architecture and sculpture across the World (to name a few Great Pyramid, Egypt, Courthouse building, Manhattan, USA, Golden Fort, India), historical monuments and buildings. It is the primary raw material for the manufacture of quicklime (calcium oxide), slaked like (calcium hydroxide), cement and mortar, as flux in the blast furnace in iron industry, soil conditioner, aggregate, glass making, paper, plastics, paint, tooth paste, medicines and cosmetics. The fossil bearing (cyanobacteria algae, skeletons and shells) limestones are potential sources of phosphate, petroleum and gas.

5.7.2. Dolomites

Dolomite is a carbonate mineral composed of calcium magnesium carbonate $[CaMg(CO_3)_2]$

FIGURE 5.63 High N–S trending low-dip hills at the east bank of river Nile stands for an vast resource of commercial quality limestone (top of the picture). The river cruse near Esna town, Egypt, presents a typical landscape with luxuriant growth of date palm trees in the middle and the flowing Nile water in the foreground.

FIGURE 5.64 Massive dolomites, sedimentary carbonate rock hosting zinc–lead mineralization from Zawar Mine, Rajasthan, India. The brownish yellow color in the top right hand of the specimen indicates sphalerite (ZnS) minerals.

(Fig. 5.64). The term is also used to describe the sedimentary carbonate rock "dolostone/dolomite rock" predominantly composed of the mineral dolomite ≥50% magnesium.

5.7.2.1. The Origin of Dolomite

Dolomites are carbonate rocks mainly composed of the mineral dolomite. The mineral is stable in seawater in nature and not known examples of its direct extraction from seawater in large quantities required for origin of dolomite rocks. Dolomite occurs by suppression of aragonite or calcite, i.e. dolomitization process. Dolomitization occurs in nature or even in untied limestone deposit or in already solid limestone rocks. Dolomites that occur in the untied deposit are called early diagenetic or sin-sedimentary. It also forms by dolomitization of limestone as late-diagenetic or postsedimentary events.

The origin of large amounts of dolomite is by suppression of calcite or aragonite through direct secretion from the seawater as consequence of the strong hydration of Mg-ions. The tendency Mg-ions is to be dissolved in seawater, and not in the crystallized state. Mineral dolomite is a two salt of Ca-carbonate and Mg-carbonate with a crystal lattice in which the properly sorted layers of $CaCO_3$ of calcite structure and layers of $MgCO_3$. The main obstacle for producing minerals dolomite from seawater requires complex arrangement of its crystal lattice with respect to calcite crystal lattice, aragonite and high-magnesium calcite due to much easier excretion from seawater than dolomite. The dolomitization in normal seawater may occur as special cases (Section 5.7.2.3).

The process of transforming calcite or aragonite to dolomite by bringing in Mg-ions can be shown by the following chemical reaction. The calcite (or limestone) joins Mg-ion in solution and removes free Ca-ion:

$$\underset{\text{calcite}}{2CaCO_3} + Mg^{2+} \longrightarrow \underset{\text{dolomite}}{CaMg(CO_3)_2} + Ca^{2+}$$

The process of dolomitization by this reaction can be achieved only in the presence of solvent, that adds Mg-ions to the new formed rock, and excludes free-Ca ions. In nature, dolomitization is actually happening by the following:

1. Bringing in Mg-ion
2. Releasing of carbonate anions CO_3^{2-}
3. Consumption of all available Ca-ions and not needed thereafter to remove Ca-ions from calcite or aragonite.

It is a process of dolomitization according to the following reaction and equation:

$$\underset{\text{calcite}}{CaCO_3} + Mg^{2+} + {CO_3}^{2-} \longrightarrow \underset{\text{dolomite}}{CaMg(CO_3)_2}$$

Sea, as freshwater, may be saturated with respect to dolomite and calcite. Mixed marine and freshwater (5–50% seawater) is unsaturated with respect to calcite and saturated with respect to dolomite. Dolomitization is suitable with pore solution of mixed groundwater zone, if they include a mixture of freshwater with 5–50% seawater. Dolomitization is particularly intense in the sabkha conditions, if the molar ratio of Mg/Ca compared to that in normal seawater that amounts to 5.26, strongly increased at ~10–30 (Section 5.7.2.2).

5.7.2.2. Early Diagenetic Dolomites

Early diagenetic dolomitization or singenetic/sin-sedimentary origin of dolomite occurs, as shown by numerous studies of recent examples, in untied deposit in the following:

1. Supratidal zone, the coastal saline (sabkha conditions) and salt lakes
2. In the zone of mixed marine and freshwater.

Early diagenetic dolomitization in supratidal zone is possible, in case, the limestone deposit is drenched with seawater during high tidal waves and in coastal saline (sabkha) and salt lakes with strong evaporation of residual sea/salt water. Early diagenetic dolomitization also occur at annual temperature of >30 °C, as well as an increase of molar Mg/Ca ratio in pore water or saline water between 15 and 30, as opposed to 5.26 in normal seawater. There is no early diagenetic dolomitization of only sedimented carbonate deposits on the bottom of the sea at normal or slightly elevated salinity and normal temperature of seawater.

In cases of early diagenetic dolomitization in sabkha conditions key factor is the evaporation of seawater at high tides and storm waves soaking carbonate deposit or evaporation of water that is lost in the recesses of high-tide zone in the form of salt lakes, saline and sabkha.

Dolomitization in the zone of mixed marine and freshwater may occur in early diagenetic or transitional stage from early to late stage diagenetic and even in late-diagenetic stage. This process of dolomitization is based on the fact that dolomite easily takes place from a mixture of marine and meteoric freshwater than from the sea or freshwater. Dolomitization in the mixed zone of marine and freshwater (brackish) will particularly be intense when the mixture contains between 5% and 50% seawater. The water mixed with a ratio of 5–50% seawater and 95–50% of meteoric water is oversaturated with dolomite, and unsaturated compared to calcite, which allows the formation of dolomite at the expense of calcite.

Dolomitization in the zone of mixed marine and freshwater begins in early diagenetic phase and the deposition lasts for a few hundred thousand years (>200,000 years), as the process is very slow. The mixed zone of seawater and freshwater, during an extensive period of time, moves simultaneously with the lowering or rising of sea level. Dolomitization will be weaker or stronger be affected large parts of coastal limestone deposits in the case of lowering of sea level (Fig. 5.65).

Early diagenetic dolomites are characterized by all the textural–structural features and layer forms as of original deposit. There are pellet, oncoid, micrite and intraclastic dolomite often with desiccation cracks or traces of erosion on the upper surfaces caused by storm tides and waves. The early diagenetic process will dolomitize all existing soft limestone sediment to a certain depth. These are, in general, pure dolomite rock without undolomitized remains, relics of limestone and contain small dolomite crystals, typically <0.01 mm. The early diagenetic dolomitization occurs specifically at a relatively high concentration of solutions that are near saturation or saturated with respect to dolomite. The process also begins to crystallize many crystals covering all the ingredients of limestone deposits, regardless of the primary mineral composition, crystal size and primary structural components (Fig. 5.66). Early diagenetic dolomitization completely transforms all the ingredients of limestone deposits to cryptocrystalline dolomite.

5.7.2.3. Late-Diagenetic Dolomite

Late-diagenetic or postsedimentary dolomite occurs by dolomitization of limestone with circulation of pore water and cold seawater, the source of Mg-ions, through the permeable limestone, at greater depths in groundwater zone of mixed freshwater and seawater. In late-diagenetic

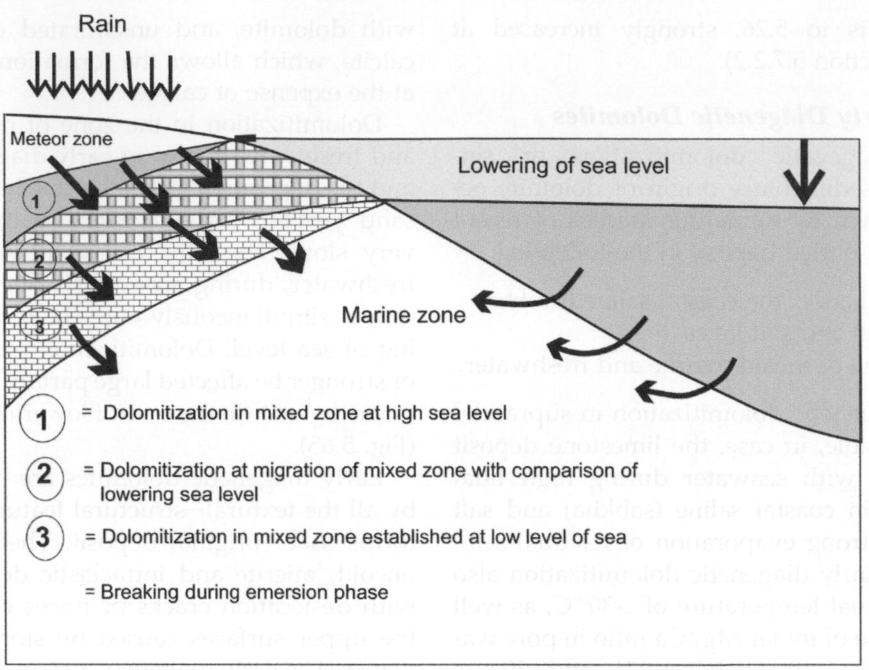

FIGURE 5.65 Dolomitization in a mixed zone of marine and freshwater in wet conditions (humid) climate with migration of mixed zone in the direction of the sea due to the relative lowering of sea level.

dolomitization at greater depth of covering the main problem is insufficient flow or circulation of pore water and bringing of Mg-ions from large distances. In late-diagenetic conditions, 35,000 m^3 of pore water at 80 °C temperature with the usual medium composition (molar ratio Mg/Ca = 0.25 and magnesium concentrations of 0.1–1000 mol of water) is required to fully dolomitize 1 m^3 of limestone. Such circulation of pore water in limestone can be achieved only through a very long time, during the whole geological periods. Dolomitization at greater depth of overlay occurs with the presence of fluid pores at relatively high temperatures (ranging between 60 and 160 °C) and very variable chemical properties during the entire process of dolomitization.

As the concentration of Mg-ions in the pore water is very low, full dolomitization of limestone will require large amounts of pores water

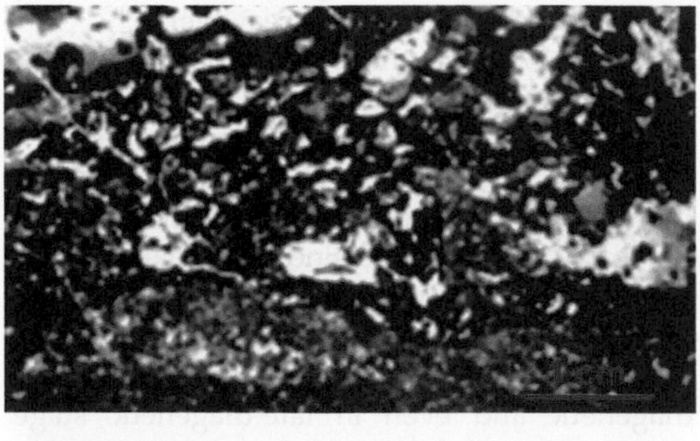

FIGURE 5.66 Microscopic glass slide of early diagenetic stromatolite dolomite, characterized by cryptocrystalline structure—dolomite crystals <0.004 mm (dark), and fenestra filled with microcrystalline dolomite cement (white) from Upper Triassic formation, Medvednica Mountain, central Croatia.

flowing over long time. Growth of dolomite crystals is slow and starts start with small number of crystallization embryos followed by large

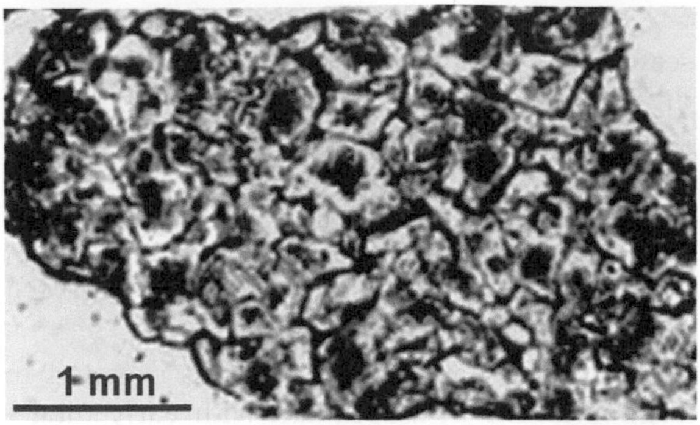

FIGURE 5.67 Late-diagenetic dolomites with macrocrystalline mosaic structure hypidiotype to idiotype dolomite crystals which in the center contains zonic distributed impurities of kerogen, from Wells Jaddua-1, Syria between depth of 450 and 460 m.

crystals (0.1–0.4 mm) with idiomorphic contours (Fig. 5.67). The result of such dolomitization is a complete change of texture–structural features of limestones and origin of dolomite rocks with macrocrystalline or microcrystalline idiotype or hypidiotype mosaic dolomite structure. Late-diagenetic dolomite texture and structure are significantly different from early diagenetic dolomite.

Limestones are dolomitized with different intensity, from place to place, due to different permeability, capillary force and the degree of tectonic cracks of certain parts of limestone layers affected by late-diagenetic dolomitization level of pore waters with variable intensity of circulation. The proportion of dolomite usually decreases from the center toward the edges of dolomite body with undolomitized remains of limestone.

Kohout (thermal) convection model flow of seawater is probably the most significant factor in late-diagenetic dolomitization of limestone at greater depths of overlay in the area of marine carbonate slopes. It is known that water in the special physical and chemical conditions can circulate through the limestone slopes of subsea carbonate by mechanism called by the author as "Kohout thermal convection". Kohout (1967) found that cold seawater from large depth may, due to convection or transfer of heat, flow and penetrate in the limestone on the outskirts of carbonate slopes containing basic water of higher temperature than the water temperature from the depths of the sea. Cold seawater is unsaturated with high magnesium calcite and aragonite, but oversaturated with dolomite, and secretion is possible from dolomite. Dolomitization by Kohout convection model is very slow and geologic time-consuming process.

2. Evaporite sediment and sedimentary rocks

5.7.3. Evaporites

Evaporite sediments or evaporites are sediments and rocks created by chemical secretion from extremely concentrated natural solution (saline) by strong evaporation of water. Evaporite deposits start forming from the edges of salt lakes in the coastal saline (sabkha), enclosed lagoons and bays in areas with arid, i.e. dry and hot climate. It is necessary to boost faster rate of evaporation of water flow in order to constantly increase the concentration of salt in water. The common examples of evaporites deposition are gypsum, anhydrite, halite and K–Mg salts (polyhalite, silvine, kizerite and karnalite).

5.7.3.1. Mineral Composition, Origin and Classification of Evaporites Rocks

In the initial stages of evaporation and concentration of seawater, that allows the secretion of Ca-carbonate in the form of aragonite, high-magnesium calcite or calcite and ends with dolomite. The dolomite then suppresses the Ca-carbonates, as explained in Section 5.7.2.2. The salt concentration increases to about 3.5 times with the succeeding evaporation of water and the salinity of seawater rises to approximately 120‰. The mineral gypsum begins to crystallize at this stage at a temperature of 30 °C and continues until the concentration of salt in water does not grow to 4.8 times higher than in normal seawater salinity (Table 5.6). The secretion of

TABLE 5.6 Limit Values Necessary to Increase the Concentration of Seawater at 30 °C for Extraction of Minerals Evaporites

Mineral Excreted	Increase in the Concentration of Seawater
Calcite, aragonite, dolomite	To 3.5 times
Gypsum	3.5–4.8 times
Anhydrite	4.8–9.5 times
Halite	9.5–11 times
K–Mg salt	>60 times

Source: Ref. 16.

anhydrite commences above this concentration at temperature of 30 °C. Necessary increase of concentration in relation to the normal concentration of seawater and the sequence of secretion of singular evaporite minerals at a temperature of 30 °C displays in Table 5.6.

The secretion of Ca-sulfates (gypsum and anhydrite) can take place from solutions of small concentration at temperatures much higher than 30 °C. Gypsum, for example, is excreted at a temperature of 58 °C from the water with normal salinity, and anhydrite secretes much above that temperature. On the other hand, gypsum and anhydrite can secrete at lower temperatures, if solutions have a high salinity. For example, anhydrite begins to exude at a temperature of 60 °C from seawater with normal concentration. The same secretion begins at 20 °C from the water with 7 times higher concentrations at arid saline or sabkha. The secretion of anhydrite, gypsum, halite and K–Mg salt is directly dependent on temperature and salinity of water. Normal salinity requires high temperature, and with increasing salinity the secretion of evaporites is possible at lower temperatures, especially in saline or sabkha and salt lakes.

Gypsum, anhydrite and halite evaporite sedimentary rocks can be found much more likely than evaporites rocks containing K–Mg salt.

Gypsum is excreted in the closed shallow seawater and salt lakes in the initial stages of drying sabkha. The initial salinity might not reach the concentration suitable for the secretion of anhydrite (Table 5.6). The extract of gypsum or anhydrite depends primarily on the concentration (salinity) of water and environments in shallow-sea water or protected shallow or salt lake, and evaporites sabkha conditions.

Anhydrite (sabkha anhydrite) is excreted in large amounts in association with early diagenetic dolomite in coastal saline or sabkha at temperatures of about 25–35 °C in conditions of dry climate and strong evaporation of water, which significantly increased the concentration of Ca-sulfate and salinity at approximately 4–7 times higher than normal salinity of seawater.

Halite (rock salt, Fig. 1.9) is excreted mostly in close marine shallow waters, saline (sabkha) and occasional salt lakes which during dry periods left without water in the form of layered cyclic sequence. Such sequences often destroyed completely by diapirism, which are very prone to salt deposits. In diapirism, by plastic injection in roof sediments a significant part of the salts can dissolve. In subaquatic conditions, i.e. the closed shallow sea and salt lakes, salt crystals in evaporation grow, at the surface of water, particularly intense at the contact water-sediment and within the sediment due to the relatively slow growth of crystals and slightly elevated salinity, resulting in large crystals of halite.

Diapirism is an anticlinal fold or sedimentary layers in which a mobile core, such as, salt or gypsum, has pierced through the more brittle overlying rocks.

The primary porosity in the salt sediments is directly dependent on the dimensions of the crystal and place of their origin. Mechanical compaction can be very different compared to the thickness of salt deposits as it depends on the pore waters and the waters in surrounding sediments outside evaporite deposits. Migration of highly concentrated fluid from the salt deposits in the water layer or from the water in the residue can cause dolomitization of limestone deposits, and cementing early diagenetic salt and other deposits in the form of extraction of gypsum, anhydrite and calcite. In such cementation primary textural–structural features of evaporites deposits may preserved very well, such as lamination, nodule, stratification, and so on.

From the standpoint of geology and petroleum the evaporite complexes, especially halite and gypsum deposits, are excellent insulator rocks beneath which often found significant amount of oil and gas accumulation.

5.7.3.2. Petrology and Diagenesis of Evaporite Sediments

The common textural and structural feature of the sabkha anhydrite is nodular configuration or nodular anhydrite, thicker and/or thinner concentric layers, lamina, or just thin layers of dolomite. Anhydrite within nodule consists of different oriented, elongated to needle crystals (acicular anhydrite). Relationship of anhydrite and dolomite in nodular anhydrite depends on salinity, temperature and strength of water evaporation in saline, as well as the duration of sabkha conditions and the content of sulfate in saline water. Extraordinary examples of sabkha cycles with nodular sabkha anhydrite are found in deep exploratory oil wells anhydrite–carbonate complexes in the Long Island district, Croatia. Upper Permian evaporites and associated carbonates and the fine-ground clasts of central part of the Dinarides (central and northern Dalmatia, Croatia and north-western Bosnia) belong to the regressive sedimentary system with evaporation conditions of coastal sabkha and peritidal environment.

Playa are the coastal salt lakes that remain with very little water or without during dry periods in the peritidal environment zone. The tides and ebb, and coastal sabkha salt water are lost in the recesses of high-tide zone (Fig. 5.70). Such environments are found along border parts of the sea in younger Perm (geological period of <265–251 million years). The level of sea gradually declines with regressive tendency, and the continental environments of clastic Playa and evaporite coastal environment progress in shallow-sea sedimentary area. While the clastic rocks (Fig. 5.68) precipitate in the environment of front beach and/or playa to the salt lakes, the community of carbonate and evaporites (anhydrite, which on the surface is hydrated in early diagenetic gypsum and dolomite) is created in the coastal sabkha and supratidal zones in condition of permanent relative lowering of the sea levels and shrinking of the sea area (Fig. 5.68).

Most of the anhydrite arises directly in the sabkha conditions and part of the anhydrite can come from dehydration of gypsum in sabkha cycles that originated in the tidal zones (intertidal) or in the initial stages of drying sabkha. Large amount of gypsum arising in underwater conditions could be dehydrated in anhydrite. These anhydrites have pseudomorph based on gypsum, and often preserve beautiful contours of gypsum crystals. Dehydration of gypsum to anhydrite is reversible process that can be shown by the following reversible reaction:

$$\underset{\text{gypsum}}{CaSO_4 \cdot 2H_2O} \leftrightarrow \underset{\text{anhydrite}}{CaSO_4} + \underset{\text{water}}{2H_2O}$$

$$\xrightarrow{\hspace{4cm}}$$
increase in temperature and salinity

Gypsum dehydration takes place at increase of salinity and temperature, and hydration of anhydrite at reduced salinity and temperature. Anhydrite occurs on the surface of Earth in arid hypersalted, and gypsum in the colder and less salty environments. If the gypsum came under the influence of hypersalted fluid its

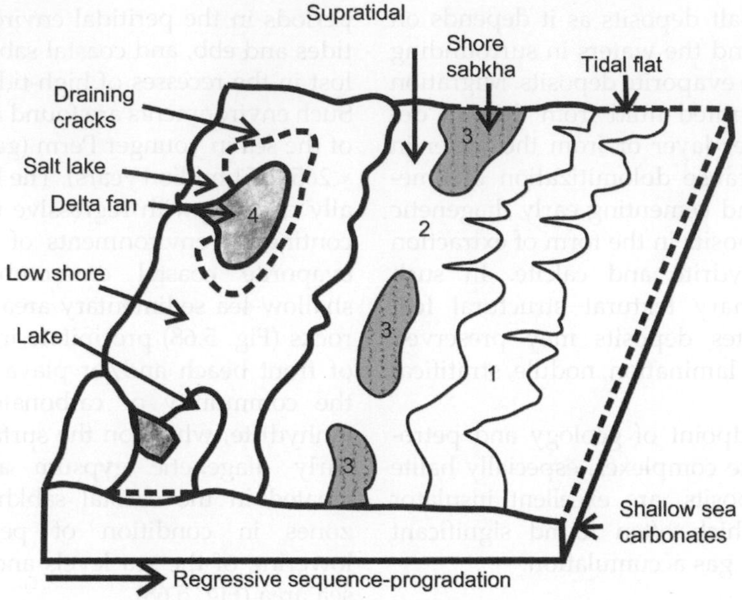

FIGURE 5.68 Environment of deposition of calcareous—evaporite complex in central Dalmatia: (1) tidal zone or intertidal environment with the deposition of limestone, (2) supratidal zones or supratidal environment with coastal sabkha, (3) early diagenetic dolomites and evaporites (evaporite-dolomite facies) and (4) salt lake in which deposit fine-ground and gypsum crystals.

conversion into anhydrite occurs at temperatures between 25 and 45 °C and the depths of the overlay of 1–2 m. The pore or meteor water, under the influence of average salinity of pore fluid and the geothermic gradient of 1 °C in each 33 m depth and if the temperature rises from ~20 °C on the surface to ~50–60 °C at depth of the overlay of >1 km, gypsum goes into anhydrite by dehydration. The relics of gypsum crystals are often preserved in the newly created anhydrite. On the other hand, under the action of pore and meteor water at normal salinity at depths cover of <1 km the anhydrite hydrates to gypsum increasing its volume by about 38%.

The conversion of gypsum to anhydrite causes other diagenetic effects at greater depths, especially melting, cementing, dolomitization and anhydratization, secretion of secondary anhydrite in cracks, tectonic cracks, cavern and melt holes in limestones and dolomites. These anhydrites and rocks in which they appear do not belong to evaporates, or be considered evaporite deposits.

Hydration of anhydrite into gypsum usually begins along cracks of cleavage anhydrite crystals and along tectonic crushed and fissured zones in the form of veins microcrystalline and macrocrystalline aggregate of gypsum or fibrous gypsum cluster.

At a higher degree of hydration of anhydrite to gypsum, larger crystals of newly created gypsum fit relic anhydrite or centripetal push anhydrite crystals. Homogeneous, fibrous gypsum occurs at complete hydration of anhydrite, which contains the remains of a rare anhydrite. Hydration of anhydrite to gypsum depends on the reduction of temperature and salinity of pore water contained in evaporates, especially by long effect of fresh rainwater on anhydrite surface. This process is particularly intense in tectonically fissured and crushed anhydrite, as

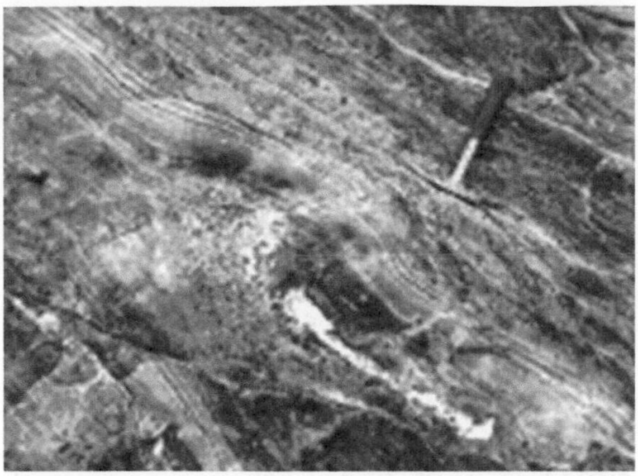

FIGURE 5.69 Gypsum with laminae and layers of early diagenetic dolomite rich in organic matter (black). Gypsum appears only in the surface of the evaporite complex, where it is originated by hydration from anhydrite.

it is the case of the upper Permian evaporites surrounding of Sinj and Knin, Croatia (Fig. 5.69).

Hydration process of anhydrite to gypsum and dehydration of gypsum in the anhydrite can be in the same rocks frequently repeated. While the thin lamina of early diagenetic dolomite, and organic matter in evaporites, due to repeated expansion and contraction, forming terminates or pleated lamina, nests and bent strips (Fig. 5.51). That is so-called enterolithic folding like tectonic folding, which are often interpreted like it. Although it is similar to tectonic deformation, it is not due to tectonics, but recrystallization and chemical changes in the volume of Ca-sulfate for shrinkage and expansion during hydration and dehydration, reversible transitions anhydrite in gypsum and gypsum in anhydrite.

The volume of evaporite rocks increases during hydration of anhydrite to gypsum by about 38%, which is accompanied by strong stresses and diapirism, injecting gypsum into the surrounding rocks, usually in the roof of evaporites. In the process enterolithic folding occurs on account of significant differences in the degree of plasticity of gypsum or lamina in dolomite which contain sabkha anhydrite, and typically leads to cracking or complete destruction of thick dolomitic lamina. In this way, along with actual tectonic crushing, result in dolomitic-gypsum breccias composed of fragments of dolomite and gypsum binder or matrix. The share of dolomite in the sabkha anhydrite is considerably smaller than the share of what was originally, because it is suppressed by early diagenetic anhydratization processes already in sabkha, and also in the hydration and dehydration processes in enterolithic folding and diapirism. Diapirism is mainly moving plastic gypsum and dolomite remains.

3. Silicon sediment and sedimentary rocks

5.7.4. Siliceous Sediments and Rocks

Silicon sediment and sedimentary rocks are composed of autogenous non-detrite silicon oxides or oxides with water. Siliceous rocks are composed of opal-A, cristobalite or opal-CT, chalcedony, quartzite and cryptocrystalline or microcrystalline quartz, which are in biochemical or inorganic chemical processes extracted from aqueous solutions enriched in silicic acid, H_4SiO_4. They can occur by deposition of opal skeletons and inorganic secretion of these minerals in solution containing silica acid. In chemical secretion silicon minerals suppress the original minerals in deposits and rocks in the process of silicification. This leads to silicified limestones, dolomites and tuffs, as well as nodule, lenses and lumps of chert in carbonate rocks.

5.7.4.1. Mineral Composition, Origin and Classification of Silicon Sediments and Sedimentary Rocks

Opal-A and opal-CT are primary constituent of the silicon sediments, and the silicon rocks have in addition of quartzite, quartz and chalcedony.

Opal ($SiO_2 \cdot nH_2O$) is an amorphous mineral flint, mixture of amorphous SiO_2 and opal-CT, i.e. cryptocrystalline cristobalite with 8–10% of water. In silicon sediments occurs or as an ingredient of opal skeletal diatoms, radiolaria and spiculae sponges diatom (biogenic opal or opal-A) or its origin is related to processes of surface-silicification as well as processes of early diagenetic and late-diagenetic silicification of deposits or solid rocks, with the participation of solutions enriched in silicic acid (H_4SiO_4).

Besides biogenous origin (opal-A of which are built houses of radiolaria and diatoms or needles of diatom sponge) and diagenetic origin (as a product of secretion of aqueous solutions of the normal temperature), opal may be hydrothermal in origin (as a product of secretion from the hot solution), as for example, in the case for geyserite.

Low-temperature cristobalite or opal-CT is a low temperature tetragonal modification of SiO_2, stable below 270 °C which is generated in silicon sediments during diagenesis by transformation from opal-A, i.e. opal skeleton. This transformation occurs in a mild increase of temperature at depths of covering a few hundred meters. Thus, opal-CT is a transitional structural form between biogenous opal (opal-A) and quartz.

Quartzite is cryptocrystalline short-fiber variety of incompletely recrystallized of chalcedony into quartz whose fibers are elongated in direction of crystallographic axis c. It occurs by recrystallization in diagenetic processes from opal-A, mainly in the recrystallization of radiolarian shells, in the presence of pore fluid saturated with quartz and rich in magnesium and sulfate. It also occurs in association with elongated fibrous chalcedony which is difficult to distinguish in micrographic thin section.

Quartz in silicon sediments, especially in layered and striped chert, appears in the form of microcrystalline and cryptocrystalline isometric clusters, and in flint in the form of dense cryptocrystalline clusters. Quartz is found in the silicon sediments made by silicification of carbonate sediments and rocks (chert concretions, nodules, bumps and lenticulares):

1. As microquartz in the form of isometric idiomorphic, mosaic crystals of diameter <35 μm, which intrude into each other, and often contain many small impurities.
2. As megaquartz in the form of mosaic equidimensional crystals of diameter from 50 to 300 μm.

Chalcedony is microcrystalline fibrous variety of quartz with small shares of water (1–2% H_2O probably built in the form of SiOH layers). Chalcedony fibers occur in parallel with each other or radially arranged ball, kidney or irregular clusters. Chalcedony is found also in the form of veins, and botryoidal mass within cherts.

The criteria for the classification of silicon sediments and silicon rocks are based on organic or inorganic origin, degree of lithification, diagenetic change and the textural and structural characteristics (Table 5.7).

5.7.4.2. Siliceous Sediments and Siliceous Rocks of Biogenic Foundation

The most important organisms and its opal skeletons that participate in the formation of siliceous sediments are diatoms, radiolarians, spicule of sponge and silicoflagellate. In the group of biogenic silicon sedimentary rocks (Table 5.7) of petrographic important sediments are diatomaceous, radiolarian and spicule muds as loose deposits, and diatomaceous and radiolarian earth and porous spiculite as weak lithificated and porous rocks, and as firmly lithificated, thick rocks without porosity: diatomite, radiolarite and spiculite.

Diatomaceous mud, diatomaceous earth and diatomite are silicon sediments mostly built of skeleton of diatomaceous algae, and with them often and skeletons of silicoflagellate, radiolaria, clay minerals, Fe oxides and Fe hydroxides. Recent diatoms are widespread in the cold sea around the South Pole and the northern Pacific

TABLE 5.7 Distribution of Silicon and Silicon Rock Sediments by Way of Origin, Structure and Mineral Composition

A—Silicon Sediment of Biogenic Origin			
Prevailing Silicon Ingredient	**Untied Sediment**	**Poor Lithified and High Porosity**	**Solid Rock without Porosity**
Diatoms skeletons	Diatoms mud	Diatomaceous earth	Diatomite
Radiolaria skeletons	Radiolaria mud	Radiometric earth	Radiolarite
Spicule sponge	Spicule mud	Porous spiculite	Spiculite
B—Silicon Sediment of Diagenetic or Other Origin			
Authigenic quartz chalcedony and/or opal		Geyserite porcellanite	Layered cornea flint, novaculite jasper nodular cornea

Ocean. There is also the largest deposit amount of recent deposition—diatomaceous mud—with predominant ingredient of diatoms skeletons. With the marine environment diatoms were and still today, adapted for life in marine and lacustrine sweetened water, so that within lake sediments often form a lamina or layers exclusively composed of opal skeleton, (diatomite lamina). The lake water poor in earth-alkalic ions (lake with "soft water"), and rich in dissolved silica, nitrates and phosphates, ecologically are favorable environment for the development of diatomaceous which deposit diatomaceous mud after dying and gradually turns in diatomite in lakes with areas of cold climate.

Diatomite, unlike diatomaceous mud and diatomaceous earth, is very hard, dense low-porosity rock, of light gray or white color, composed opal skeletons of diatomaceous, amorphous opal-A, cemented together with opal or in older rocks, microcrystalline or cryptocrystalline quartz or fibrous chalcedony cement. Chalcedony fills the pores of the largest dimension.

Radiolarian muds, radiolarian earths and radiolarites are predominantly composed of radiolaria skeletons built of opal-A, diagenetic transformed chalcedony and quartz. Radiolarian belongs to protozoa—marine plankton (zooplankton) organisms that float near the sea surface. These are single-celled organisms whose homes consist of $SrSO_4$, some organic silicate, or opal-A. The only preserved fossils are opal homes (opal-A) of spherical shapes, diameters between 50 and 250 μm, usually about 150 μm. The rocks consist of fungous glassy membranes with many regular radially spaced thorns.

Radiolarites are thick rocks of glassy shine, have microgranular structure, mainly composed of radiolaria skeleton and fibrous chalcedony aggregates. Black radiolarites are rich in organic matter and are called *touchstone*.

Spicule muds and spiculites are mainly composed of spicule silica sponge, built of opal-A. Opal-A in spiculites, during diagenetic processes, as one of the skeletons built and the one excreted as cement, recrystallized in opal-CT and cryptocrystalline quartz, and chalcedony. Spiculites also contain carbonate, clayey and powder matrix as essential ingredients and glauconite. Spicule of sponges is constructed from opal or chalcedony, and an elongated central channel of spicule filled with opal cement.

Biogenic silicon deposits occur in areas with high primary production of opal and radiolarian skeletons or spicule with low influence of terrigenous material and low carbonate, or the CCD boundary line is located at short depth. All or most of the carbonate in such environment is dissolved in water column before it sediments at the bottom, or melting occurs within sediment located below the CCD boundary line. The solubility of opal in seawater is relatively good and enhanced with increasing temperature and depth, and therefore deep sea silicon sediments are good indicators of diagenetic processes. Silicon muds have high porosity (75–90%) and remain diagenetic unchanged near the seabed, up to several hundred meters of deposits.

5.7.4.3. *Siliceous Sediments and Siliceous Rocks of Diagenesis Origin*

The silicon rocks are divided into two groups based on appearance, texture characteristics, form of occurrence and origin: inorganic and/or diagenetic origin (Table 5.7).

1. Weak lithificated and high porous rocks; geyserite, porcelanite, tripoli.
2. Firm lithificated, dense rocks with no porosity: layered cherts, flint, novaculite, jasper, nodular and lenticular cherts.

Geyserite is a siliceous sedimentary rock which occurs by secretion of opal from hot water in geysers and hot springs at its outbreak to the surface of the Earth. Geyserite and rocks of similar origin are known as *silicon sinter*. Geyserites are incrustation of fibers and pearls (silicon sinter) resulting in the secretion of opal in evaporation of silica-rich hot springs fiorite, according to the locality of Santa Fiora from Tuscany in Italy where the "sintered silicon" is extracted as stone for making ornaments.

Porcelanite is a chert variety cryptocrystalline in structure, blurry shine of white color very similar to unglazed porcelain. It consists of opal-A, opal-CT (cryptocrystalline cristobalite) and sometimes tridymite. It is more porous and softer than chert. It occurs by diagenetic processes of recrystallization of opal-A into opal-CT from radiolarian muds and secretion of tridymite from pore solution in a greater depth of the overlay.

Tripoli is white or light gray, porous silicon sediment composed of microcrystalline quartz. It occurs by partial silicification of carbonate rocks, where the quartz component stays after the weathering, while the carbonate component almost entirely secreted from the rock.

Cherts appear in two ways:

1. As layered rocks.
2. As nodules, lenses and irregular clumps.

These rocks are known under the general name of chert, and some call it hornfels. Cherts include all solid-silicon rocks, regardless of origin of silicon mineral, consisting of cryptocrystalline or microcrystalline quartz of non-detrite origin and/or chalcedony (Fig. 5.70), and opal-CT. These are dense, very hard rocks, sharp fracture silicon minerals contain Fe-oxides or hydroxides and organic matter (Fig. 5.70). The silicon minerals within cryptocrystalline to microcrystalline mass composed clearly preserved skeletons, or their unclear remains, "ghosts" due to intensive recrystallization. It applies to radiolarians that from the entire silicon skeleton most resistant to recrystallization.

The dimensions of quartz crystals in the cherts rarely exceed 10 μm (Fig. 5.70), and for further research of cherts structure needs electronic microscope.

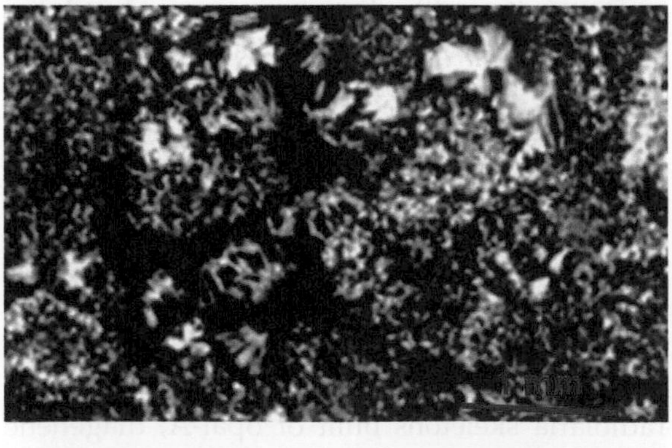

FIGURE 5.70 Microscopic recording of chert composed of microcrystalline quartz cluster and fiber chalcedony.

Bedded cherts (layered hornfels) appear in the form of thin layered (3–8 cm) or tens of meters thick chert deposits within dark gray to black shale or fine-grained graywacke sandstone, layered iron sediments, and striped cherts. Bedded cherts are common in preCambrian, Paleozoic and Mesozoic. The Jurassic and Cretaceous deposits occur almost everywhere in the World. The chert family can be described as the following:

1. *Cherts*, in the wider sense, are known as flint, novaculite and jasper, porcelanite and tripoli.
2. *Flint* is the chert with cryptocrystalline texture used as artifacts, weapons and tools of prehistoric man.
3. *Novaculite* is a variety of light or white, dense and solid chert, laminated, Devonian age, characterized by homogeneous cryptocrystalline to microcrystalline anhedral texture, isometric tiny quartz crystals and over fibrous chalcedony.
4. *Jasper* is a dense, opaque variety of cryptocrystalline chalcedony, contain significant proportion of Fe-oxides and hydroxides namely hematite, limonite and goethite so that reddish characteristic come from hematite and brownish color from goethite.
5. *Tripoli* is a naturally-occurringfine-grained microcrystalline mineral from chert family. It has special application abrasive mineral used in a variety of industries for sharpening, buffing and polishing end uses.
6. *Novaculite* is a form of microcrystalline or cryptocrystalline quartz of chert family. The color varies from white to gray-black and the specific gravity shows an increase from 2.2 to 2.5. The very hard dense rock is used as abrasive machining for steel tools, sharpening and grinding blocks or whetstone.

Nodular and lenticular cherts occur within limestones and dolomites, and significantly less within pelite and sandy sediments. It consists of microcrystalline and/or cryptocrystalline quartz, equidimensional quartz crystals, with smaller/larger amount of quartzite and chalcedony, and opal-CT. Microcrystalline quartz in limestone suppresses micrite mass, typically contains numerous small impurities of calcite, evaporite minerals and/or the primary structural components (micrite, pellets, ooids, and fossils), chert concretions in limestone of southern Istria (Croatia). In general, nodular and lenticular cherts are egg or spherical shapes regardless of whether they appear inside layer or between two layers (Fig. 5.71).

Most bedded cherts occur mainly by processes of recrystallization of diatomite, radiolarites and spiculite or acid tuffs and volcanic glasses. Nodular and lenticular cherts generally occur in the processes of suppression of carbonate sediments or some other rock with opal, chalcedony or quartz, with the participation of silicon acid, i.e. processes of silicification (Section 5.7.1.4.3).

Sediments and sedimentary rocks that originate from igneous, metamorphic and older sedimentary rocks, is ever forming new sedimentary deposits through the millions of geologic age. Erosion of preexisting rocks is the primary source of natural embryo-grains, that move, deposit, lithify and form the sedimentary rocks. The process continues for eternity (Fig. 5.72).

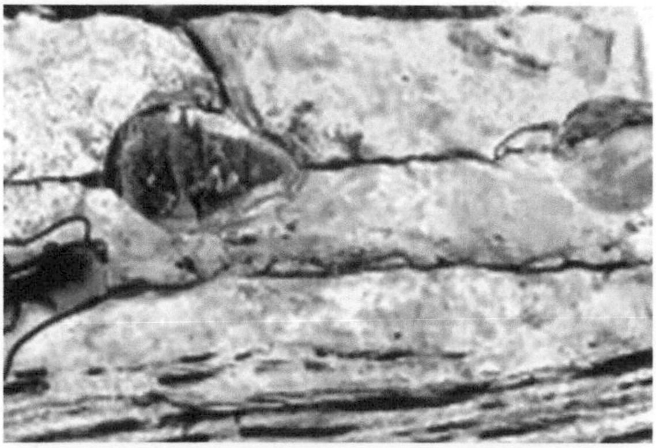

FIGURE 5.71 Egg-like chert nodule and spherical impression in limestone from which was removed chert nodule. Southern Istria, Croatia.

FIGURE 5.72 Erosion continues to sculpt the extremely water saturated sandstone accelerated by flash flood of Virgin River at Zion Canyon, Utah State, USA. The natural erosional caves and holes at about 4600 feet (1400 meter) above the Mean Sea Level fascinate scientists and nature lovers, like these kids (Srishti-Srishta) to rest for a while. The erosional fines and coarse materials move near and far distances to form new sedimentary rocks over geological time. *Source: Soumi, September 1, 2013.*

FIGURE 5.73 Visit to awesome Grand Canyon, a steep-sided gorge in Arizona State, is a dream for any geologists, trackers and nature lovers. The Canyon is 446 km long, up to 29 km wide and attains a depth of over 1800 m. It is carved by the Colorado River exposing over 600 million years of Earth's geological history through layer after layer of igneous, sedimentary and metamorphic package of limestone (Ls), sandstone (Ss), shale (Sh), conglomerate, schist over granite basement. The South Rim limestone mined for rich lead ore. Photograph: September 30, 2005 from the South Rim at ~2,100 m above mean sea level.

The sedimentary depositional system preserves its own history for the students of mineralogy, stratigraphy and exploration geology (Fig. 5.73).

FURTHER READING

Tucker et al.,[54] Tucker,[55] Bathurs,[1] Engelhardt,[15] Füchtbauer,[19] Flügel,[20] Pettijohn et al.,[32] Potter, et al.,[35] Slovenec,[43] Slovenec et al.[42] and Sam Boggs[10] provide a detailed description on sedimentary rocks and their origin. Moore[31] and Boggs (Jr)[3] will be an interesting reading on the subject. Academicians and students knowing Croatian language will be benefited reading Tišljar[47–51] on sedimentary rocks.

CHAPTER

6

Metamorphic Rocks

OUTLINE

6.1. Origin and Structure of Metamorphic Rocks 213

6.2. Types of Metamorphism and Classification of Metamorphic Rocks 219

6.3. Rocks of Dynamic Metamorphism 220

6.4. Rocks of Contact Metamorphism 220

6.5. Rocks of Regional Metamorphism 223

6.5.1. Schists of Low-Grade Metamorphism 223

6.5.2. Schists of High-Grade Metamorphism 225

6.6. Rocks of Plutonic Metamorphism 230

Further Reading 232

6.1. ORIGIN AND STRUCTURE OF METAMORPHIC ROCKS

Metamorphic rocks (from the Greek "meta" means "change" and "morphé" means "form") result in transformation or metamorphism (change) or solid-state recrystallization of existing igneous and sedimentary, and even metamorphic rocks. The changes occur in physical and chemical conditions, principally heat (temperature), pressure and introduction of chemically active fluids and gases. The conversion may alter the mineral composition including formation of new minerals (garnet, kyanite, chlorite, sericite, staurolite, analusite, etc.), the texture (granite to granite gneiss) and the both. Such changes or alterations are called *metamorphism*. The preexisting original rocks, affected by metamorphic changes, are called *protolith*. Increasing the temperature in the interior of the Earth can be so high that a part of rock or the whole rocks melt (or friction melt). The metamorphic changes of original igneous and/or sedimentary rocks may be partial, gradational or complete. There may not exist any sharp line between igneous and sedimentary rocks and their metamorphic complements. The dolerites may merge into hornblende schist and limestones grade into marble, without sharp line of demarcation.

The primary metamorphic agents in the Earth's crust are pressure, shearing stress, increase in temperature, and chemical effects of liquid and gases.

Introduction to Mineralogy and Petrology
http://dx.doi.org/10.1016/B978-0-12-408133-8.00006-7

The weight of the overburden sediments or upper part of the crust will have little effect in transformation, other than compaction and lithification including secretion of liquids by bringing and binding the grains close together in sedimentary rocks. The compressive forces, due to orogenic movements or mountain-making process, act as lateral thrust on the crust. The entire mass of geological strata including intrusive, extrusive and fragmented volcanic igneous rocks are folded, closely deformed and faulted by this powerful pressure and tend to produce shearing stresses. The deformation by shearing stresses would be the potential agent in producing metamorphism. The large crystals and pebbles will be flattened, elongated and deformed.

Heat is a powerful metamorphic agent. The increase of temperature in the lithosphere is driven by three factors:

1. Geothermic gradient gradually increasing the temperature of the surface toward depths of the earth (~25–30 °C for each kilometer of depth or 1 °C for every 33–40 m depth).
2. Effect of heat by magmatic body that is printed in the interior of the lithosphere.
3. Friction in rock masses by tectonic movements in the lithosphere.

The water, under the influence of heat and pressure, acts as a powerful chemical agent. It operates as solvent and promotes recrystallization.

The common rock forming minerals in metamorphic groups are kyanite, zoisite, staurolite, garnet, biotite, talc and graphite representing as characteristic constituents. Chlorite and sericite are attributes of low-rank (mild) metamorphism. Kyanite and staurolite are high-rank (more intense) metamorphism. Some minerals like quartz can be formed under all conditions of metamorphism or be present as relicts from the original rocks.

The rocks that plunge into the deeper parts of the lithosphere by geological processes are subjected to remain under increasing high pressure. In the shallower parts of the lithosphere, directed pressure (stress) causes the crushing of minerals and rocks. The directed pressure (static stress) can cause partial melting and recrystallization of minerals by Ricke principle at greater depths. This is because the melting point of minerals decreases with increase of pressure at that place and also increases its solubility. In places of lesser pressure, minerals are often recrystallize. The plate-, leaf- and stick-like minerals occur in parallel rows under the influence of unidirectional pressure during metamorphism. The newly formed metamorphic rocks attain schistose texture and lepidoblastic (leaf-like) or nematoblastic (stick-like) structure.

The rocks of metamorphic family are crystalline and resemble sedimentary rocks in having parallel lamellar structure with well-defined foliation that simulates bedding. The varieties of foliation are recognized as (1) gneissic, (2) schistose and (3) slaty.

1. *Gneissic* structure is applied to perfect to roughly foliated rocks and characterized as banded and lenticular layers of different composition (say quartz and hornblende) with parallel alignment make the rock with distinct structure (Fig. 6.1) that resembles

FIGURE 6.1 Banded or layered granite gneiss as illustrated by alternate layers of crystalline feldspar and quartz (white), and biotite and hornblende (black) resembling unique sedimentary bedding structure.

sedimentary bedding. In lenticular gneiss, some of the layer components are represented by aggregates of thicker and thinner lenses wrapped by other minerals (Fig. 6.11). The coarse and fine lenticular structure resembles lumpy or stretched lenses aligned along the layers.

2. *Schistose* or *schistosity* is well-foliated rock that tends to cleave into thin flakes made up of flaky minerals like mica. Schistosity is the most significant macroscopic feature encompassing the majority of the metamorphic rocks, particularly those whose metamorphism was carried under one-way pressure and stress. Schistosity manifests itself in a parallel or subparallel agreement of slips, flaky and sticky minerals (mica, chlorite, and amphibole). The elongated crystals of feldspar, quartz and Al-silicates (andalusite, sillimanite, kyanite, and cordierite) are oriented in parallel with the schistosity. Schistosity can be manifested by a parallel or a cluster of striped orientation of bright (leucocratic) and dark (melanocratic) minerals, or alternately stacking lanes of coarse and fine crystalline mineral aggregates (Figs 6.2 and 6.3).

FIGURE 6.2 Schistose, basic feature of crystalline schist, is caused by parallel bending of slips mineral around the grains of quartz and feldspars. The development of small- to medium-scale crenulated structure is prominent feature.

FIGURE 6.3 Typical mica schist mainly composed of sericite with satin color and silky golden shine. Mica schist is often folded showing microcrenulation cleavage texture and wavy structure. *Source: Prof. A.B. Roy.*

3. *Slaty* rocks are well foliated, aphanitic and composed of extremely fine-grain minerals. The slaty rocks have remarkable property of splitting into thin slabs. The foliation is so perfect that it can be termed as *slaty cleavage*.

In the great depths, the pressure acts on all sides (called hydrostatic pressure) and that results in formation of new minerals with higher density than the density of the original minerals. The new minerals occur in isometric shapes and oppose the pressure striving to be as close to a sphere shape (forms in cubic system, hexahedron, octahedrons, and orthorhombic dodecahedron) or monoclinic and triclinic minerals. The newly created isometric minerals of metamorphic rocks (such as calcite in the marble and quartz in the quartzite) have no preferred orientation or schistosity. Usually these are mineral deposits of mosaic type (Fig. 6.4(A)). The primary texture of the resulting metamorphic rocks is granoblastic (Fig. 6.4(A)). Granoblastic is an anhedral phaneritic equigranular metamorphic rock texture. The characteristics granoblastic textures include grains visible to the unaided eye, sutured boundaries and approximately equidimensional grains. The grain boundaries intersect at 120° triple junctions

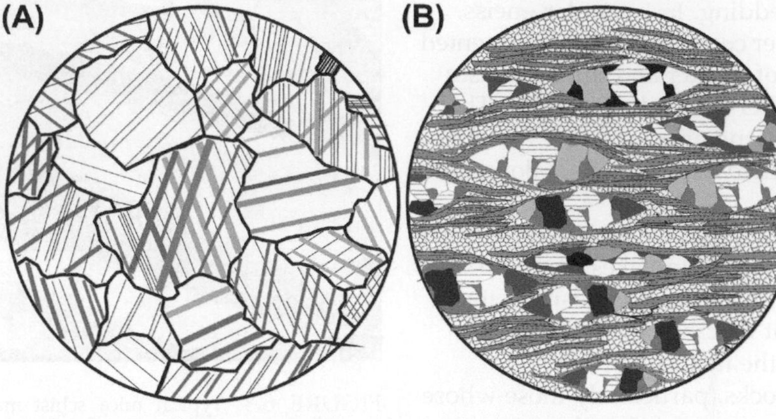

FIGURE 6.4 Conceptual diagrams showing (A) granoblastic marble texture characterized by large granoblastic calcite, and (B) structure of gneiss such as network of eye-like large granoblastic quartz and feldspars clusters surrounded by major slips lepidoblasts mice in finely granulated quartz–feldspar–mice mass.

under ideal conditions. The schistose textures are variation resulting from the ideal results from stress-produced foliation during crystallization.

The textures of metamorphic rocks are the result of recrystallization of the original minerals subjected to metamorphic processes. Minerals formed from metamorphic processes are called *blasts* (Greek meaning "lump"). Many types of structures of metamorphic rocks have a suffix of blast. Metamorphic recrystallization forms main mineral components of metamorphic rock, which can be large, and named as *granoblasts* and *porphyroblasts*. The minerals without correct crystal form are called *xenoblasts* (Fig. 6.4(A)). The smaller share of minerals which have a typical crystalline form are called *idioblasts*, for example, magnetite, rutile, garnet, andalusite and staurolite.

If metamorphic rocks contain predominantly of isometric (equidimensional) mineral grains and it is known as *granoblastic texture* (Fig. 6.4(A)). A rock composed of leaf-like mineral is called *lepidoblastic* texture (Greek, "lepis" = shell). Porphyroblast texture is typical of rock that contains some large crystals (porphyroblasts) and granular mass. *Nematoblastic* texture represents (Greek, "nema" = fiber) for rock composed mainly of rod, needle or fibrous minerals. Eye structure is a special kind of porphyroblastic structure in which individual porphyroblasts (quartz and feldspars) accumulate in clusters like eye wrapped with flakes inside finely crystalline granoblastic or lepidoblastic primary mass (Fig. 6.4(B)).

TEXTURE AND STRUCTURE OF METAMORPHIC ROCKS

Textures are the relationships of crystals at the smallest scale such as fine-, medium- and coarse-grained poikiloblastic, granoblastic, xenoblastic, idioblastic, nematoblastic, porphyroblastic and helicitic or snowball.

Structures are larger scale features, at times requiring a whole outcrop to fully describe such as foliated, lineated, cleavage, layered or banded, schistose, gneissose, slaty, crenulated or small-scale regular folds.

An extremely significant factor of the metamorphism is the presence of water in rocks, especially the water in the form of constitutional water (OH groups) and crystalline water (H_2O), that exists and associates within hydrosilicate minerals by solid molecular bonds. The water squeezes out of hydrosilicate minerals and change to anhydrous silicate minerals by increasing the pressure and temperature to particular level. For example, the clayey sediments containing chlorite, kaolinite, and smectite–montmorillonite with increasing temperature will transform into a cluster of and alusite, quartz, cordierita, biotite and Na-plagioclase, i.e. the rock hornfels (Section 6.5).

The sequence of transformation or metamorphism of minerals and rocks occurs at a constant and gradual increase of temperature and pressure conditions. The existing mineral assemblage of lower temperatures and pressures gradually transform into a stable mineral assemblage at higher temperatures and pressures. This process of transformation is called *progressive or prograde* metamorphism. The conversion of clay minerals in the clay sediments into sericite and chlorite is an example of progressive metamorphism. The sediments itself may transform to phyllites, and phyllites may transform into mica schist under further increase in temperature and pressure (Table 6.1). Prograde metamorphism results in higher density and generally larger crystals.

On the other hand, the metamorphic processes that occur in gradual lowering of temperature and pressure leads to *retrograde* metamorphism. The quartz-amphibole rock will be metamorphosed to quartz-amphibolite (Fig. 6.5) or amphibole quartzite depending on the ratio of quartz and amphibole in the parent rock, formed in great depths at high temperatures and pressures. The same rock may ascend up close to the surface by tectonic movements and keeps on in terms of low pressure and temperature. The mineral garnet, Ca-plagioclase and amphibole will be metamorphosed into chlorite, actinolite, zoisite, and epidote $\pm$ garnet and saussurite schist. Therefore, the quartz-amphibolite will transform into the green schist or garnet-chlorite schist by retrograde metamorphism (Fig. 6.6). The primary minerals of basalt and gabbro will be metamorphosed to green schist under similar condition and process of retrograde metamorphism. The processes of retrograde metamorphosis are extremely slow in geological time scale and will take millions of years.

In prograde/progressive and retrograde metamorphism, part of mineral in the parent rock that undergone a metamorphic process can remain preserved and unaltered as *relict minerals*. Therefore, a metamorphic rock usually contains two or more generations of minerals, and can include old and new community of minerals as assemblages. Specifically, the mineral composition of metamorphic rocks is a community of all minerals in the rock, as those of stable and unchanged at a given temperature and pressure, as well as, those emerging in the new changed conditions. This is due to gradual metamorphic changes during the geological history of the afflicted rock. Some assemblage includes only those minerals in metamorphic rocks that are formed under the same conditions of pressure and temperature, which determine the degree of metamorphism and metamorphic facies. The same metamorphic facies can contain different mineral assemblages as they are a direct product of different mineral and chemical composition of the original rock, i.e. *protolith*. A protolith is the original, unmetamorphosed rock from which a given metamorphic rock is formed, e.g. shale or mudstone is the protolith in slate.

Metamorphic facies indicates all the different mineral assemblages of equal and similar conditions with regard to the origin of pressure and temperature, irrespective of origin of some minerals in metamorphic rocks. In this book, metamorphic rocks are systematized according to metamorphic facies.

Recrystallization of preexisting original mineral components during metamorphism is the principal phenomena. Calcites of different grain sizes and matrix that make up limestone are converted to the aggregate coarse crystalline calcite during progressive metamorphism, resulting in

TABLE 6.1 Primary Distribution and Mineral Composition of Metamorphic Rocks

Metamorphism	Metamorphic Rocks	Mineral Ingredients	Premetamorphic Originating Rocks
Dynamic or kinetic	Mylonite	Finely crushed quartz-feldspars and other minerals	Sandstones, pelite sediments, limestones, dolomites, intrusive and effusive igneous rocks
	Flazer cataclasite	Rock bits and numerous finely crushed minerals	
	Eye gneiss	Feldspars and quartz in eye-shape mica and other silicates	
Contact	Hornfels	Pyroxene, garnet, mica andalusite, cordierite, carbonates	Shales, sandstones, limestones, basalt and tuff
	Skarn	Garnet, hedenbergite, epidote diopside, wollastonite, quartz	Limestones and dolomites
Regional or contact	Marble	Calcite or dolomite, forsterite tremolite, wollastonite, diopside	Limestones and dolomites
	Quarzite	Quartz, mica, sillimanite, garnet feldspar, andalusite, corundum	Sandstones, siltstones, silicic sediments
SCHISTS OF LOW-GRADE METAMORPHISM			
Regional low grade	Argillaceous and argillite	Cryptocrystalline quartz, feldspar, chlorite, illite	Shale, siltstone pelite tuff, clay tuffite
	Phyllite	Microcrystalline quartz, mica chlorite, graphite	
	Green and chlorite schist	Chlorite, actinolite, quartz, epidote, albite, talc	Basic magmatite
	Glaucophane schist	Glaucophane, garnet, mica quartz, calcite	Basalt, diabase feldspar, sandstone
	Talc schist	Talc, Mg-silicates, magnesite calcite, dolomite	Ultramafic igneous rocks
SCHISTS OF MEDIUM- AND HIGH-GRADE METAMORPHISM			
Medium–high	Mica schist	Mica, quartz, garnet, feldspar staurolite, sillimanite, corundum	Shale, siltstone, graywacke acid and neutral magmatite
	Amphibole schist	Hornblende, feldspar, garnet quartz, biotite, magnetite	Base magmatite, clay limestone, marl
Regional high degree	Garnet schist	Garnet, mica, chlorite feldspar, quartz, hornblende	Nearly all igneous and sedimentary rocks
	Distended sillimanite, cordierite, staurolite schist	Kyanite, sillimanite, cordierite, staurolite, garnet, mica, quartz	Shale, siltstone, clay
	Graphite schist	Graphite, quartz, mica feldspar, chlorite	Pelite sediment rich in organic matter

TABLE 6.1 Primary Distribution and Mineral Composition of Metamorphic Rocks (*cont'd*)

Metamorphism	Metamorphic Rocks	Mineral Ingredients	Premetamorphic Originating Rocks
	Gneiss	Quartz, feldspar, mica garnet, kyanite, staurolite, sillimanite	Acid magmatite and feldspar arenite
	Amphibolite	Hornblende, plagioclase, garnet, mica, epidote	Base and neutral magmatite and clay limestone
Plutonic	Granulite	Quartz, feldspar or quartz pyroxene, garnet, kyanite	Acid magmatite and graywacke sandstone
	Eclogite	Omphacite, garnet, glaucophane plagioclase, kyanite	Base magmatite

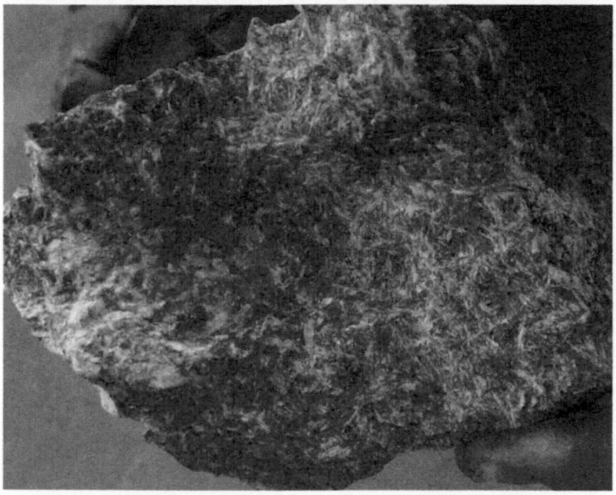

FIGURE 6.5 Quartz-amphibolites or amphibole quartzite is metamorphosed at great depth under high temperature and pressure condition by prograde metamorphism (pictured from Kolihan Section, Khetri copper project, India). The quartz and amphiboles are recognizable in the left and right portions, respectively.

FIGURE 6.6 The quartz-amphibolites is transformed to garnet amphibole chlorite schist under retrograde metamorphism. The sample is taken from Kolihan Section of Khetri copper project, India and is one of the host rock types for copper mineralization.

metamorphic rock marble with granoblastic or porphyroblastic texture (Fig. 6.4(A)).

Metamorphic rocks formed from parent sedimentary rocks are called *parametamorphite* (e.g. formed from sandstone). Similarly, those resulting from original igneous rocks are known as *orthometamorphite* (say, formed from granite and granodiorite).

6.2. TYPES OF METAMORPHISM AND CLASSIFICATION OF METAMORPHIC ROCKS

The type of metamorphism can broadly be grouped in three categories: dynamic, contact and regional. These can be further expanded to five main types of metamorphism (Table 6.1):

1. Dynamic (kinetic) metamorphism
2. Contact metamorphism
3. Regional low-grade metamorphism

4. Regional high-grade metamorphism
5. Plutonic metamorphism.

Each of these metamorphic types belong to metamorphic rocks that are formed by corresponding processes.

6.3. ROCKS OF DYNAMIC METAMORPHISM

Dynamic or kinetic metamorphism is caused by mechanical deformations without significant changes in temperature. Dynamic metamorphic rocks are formed by mechanical crushing of existing rocks during tectonic movements or actions of directed pressure, i.e. the dynamic stress. An important feature of this type of metamorphism is profoundly changing textural and structural characteristics of the rocks without significant recrystallization, and/or changes in chemical reconstitution to form new minerals. The mineral ingredients are first compacted, bend and lengthened by the action of strong unidirectional pressure. The minerals break, crush or crumble in a granular mass when the pressure exceeds the size limit of their elasticity. The completely crushed rocks become compact, solid and very hard with mutual compaction of the fragments due to the high pressure. Dynamic metamorphism is associated with zones of high to moderate strain such as fault and thrust zones. Cataclasis, crushing and grinding of rocks into angular fragments occur in dynamic metamorphic zones, giving cataclastic texture. The textures and structures of these rocks are similar to breccias and consist of crushed square grains or powder of crumbled rock fragments.

The rocks, resulting from dynamic metamorphism, are divided into mylonites, flazer, cataclasite and eye-gneisses (Table 6.1) depending on the textural–structural changes caused by crushing.

Mylonite (from the Greek, "myle" = "mill") caused by intensive grinding and crushing of various rocks along tectonic zones influenced by strong pressure or a dynamic stress. In fact, there will not be any significant recrystallization or formation of new minerals. The minerals represent tiny microscopic grains and crushed rocks formed in the zone of major tectonic movements under the influence of intense stress. The mylonites can be formed by converting many kinds of rocks, particularly those predominantly composed of quartz, feldspar, calcite and dolomite. The likely rocks are granite, gneiss (Fig. 6.7), granodiorite, diorite, sandstone, limestone, marble and dolostone. Finally, the granular mylonites are composed of quartz, feldspar, sericite and chlorite, similar to phyllites, and generally occur on planes of movement. The crushed quartz-feldspar rocks are known under the name *phyllonites* (from phyllite and mylonite).

Flazer cataclasite are characterized by the appearance of a textured bar strip curved around completely disintegrated rock material and between the elongated grains and fragments (eye) rocks.

Augen gneisses form by the strong crushing of coarsely crystalline gneisses. The augen gneisses contain a large amount of fine-grained mylonite matrix in which the residual buildup of lenses (the eye) or the larger crystals of fractured feldspar and/or quartz embedded (Figs 6.4(B) and 6.8). The matrix is completely fragmented (rock flour), finely crushed and more or less recrystallized rock mass. The name augen gneiss is derived from German ("augen" = "eye").

6.4. ROCKS OF CONTACT METAMORPHISM

Contact (thermal) metamorphism occurs in a large range of temperatures caused by injection of magma and lava into the cooler country rocks of lithosphere at relatively low pressure. It is limited to the area of contact of rocks with the igneous body (Fig. 6.9) and that is why it is called *contact metamorphism*. The effect

FIGURE 6.7 Mylonite rock comprises biotite–quartz–feldspar and displays strong fluxion banding (flow layers) with numerous white porphyroblasts and porphyroclasts of feldspar that show strong component of flattening into the fabric implying blastesis during the mylonitization. This rock type separates the biotite–garnet–feldspar–quartz "gneisses" hosting the rich sulfide ore from intensely metamorphosed partially melted biotite–garnet–feldspar gneisses at the footwall at Rampura-Agucha zinc–lead–silver mine, India.

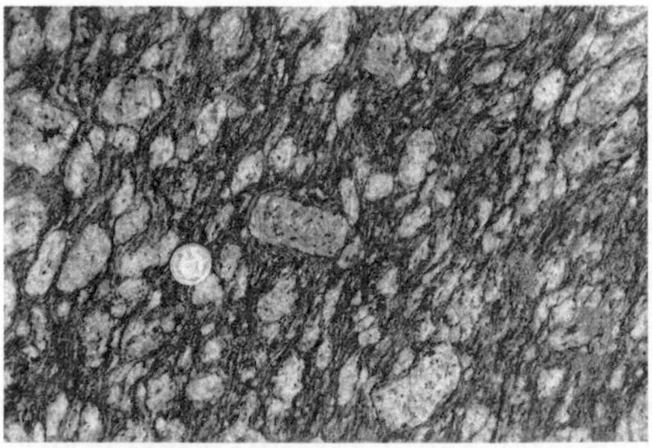

FIGURE 6.8 Typical texture of "Augen gneiss" or "Eye gneiss" representing large lenticular eye-shaped mineral grain, usually, feldspar and distinctly arranged in linear or banded structure. Alternate foliated layers are composed of biotite. Mica flakes wrap around the rectangular porphyroblast of feldspar due to stree free metamorphic crystallization, Udaipur region, India. *Source: Prof. A.B. Roy.*

of heating on the surrounding rock is predominant at the contact of magma or lava and propagates in several concentric rims or the contact metamorphic zones or metamorphic *aureole*. The size of the aureole depends on the heat of the intrusion, its size, and the temperature difference with the wall rocks. Specifically, the rocks at the contact with the magma or lava are exposed to significantly higher temperatures than the rocks away from the magmatic bodies, and in each of the zones form specific mineral assemblages. The pressure does not substantially change from zone to zone for the newly created contact metamorphic minerals. The contact-metamorphic rocks are usually known as *hornfels*. These rocks may not present signs of strong deformation

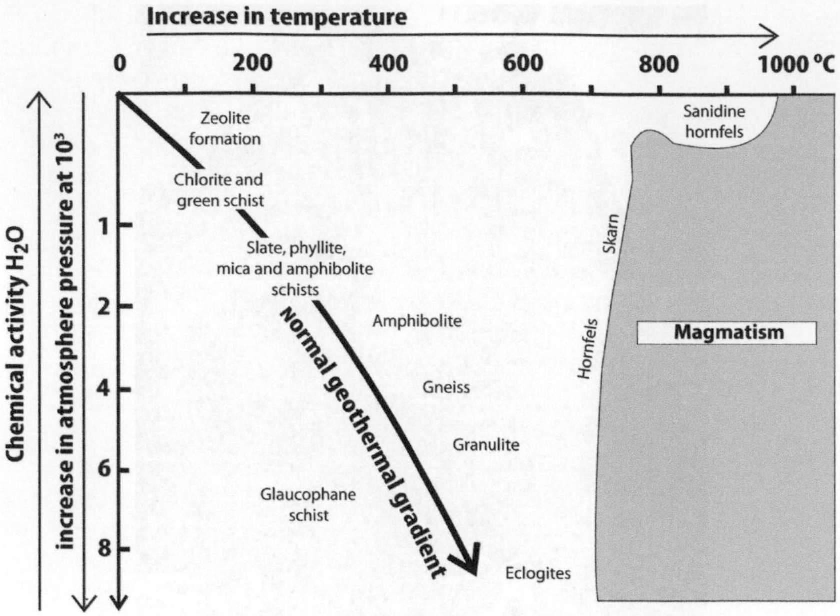

FIGURE 6.9 Stabile areas of major mineral community (metamorphic facies) with respect to the increase in pressure and temperature with increasing depth of the overlay and by heating with magma.

and are often fine grained. The grains are of equidimensional contours and are relatively small in size due to short duration of metamorphic processes of cooling. The newly formed metamorphic rocks will contain minerals of smaller dimensions and less metamorphosed further from the contact. The vapor and gases play an important role near the contact with the magma or lava. If the surrounding rocks are metamorphosed by hot water vapor, gases, or pneumatic gas released from magma or lava, it is *pneumatolytic metamorphism*. The surrounding rocks can be metamorphosed with hot solutions that originate from the magma, and such a metamorphism is called *hydrothermal metamorphism*.

The intensity of contact metamorphism depends on the size of intrusions (magmatic body), the cooling rate of magma or lava and the quantity of gases and vapors emitted from the magma. It is clear that due to higher temperatures and slow cooling of magma injected into the rocks at greater depths, the intensity of change will be stronger in large igneous intrusions of the body than in the subsurface injection or the outpouring of lava. The contact metamorphic rocks commonly occur in large masses at the edge of the intrusive igneous massif (Fig. 6.9), and smaller crowds at the contact of volcanic rocks.

The most sensitive rocks to contact metamorphism are clay and carbonate sediments, and are subjected to the highest degree of metamorphic change and metamorphosed to "hornfels" and "skarn" (Fig. 6.9).

The rocks of contact metamorphism are hornfels created from clay and pelite sediments and/or tuffs. The skarn rocks are created from marl and clay limestone and dolomite (Table 6.1). The contact metamorphic marbles occur at the contact of the magma or lava with limestones and dolomites. The contact metamorphic quartzites occur at the contact magma or lava with quartz-rich sedimentary rocks (sandstone, siltstone and chert). The resultant marbles and quartzites do not differ significantly with respect to the structure and composition from those generated in high-degree regional metamorphism (Section 6.5.2).

Hornfels are medium to coarse crystalline rocks, dark color and rich in silicates with granoblastic and porphyroblastic texture. They consist of andalusite, garnet and cordierite as major minerals and quartz, feldspar, biotite, muscovite and pyroxene as typical minerals. Hornfels often include epidote, diopside, actinolite, or wollastonite and sometimes titanite and tremolite.

Skarns arise from pure limestone or dolomite at the direct contact with the magma from which large quantities of Si, Al, Mg and Fe infiltrate. Skarns are characterized by a granoblastic to porphyroblastic texture. Skarns usually contain one of these minerals of hedenbergite ($CaFeSi_2O_6$), grossular ($Ca_3Al_2(SiO_4)_3$), diopside ($FeCaSi_2O_6$) or wollastonite ($CaSiO_3$) as an essential ingredient (Table 6.1). The marbles and quartzite are formed from limestone and dolomite and quartz sandstone and siliceous sediments, respectively, by contact and regional metamorphism (Table 6.1). They are presented in detail in Section 6.5.2.

6.5. ROCKS OF REGIONAL METAMORPHISM

Regional or *Barrovian* metamorphism is caused by the general increase in temperatures over large areas of the continental crust. It is typically associated with mountain ranges, particularly in the subduction zones with a geothermal gradient. The geothermal gradient gradually increases the temperature by about 1 °C from the Earth's surface to the vertical depth of every 33–40 m. The pressure will also increase proportionately in depth by the overlying thickness of rock strata. Therefore, both the temperature and pressure will simultaneously increase in depth on regional basis. Differences in pressure at greater depths of the Earth's crust is not only a consequence of the depth of covering, i.e. the weight of overlying rocks, but are the result of tectonic movements, such as the collision of continental masses or between two tectonic plates (Section 3.4.2). This will cause strong compression and folding of rocks.

The intensity and degree of metamorphic change increases with depth, which is manifested by the various new communities of metamorphic minerals, characteristic for different conditions of temperature and pressure. The effect is changes not only in the mineral composition of the original rock (protolith) but also in the texture and structure of the newly formed metamorphic rocks, such as schistosity, lepidoblastic, nematoblastic and granoblastic texture (Section 6.1).

Increasing temperature and directed pressure cause distinct schistosity of metamorphic rocks that manifests with schedule from slips and stick mineral substances parallel with the plane perpendicular to the target pressure (Figs 6.2 and 6.3). Rocks of such textures and structures are called *crystalline schists*, and may occur at low and high degree of metamorphism.

6.5.1. Schists of Low-Grade Metamorphism

Regional low-grade metamorphism takes place with a small increase in temperature (above 200 °C) at significantly increased directional pressure. The directed pressure or stress, as opposed to hydrostatic pressure, operates only in one direction. This type of high pressure under regional metamorphism affects recrystallization of mineral constituents of rocks resulting in the formation of new plate, e.g. leaf-like and elongated minerals. The longest surface of the new minerals is oriented perpendicular to the direction of pressure (Figs 6.2 and 6.3). Such orientation of minerals or *schistosity* is the significant feature of schist of low-grade metamorphism, and their slips and stick form is the basic feature of lepidoblastic and nematoblastic structure.

Typical low-grade metamorphic minerals are albite, muscovite, chlorite, actinolite and talc. The main and most widely spread metamorphic rocks from the group of low-grade schist metamorphism are argillaceous rocks namely slate, phyllites and schists as shown in Table 6.1.

Slate is an extremely dense, fine-grained metamorphic rock form under low-grade regional metamorphism emerged from pelitic sedimentary rocks such as shales and fine-grained tuffs (Table 6.1). The level of precrystallization of primary minerals witness very weak changes under the increase of only directed pressure and low temperature. These metamorphic rocks largely retain the primary mineral composition and structure of the original rocks, except a smaller or larger part of typical clay minerals, partially or fully recrystallized as illite, sericite and chlorite. Slate is characterized by an excellent flat schistosity, which is usually difficult to differentiate from sheet pelitic sedimentary rocks, due to the flat surfaces of schistosity split into sheets or thin plates. This structure is caused by parallel orientation of microscopic tiny leaves of illite and chlorite. The essential mineral ingredients of slate are cryptocrystalline quartz, feldspar (albite), chlorite, calcite and illite, and organic matter. These unique properties enable it useful for roofing, inner walls and as small board (slate) for small school children in the class room in the olden times.

The pelite sediments (claystone, mudstone and shale) turn into slate passing through rocks from the transitional stage between the diagenetic changes and the lowest degree of metamorphism. Such rock may be called as *argillite*. The slate contains a higher proportion of uncrystallized clay minerals, and instead of schistosity shows sheet or thin horizontal lamination, unlike argillite.

Phyllite is foliated metamorphic rock rich in tiny sheets of sericite mica. It presents gradation in degree of metamorphism ranging between slate and mica schist. The color varies between black and gray to greenish gray. Phyllite forms from pelitic sediments (shale and mudstone) at slightly higher degree of regional metamorphism from slate (Table 6.1). It may also result from finely grained tuffs and clayey tuffite. Phyllites have excellent fissility with tiny sheets showing thin schistosity due to the high content of mica and chlorite. The schistosity surfaces display silvery shine. The essential mineral ingredients of phyllites are microcrystalline quartz, fine-grained micas (sericite and muscovite) and chlorite. The quartz is usually in the form of elongated thin lenses or veins within the sheets, which contain mainly fine-grained mica and chlorite. Phyllites no longer include clay minerals, unlike slate, because of higher level of metamorphism and fully recrystallized into a fine-grained mica and chlorite.

Sericite schist is a variety of fine-sheet phyllites in transition to mica schist. It consists mainly of sericite ± quartz, i.e. clusters of very small slips of illite, muscovite and other mica that cannot be accurately determined by petrographic microscope (see sericite, Section 2.5.8.5.2). Sericite schist exhibits shining silver, gray, brown color with excellent fissility.

Green schists and *chlorite schists* are fine-grained to medium-crystalline schist of low-grade metamorphism resulting from basic igneous rocks at relatively low temperature and pressure (Figs 6.6 and 6.9). Some green schist may also occur in progressive regional low-grade metamorphism of calcite-rich pelitic sediments. Green schists are named after their characteristic green color, caused by the high content of green minerals like chlorite, epidote, actinolite and zoisite. In addition to these minerals, it includes quartz, acid plagioclase, tremolite, calcite, dolomite, magnesite and hornblende.

Glaucophane schists are formed in a regional low-grade metamorphism of basalt, diabase, feldspar arenite, graywacke sandstone and marl sediments at relatively low temperatures and high directed pressure (Fig. 6.9). The glaucophane schist includes a high content of Na-amphibole (glaucophane) associated with albite, epidote, garnets, mica, quartz and calcite.

Talc schist originates in regional low-grade metamorphism by transformation of ultrabasic igneous intrusive magmatic rocks (peridotite, dunite, and olivine gabbro) and also serpentinite that occurs by hydrothermal metamorphism from olivine-rich ultrabasic rock. They consist of Mg-silicate (talc, antigorite, and chlorite), actinolite, magnesite, calcite and dolomite. The talk is usually in the form of thin lenses or zones along the surface of schistosity. The calcite, magnesite and dolomite exist as irregular masses or veins.

6.5.2. Schists of High-Grade Metamorphism

The physical and chemical properties of minerals, in the deep parts of the lithosphere, change on account of interaction of high pressure and high temperature during high degree of metamorphism. The process causes almost complete primary recrystallization of mineral ingredients in rocks that are affected by such high degree of metamorphism. The primary minerals become unstable in condition of very high pressure and high temperature and recrystallize into minerals that are stable in condition of higher degree of metamorphism. The general characteristics of these minerals are decrease in volume and higher packing density of ions in the crystal lattice. The extremely high hydrostatic pressure and high temperature cause the formation of minerals from the group of nesosilicates, sorosilicates, inosilicates and tectosilicate. The new minerals mainly crystallize in the cubic, rhombohedral and monoclinic system and thus constitute schistose rocks with granoblastic, porphyroblastic and nematoblastic textures. This forms a brand-new mineral composition with entirely new structures and textures.

The distinctive minerals of higher degree regional metamorphism are sillimanite, staurolite, cordierite, andalusite and biotite. The characteristic rocks of high-degree regional metamorphism are various schists, such as mica, amphiboles, cordierite, sillimanite, staurolite and graphitic schists, and also amphibolites and gneisses.

Mica schists are distinct schistose metamorphic rocks (Fig. 6.3). The schists originate from clay, shale, mudstone, graywacke sandstone, and acidic and neutral igneous rocks, fine-grained basalts and tuffs at medium and high degree of regional metamorphism and at significantly higher level of phyllite and sericite schist (Fig. 6.9). They follow by gradual steps to the highest degree of progressive regional metamorphism such as slate → phyllite → sericite schist → mica schist (Fig. 6.10). Specifically, the intensity of the recrystallization at higher pressure and temperature is significantly increased in relation to the conditions of phyllites. Recrystallization causes many larger crystals in particular to mica (muscovite and biotite), quartz, feldspar, garnet, staurolite and sillimanite and is well observed macroscopically. In addition to sericite, the mica schist contains quartz, acid plagioclase, chlorite, and typical metamorphic minerals such as graphite, garnets, kyanite, andalusite and staurolite. The rock can be designated as sericite, muscovite, biotite, graphite, staurolite and andalusite mica schists.

Amphibole schists are schistose metamorphic rocks that are formed under conditions of high-degree metamorphism from basic magmatic, clay limestone and marl, and are mainly composed of hornblende and feldspars and minor amounts of quartz, garnet, magnetite and biotite (Fig. 6.6).

Disten or sillimanite, cordierite and staurolite schists occur in conditions of high-degree metamorphism from pelite sediment, i.e. shale,

FIGURE 6.10 Chlorite quartz schist ± garnet (top), hosting foliation parallel or statiform type of late veins of chalcopyrite (brassy golden) and pyrrhotite (brown) in transition zone at Kolihan Section, Khetri Cu Mine, India. *Source: Prof. Martin Hale.*

mudstone and marl. The essential mineral components of medium to large crystalline schist are aluminosilicate (or disten), sillimanite, staurolite and cordierite, and in association with quartz, K-feldspar, Na-plagioclase, garnet, mica, graphite, hematite and magnetite. The aluminosilicates occur in the form of larger or smaller porphyroblasts and granoblasts. The resulting schist receives appropriate name with prefix of the individual aluminosilicate, for example, disten schist, sillimanite schist, staurolite schist, etc. These groups of schistose rocks are the product of high-degree regional metamorphism from organic matter rich in clay sediments (or mudstone and shale).

Graphite schists are black in color, and contain graphite between 5% and 10% in the form of elongated lenses, lines, layers or veins within the cluster of mica, chlorite, quartz, K-feldspar, garnet, sillimanite and magnetite. Graphite schist is the product of high-degree regional metamorphism from organic matter-rich clay sediments like mudstone and shale (Fig. 8.11).

Gneisses are common and widely distributed medium- to coarse-grained rocks formed by high-grade regional metamorphic process and plutonic metamorphism under high pressure and temperature from a variety of preexisting igneous and sedimentary rocks. If the original rocks contain quartz and feldspars, the resultant product will be "orthogneiss" and is case of quartz and clay minerals are known as "paragneiss". In other words, orthogneiss designates gneiss derived from an igneous rock, and paragneiss originates from a sedimentary rock. Gneiss resembles structure to describe appearance of alternate layers of unlike mineral composition and colors (Fig. 6.1). The gneiss and gneissose are very similar to schist and schistose in metamorphic rocks with structure presenting gneissosity (alternate layers or bands) and schistosity (thin lamination). The term *gneiss* is used as suffix with the original rocks that have been transformed by metamorphism. Granite gneiss is the most common type and formed by transformation of granite. Similarly, presence of excess hornblende in the rock will be designated as hornblende gneiss or amphibole gneiss or simply amphibolites after metamorphism.

Orthogneissic rocks are medium to coarsely foliated and largely recrystallized. Mineral composition of orthogneiss is similar to the composition of granites (Figs 6.11 and 6.12). The essential minerals are quartz and feldspars (orthoclase, microcline, perthite, and

FIGURE 6.11 Granite gneiss (orthogneiss) composed of large porphyroblasts of feldspar (pink) arranged in a gneissic or layered structure, roughly along north east–south west direction, with alternate fine-grained matrix. *Source: Prof. Arijit Ray.*

FIGURE 6.12 Photomicrograph of thin section of granite gneiss (orthogneiss) composed of large porphyroblasts of feldspar (white) and quartz (blue) in a fine-grained matrix of feldspar, quartz, biotite (yellow brown) and ilmenite (black). All mineral grains display an excellent gneissic structure. *Source: Prof. Arijit Ray.*

Na-plagioclase), and with significant amount of biotite and/or muscovite.

Paragneiss results from metamorphism of fine-grained clastic pelitic sediments (mud, mudstones and other argillaceous rocks) that recrystallized in the deep zones of the earth's crust in an amphibolite facies of metamorphism. Paragneiss contains quartz and feldspar and amply mixed up with other minerals typically sillimanite, kyanite, cordierit, andalusite and staurolite. A certain excess of alumina, from the content of clay material in the primary sediments is typical of paragneiss. The varieties of paragneiss are distinguished according to the admixture content. There are numerous varieties of gneiss depending on the different mineral composition. The most prevalent are the quartz-feldspar, muscovite, biotite, sillimanite, andalusite and cordierite gneiss. Gneisses are characterized by generally "eye" or porphyroblastic schistose texture, and often granoblastic or granolépidoblastic texture along with development of characteristic gneissic structure (Figs 6.4(B), 6.13 and 6.14).

Amphibolites are the most common metamorphic rocks formed by regional metamorphism under high pressure and high temperature (Fig. 6.5). Amphibolites usually occur along with the mica schist and gneiss. Schistosity is considerably less pronounced in amphibolites than those in the amphibole schists and is mainly influenced by more or less parallel orientation of prismatic crystals of black hornblende. The rocks are often striped due to semiparallel sort of mutual narrow bands, predominantly of rich black hornblende and light plagioclase. Mineral composition of the amphibolites is simple and mostly contains hornblende and plagioclase, with variable amounts of anthophyllite, garnet, mica, quartz and epidote. The rocks may originate from pelitic sediments, with amphibole (hornblende), plagioclase and typically include green pyroxene. Amphibolites are characterized with nematoblastic or granoblastic texture Amphibolites are formed from more basic magmatites usually containing more magnesium amphibole (anthophyllite), and the ones crops up from weak basic and neutral magmatites containing nearly equal amounts of hornblende and

FIGURE 6.13 Orthogneiss or granite gneiss rock showing alternating sorting of strips/bands composed of leucocratic quartz-feldspar (light) and melanocratic mica-sillimanite (dark) minerals. The rock has been deformed showing development of micro structure like fault in the upper middle part and folds in the left hand bottom corner. *Source: Prof. A.B. Roy.*

Ca-plagioclase. There will be a considerable amount of quartz in the amphibolites are formed from tuffs.

Amphibolites may also originate from a variety of neutrals and basic igneous rocks (orthoamphibolite), and out of marl, pelitic sediments, and clay limestone (paraamphibolite). Orthoamphibolite includes mainly amphibole (hornblende) and albite, and even small amounts of epidote, zoisite, chlorine and quartz. The accessory ingredients are leucoxene, ilmenite and magnetite. It often contains incomplete metamorphic remains of protolite igneous rocks. Paraamphibolites are more balanced in composition than orthoamphibolites. Paraamphibolites contain biotite, more of quartz, albite, wollastonite and calcite, in addition to hornblende and plagioclase, unlike orthoamphibolite. It contains less of protolith (incomplete metamorphosed relics of sedimentary rocks) than orthoamphibolite.

FIGURE 6.14 Biotite–garnet–feldspar–quartz gneisses display strong fluxion banding (flow layers) and extensively microfolded replicating regional structure with numerous white porphyroblasts and porphyroclasts of feldspar showing varying degrees of flattening into the fabric with psammitic bands, and are locally overprinted by large feldspar porphyroblast growths, exists close to the host mineralization at Rampura-Agucha zinc–lead–silver mine, India. *Source: Prof. Martin Hale.*

Quartzite is composed almost exclusive of monomineral, medium to coarse crystalline metamorphic rocks with granoblastic texture. The isometric grain shape and their jagged contacts are a consequence of the process of recrystallization and loss of primary forms, and contacts of grain under high pressure and high temperature conditions. Pure quartzite forms from the quartz-rich sedimentary rocks (sandstone, siltstone and hornfels). The quartzite, in addition to quartz exceeding 90% share, may contain minor amounts of mica (mica quartzite), feldspar (feldspar quartzite), chlorite (chlorite quartzite), garnet (garnet quartzite) and amphibole (amphibole quartzite (Fig. 6.15)).

Quartzite forms in regional metamorphic of feldspar and quartz sandstone, siltstone, chert, and rarely in quartz-rich (pegmatites) veins. The metamorphic transformation of sandstone into quartzite is usually caused by high temperature and high pressure, with the participation of tectonic movements, especially in orogenetic tectonic compression zone. It may also occur by contact metamorphism of quartz-rich sandstones. The quartzite often displays relict structures, i.e. remnants of sedimentary textural–structural features.

FIGURE 6.15 Amphibole quartzite composed mainly of massive hard conchoidal quartz (greasy bluish) and needles of amphibole (dark) in linear style and located at the footwall of main copper mineralization at Kolihan Section, Khetri Copper Mine, India. Amphibole quartzite also hosts part of copper mineralization (Lode-I). The major mineralization is hosted by garnet mica schist and graphite mica schist.

Quartzite is extremely resistant to chemical weathering, and is often found in the form of morphological elevations or areas that protrude from the surrounding rocks. As very resistant to weathering appears in the form of clasts (as fragments of quartzite) in the sands or sandstones, conglomerates, schist or generally in many clastic types of sediment (Fig. 6.16).

Marble is a metamorphic rock that contains predominantly of calcite (Fig. 6.4(A)) and/or dolomite, and minor minerals like tremolite, forsterite, spinel, garnet, diopside and wollastonite. It occurs in regional and rarely contact metamorphism of carbonate sediments like limestone or dolomite, which contains a smaller or larger amount of clay. Marble forms by combined effects of high pressure at great depths of covering and high temperatures from the carbonate sedimentary rocks (limestone and dolomite). Marble may also occur with the high pressure and temperature increased at contact zones with magma and lava. Typical marble has granoblastic texture characterized by coarse crystalline aggregates of isometric calcite crystals with single larger xenoblasts of calcite (Fig. 6.4(A)).

Pure-white marble is formed from unadulterated limestone, which in addition to calcite practically does not contain other ingredients (Fig. 6.17). The colored marbles originate from limestones with mineral impurities such as clay and mica. Marbles of different nuances of colors derived from clay limestone containing organic matter (gray and dark gray marble) or contaminating iron oxides (pink and reddish-brown marble) and presence of serpentine from contact ultramafic magma (green marble).

Marble is exceptionally appreciated as a decorative stone due to the special relationship between brilliant colors and excellent polishing capabilities. Marble has been prized for its uses in historical sculptures since time immemorial due to its softness, homogeneity, and a relative resistance to shattering (Figs 6.17, 6.18 and 6.19).

FIGURE 6.16 Steep (back side) quartzite ridges standing as stony (nothing other than hard massive quartzite) barren (no trace of plants and forest) standing 50 m above valley level (500 m above main sea level). The quartzite ridge is sitting at the top of rich metal-grade deep seated ore body of 60 Mt averaging 5% Zn, 3% Pb and 215 g/t Ag that starts at a depth of 120 m from valley level. The cover quartzite does not respond to geological and geochemical exploration signature.

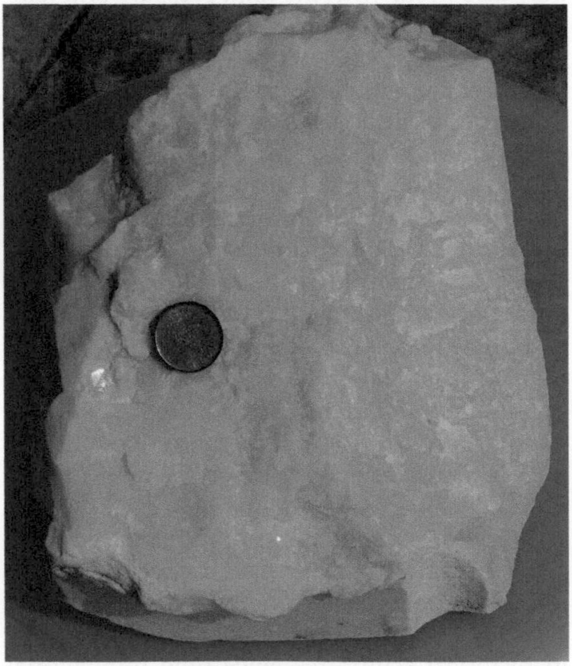

FIGURE 6.17 Pure-white marble composed of milky-white calcite having excellent polishing capabilities.

6.6. ROCKS OF PLUTONIC METAMORPHISM

Plutonic metamorphism occurs at very high temperatures and high pressure in the deeper parts of the lithosphere, usually in combination with partial remelting of rocks under the influence of high temperatures, i.e. conditions similar to igneous episode. The plutonic metamorphism of pelite-sandy sediments creates garnet, sillimanite and cordierite paragneisses. Pyroxene orthogneiss generates from basic and ultrabasic igneous rocks. Granulite and eclogites occur with plutonic metamorphism at very high pressure and high temperature (Fig. 6.9).

Granulites are massive medium- to coarse-grained metamorphic rocks. Granulites are characterized by a granoblastic texture and a large amount of quartz and feldspar or quartz and pyroxene with variable proportion of garnets. It occurs at very high pressure and high temperature (Fig. 6.9) from different rocks (mainly acid magmatites, arkose and graywacke sandstone) having a different mineral composition.

Eclogites occur at extremely high pressure and high temperature (Fig. 6.9) from basic magmatites

FIGURE 6.18 Victoria Memorial Hall, 56 m high, is monumental memorial building and garden setup dedicated to Queen Victoria, Empress of British India, 1901, costing 10.5 million Indian Rupee (present day equivalent to 210,000 US$) US$ and located in Kolkata, formar capital of British India. It currently serves as museum and a tourist attraction. The majestic structure in white marble is modeled on Taj Mahal, Agra, India.

FIGURE 6.19 Taj Mahal, the jewel of Muslim art in brilliant milky-white marble and one of the universally admired masterpieces of the world's heritage, was built between 1632 and 1653 by Mughal emperor Shah Jahan in memory of his wife Mumtaj Mahal. It is located at Agra, India.

rocks (gabbro and basalt). The optimum pressure requires for formation of eclogites is certainly much greater than the typical pressures of Earth's crust. Moreover, it forms from unusually denser rocks and minerals. The necessary depth requires for creation of eclogites is believed to be more than 4 km and temperatures higher than 500 °C, so that olivine and basic plagioclase could be transformed into alkali amphibole omphacite $[(Ca,Na)(Mg,Fe^{2+},Al)Si_2O_6]$. It also contains smaller amount of glaucophane, plagioclase, amphibole, kyanite and rutile. Eclogites have relatively small presence and share of the metamorphic rocks in the Earth's crust, but are the main ingredients with peridotite of earth layer (Section 3.1).

Migmatites are special kind of mixed igneous and metamorphic rocks with complex structure. It forms by combination of plutonic metamorphism and partial melting of preexisting rocks such as gneiss, and/or injection of bright granite magma. The granitic magma is the outcome of partial melting of mineral components of some dark metamorphic rock. Migmatites form under extreme temperature conditions during prograde metamorphism with partial melting of preexisting rocks. Migmatites often appear as extremely deformed and extensively tight micro-folded.

FURTHER READING

Principles of Igneous and Metamorphic Petrology by John D. Winter[63] will be an excellent reading in the subject with much greater details. *Metamorphic Petrology* by Mason[29] can be a good choice.

CHAPTER

7

Precipitation Systems of Major Sedimentary Bodies—Collector Rocks of Oil and Gas

OUTLINE

7.1. Introduction 233

7.2. Main Forms of Collector Sedimentary Bodies in Clastites 234

7.2.1. Alluvial Fans 234

7.2.2. Deltas 236

7.2.3. Sand Bodies in Coastal Marine Environments (Beaches and Offshore) 239

7.2.4. Debrites 242

7.2.5. Turbidity Fans 245

7.3. Main Forms of Collector Sedimentary Bodies in Carbonate Rocks 247

7.3.1. Carbonate Platforms 247

7.3.1.1. Carbonates of High-Energy Shallows 248

7.3.1.2. Peritidal Carbonates 250

7.3.1.3. Carbonates of Restricted Shoals, Lagoons, and Inner Shelf 253

7.3.1.4. Carbonate Bodies of Reef and Perireef Limestones in Carbonate Platform 254

7.3.2. Carbonate Debrites and Turbidites or Allodapic Limestones 257

7.3.3. Reef and Perireef Bioclastic Limestones Outside the Carbonate Platforms 258

Further Reading 260

7.1. INTRODUCTION

Sedimentary rocks are the most important medium in which oil and gas generates and accumulates. Sedimentary rocks are the sources of origin of oil and gas, as well as the medium in which the liquid and gaseous hydrocarbons can accumulate in large quantities forming oil or gas fields or deposits suitable for the exploitation and recovery by mining. These are the main

Introduction to Mineralogy and Petrology
http://dx.doi.org/10.1016/B978-0-12-408133-8.00007-9

insulator rocks without which no oil and/or gas field formation is possible.

Most common reservoir rocks of the oil and natural gas are the conglomerates, breccias, and sandstones, among the clastic sedimentary rocks. Similarly, the limestones and dolomites are most suitable among biochemical and chemical components. The pelite sediments and marls, i.e. evaporites, are equally favorable and significant insulator rocks in oil and gas reservoirs.

The strongly fractured igneous rocks, particularly at the peripheral parts of the igneous mass, may be the appropriate collectors of oil and gas in special conditions. This is particularly suitable at the cover of waterproof clay and marly sediments and good communication with the source rocks. The primary porosity of igneous rocks is usually poor and is mainly related to minor intergranular and oscular porosity. However, the development of secondary porosity due to intense tectonic crushing known as "cracking porosity", may often be significant.

The metamorphic rocks of the large "Pannonian basin" in East-Central Europe and the buildup of basic "highlands" of Tertiary substrate, particularly Miocene sediments, often contain oil and gas. The fissured edges of the crystalline massif consisting of metamorphic rocks, with or without the intrusive igneous rocks, may be collectors of oil and gas in an environment of sedimentary rocks under certain geological and geotectonic conditions.

The significant collector characteristics and the possibility of oil and gas deposits in sedimentary rocks depend on their manner of appearance, shape and size of sedimentary bodies and basins, as well as their relation to the insulating rocks. In such favorable relations, oil and gas may occur already in the deposition or by subsequent postsedimentary tectonics.

The main forms of sedimentary bodies and collector rocks of oil and gas, that are the direct consequence of the method and conditions of deposition, will briefly be discussed.

7.2. MAIN FORMS OF COLLECTOR SEDIMENTARY BODIES IN CLASTITES

Clastic sedimentary rocks are the main collectors of oil and natural gas. These sedimentary bodies are largely composed of gravel, conglomerate, debris, breccia, sands, and sandstones The key and most frequent forms are represented by alluvial fans, deltas, tidal plains, sandy beach, sandbank, barrier islands, underwater dunes and sand ridges, as well as debrites and turbidity fans.

7.2.1. Alluvial Fans

Alluvial fans are morphological fanlike formations of small to extremely large sizes resulting in accumulation of clastic materials at the outputs of river flows from the narrow valley of mountain chains into lowland areas (Fig. 7.1). The dimensions and shapes of

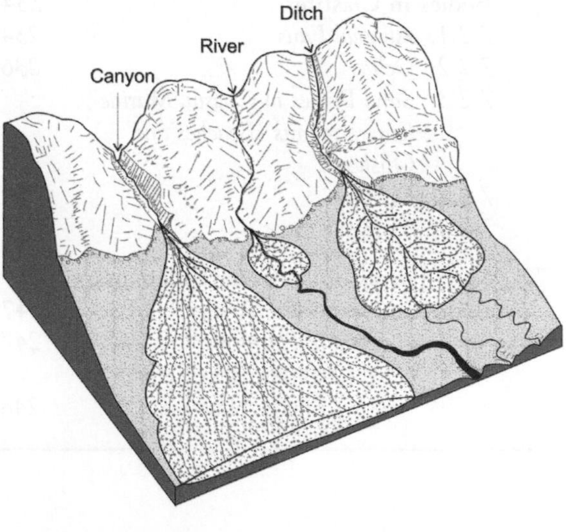

FIGURE 7.1 Deposition of alluvial fans on outputs of river flows from the narrow valley of mountain chains into lowland areas. A large amount of gravel–sand deposits accumulate in a fan shape at the ends of canyons and trenches. Migrating channels may exist and extend from top of the fan, and at its end could continue into riverbed.

sedimentary bodies of alluvial fans essentially depend on the surface spread of the catchment area of the mountain river systems, the intensity of physical weathering of rocks, and the amount of water energy in the river systems. Their surfaces typically vary from a few tens of square meters to several thousand square kilometers, a length of several hundred meters and often exceeding 50 km.

The sediments of alluvial fans are relatively easy to recognize and distinguished from other river sediments by fan-shaped sedimentary bodies consisting of coarse gravel and irregular blocks with diameter greater than 1 m (Fig. 7.2) in the upper proximal portion. The most important features of alluvial fans sediments are:

1. A large amount of gravel with the blocks in the proximal range or upper fan (Gms in Fig. 7.2) whereby gravel, particularly in the distal area fans, typically contain muddy matrix.
2. Reduction in grain size toward the lower part of fans.
3. Spreading of the marginal parts of the fans with thinning and decrease of thickness.
4. In vertical sequence of layers, clear sequence of enlargement upward occurs with several meters thickness due to progradation of younger sedimentary units with the gradual progress of alluvial fans in the valley or lake.
5. Covering of fans with younger muddy sediments of floodplains, i.e. marly lacustrine or marine sediments, as a consequence of the establishment of river, lake, or sea depositional systems in the fan areas.

There are three different types of alluvial fans in the subenvironment designated as proximal or upper fan, mid or middle fan, and distal or lower fan.

In the proximal fan subenvironment, located just at the exit of the mountain range (upper fan), coarse-grained massive gravel (Gm) and blocks which contain relatively high proportion

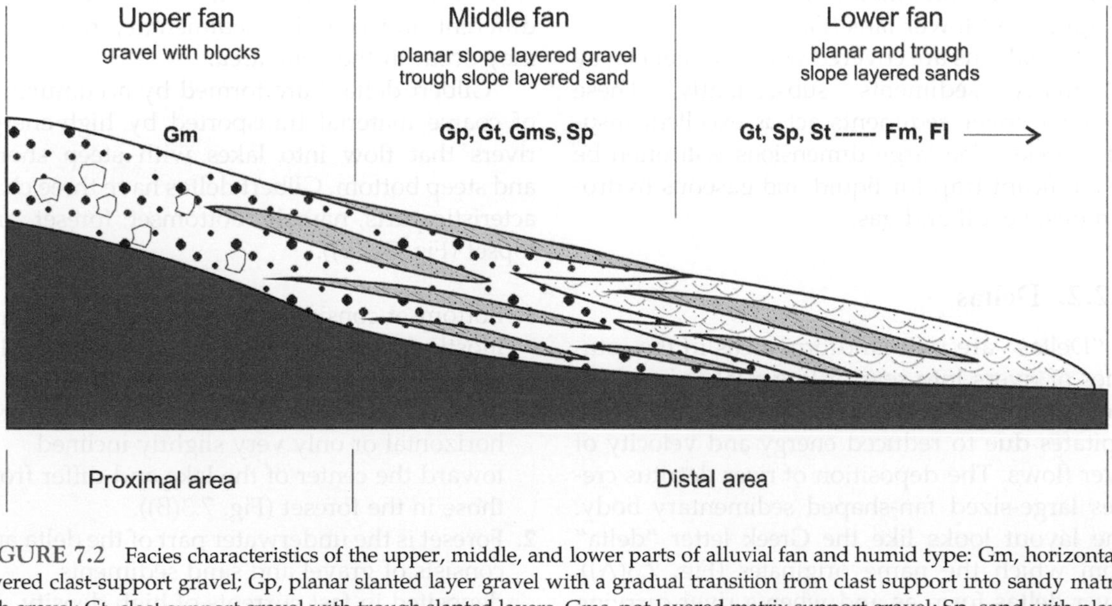

FIGURE 7.2 Facies characteristics of the upper, middle, and lower parts of alluvial fan and humid type: Gm, horizontally layered clast-support gravel; Gp, planar slanted layer gravel with a gradual transition from clast support into sandy matrix-rich gravel; Gt, clast support gravel with trough slanted layers, Gms, not layered matrix support gravel; Sp, sand with planar slanted layer; St, sand with trough slanted layers; Fm, massive sandy pelites; Fl, horizontally laminated sandy pelites and silts.

of fine-grained matrix are highly prevalent. The fan sediments gradually change to more and more sand and even powder or silty detritus further downstream in the lower parts (Fig. 7.2). The inclinations of conglomerate bodies in proximal parts are typically 10°–15°, and on the outskirts, they are up to 30°.

In the middle part of alluvial fan, the morphology depicts characteristics of frequent changes of planar inclined layered gravels (Gp), clast-supported troughed inclined stratified gravel (Gt), massive matrix-supported (muddy) gravels (Gms), and planar inclined layered sands (Sp) of different thickness and diverse granulometric composition.

The dimension of the sedimentary deposition is often extremely wide in the distal part of the alluvial fan (lower fan). There is prevalence of sands with planar (Sp) and trough slanted stratification (St) over clast-supported pebbles with trough slope stratification (Gt) and massive (Fm) and horizontally laminated silty sediments (Fl).

The characteristic features of the midfan represent the combined attributes of both sides of upper and lower fans (Fig. 7.2).

Alluvial fans are covered with younger clayey or marly sediments subsequently. These younger cover sediments act as excellent insulator rocks. The large dimensions will often be a significant trap for liquid and gaseous hydrocarbons, i.e. oil and gas.

7.2.2. Deltas

"Deltas" are vast landform areas in the estuaries of rivers into seas or larger lakes in which most of the detritus carried by these flows precipitates due to reduced energy and velocity of river flows. The deposition of river detritus creates large-sized fan-shaped sedimentary body. The layout looks like the Greek letter "delta" from which the name originates (Fig. 7.3(A)). River deltas form as and when a river carrying large sediment reaches the body of standing lake, ocean, or reservoir water. The river rapidly deposits the detritus sediments and spreading at the estuaries causing a reduction in the rate of flow. The river that cannot remove the sediment quickly due to reduction of water flow may end up forming a delta. Deltas can also form in inland region where the river water spreads out and deposits sediments. The river is no longer confined to its channel when its flow enters the standing water and expands in width causing decrease in the flow velocity. The river is no longer in a position to transport sediments further, and hence drops out of the flow and deposits (delta). The growth of a river delta farther out into the sea over time takes place by progradation. In doing so, granulometric separation of detritus occurs with the constant spreading of delta outward. Coarse-grained sediments accumulate immediately near the river mouth, followed by deposition farther from the mouth to the lake or sea bottom i.e. in the basin plain. The deposition and resedimentation of detritus, particularly in the upper part of the delta (delta plain), may significantly affect tidal currents, waves, and ocean currents, so that a very different and complex sedimentary formations may occur in the delta area.

"Gilbert deltas" are formed by accumulation of coarse material transported by high-energy rivers that flow into lakes with steep shores and steep bottom. Gilbert deltas have three characteristic parts, namely, bottomset, foreset, and topset (Fig. 7.3(B)).

1. Bottomset consists of fine-grained sediments, mostly fine-grained horizontally laminated sandy and muddy sediments that are deposited at the lake bottom. The layers are horizontal or only very slightly inclined toward the center of the lake and differ from those in the foreset (Fig. 7.3(B)).
2. Foreset is the underwater part of the delta and consists of gravel and sand sediments deposited in fast currents of high density. These sediments are primarily inclined

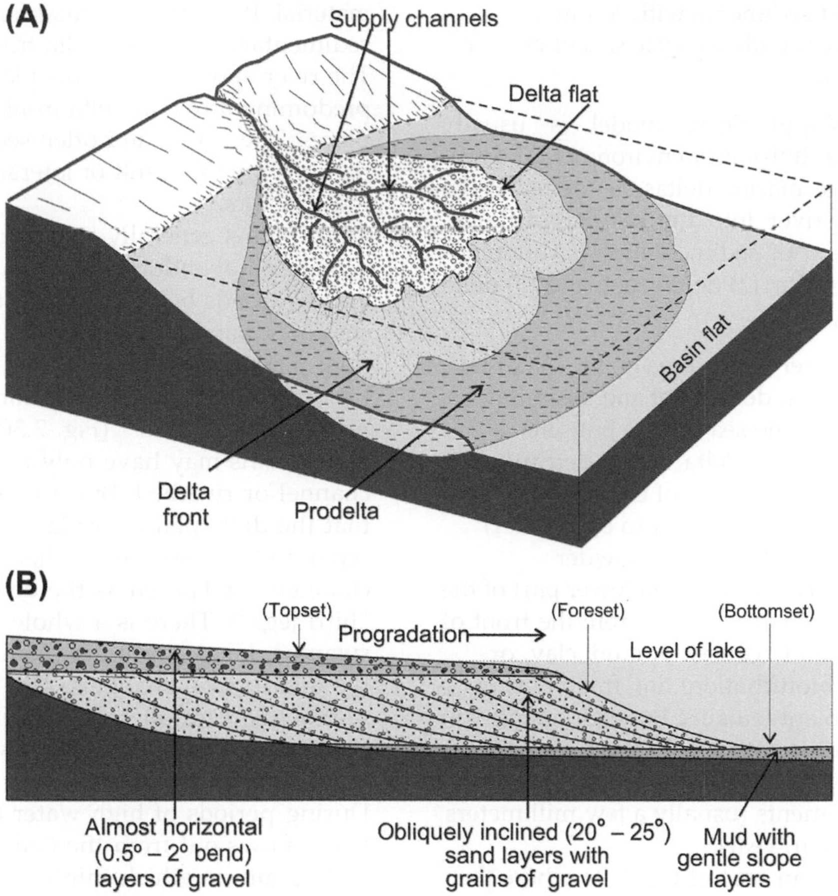

FIGURE 7.3 The conceptual diagram displays (A) sea delta fan with delta plain, steep delta front, and spacious prodelta on the seabed and (B) longitudinal profile of lacustrine fan delta of Gilbert type with the basic characteristics of sediment in topset, foreset, and bottomset of delta showing accumulation of material transported by high-energy river flow on the estuary of lake of steep shores and bottom. *Source: Ref. 22.*

toward the center of the lake at an angle of 10°–25°, due to prograde precipitation sequence as well as the morphology of the delta (Fig. 7.3(B)). Larger erosion channels can be seen sometimes in foreset part of the delta. The height of foreset is typically a few meters. Inclined layered conglomerates with angles typically varying between 20° and 25° precipitate (with heavy delta progradation of foreset) in the middle of the lake deltas with strong river flows and steeper relief of land. The lower slopes of foreset contain layers of sand with gravel pebbles which have lower inclination and gradually pass into bottomset.

3. Topset is nearly horizontal or only slightly inclined in the direction of the basin. Topset is the package of layers of sand and gravel sediments with inclination of 5–10 m/km, i.e. from 0.5°–1°/km. This part of the delta exists mostly outside and only partly underwater of the lake into which it flows (Fig. 7.3(B)). The layout is semicircular with a radius of not more than a few kilometers. It is an extension of the river flow and therefore contains only

gravel–sand sediments with common inclined layers with separate sets of clay or pelite lamina.

The Mississippi Delta model is usually applied for the division of environmental deposition in case of marine deltas, i.e. deltas at the mouth of the river into the sea. Accordingly, the delta system of sedimentation is differentiated as (1) prodelta, (2) delta front, and (3) delta plain (Fig. 7.3(A)).

1. Prodelta is a very wide area on the sea or lake bottom between delta front and basin flat, i.e. the flat part of the sea or lake bottom (Fig. 7.3(A)). In prodelta, tiniest detritus, mostly sludge; powders of clay; as well as marls with thin (millimeter to centimeter) fine-grained sand or sandy powder precipitates. The middle and lower part of the prodelta, which are farthest from the front of the delta, mainly consist of mud, clay, or marls with bioturbation; tiny fragments of shells; and plant remains. But their quantity in the sediment is very low because of rapid sedimentation i.e. deposition of large layers of muddy sediments (usually a few millimeters to about 10 cm/year).
2. Delta front is an area of the delta where the river enters into the sea or lake area and deposits most of the material passed over by a sudden expansion of flow and decreases its speed (Fig. 7.3(A)). The depositional sequence of progradation sediments has an inclination of 10° or more, depending on the morphology of the coast and sea or lake bottom, quantity of detritus, and portions of its deposition. The accumulated mass of sediment by mutual action of basin process, primarily tidal and coastal currents, is distributed in sedimentary body that consists of coarse granular onsite and finely granular sediment further in the distal part. The distal fine-grained sediments are covered by proximal coarse granular sediments by progradation of delta with constant input of material. It is a fundamental feature of sedimentary bodies of delta front deposition. The river and tidal currents play a predominant role in delta front in some places. The waves and often sedimentary patterns are the result of interaction of all these factors.
3. Delta plains generally consist of numerous active and abandoned river channels and riverbeds between which there are shallows, bays, and occasionally floodplains, tidal plains, and marshes (Fig. 7.3(A)). Its characteristics generally correspond to topset of lake delta[22] (Fig. 7.3(B)). Some delta plains may have only one distribution channel or riverbed. But it is more likely that the delta plains display two or more separate fan distribution channels (three channels are known as the delta of the "bird legs"). There is a whole tangle of swamps, lakes, and tidal plains between the distribution channels. The gravel (sand sediments) precipitate in distribution channels of delta plains with similar characteristics as in river environments. During periods of high water levels, water flows out from the bed and flood delta plain on which thinly laminated mud of floodplain consequently accumulates, which are usually intensely bioturbated.

A large and relatively rapid input of terrigenic material (derived from the erosion of rocks on land) with river flow may cause progradation of delta complex into sea. Large portions of sedimentation and constant subsidence of precipitation area led to the formation of thick, widespread delta deposits with characteristic of vertical and lateral distribution of certain types of sediment. These types of deposits are arranged in a specific vertical sequence in delta complex with the progradation of the delta and its advance into the sea. This is called delta sequence (Fig. 7.4).

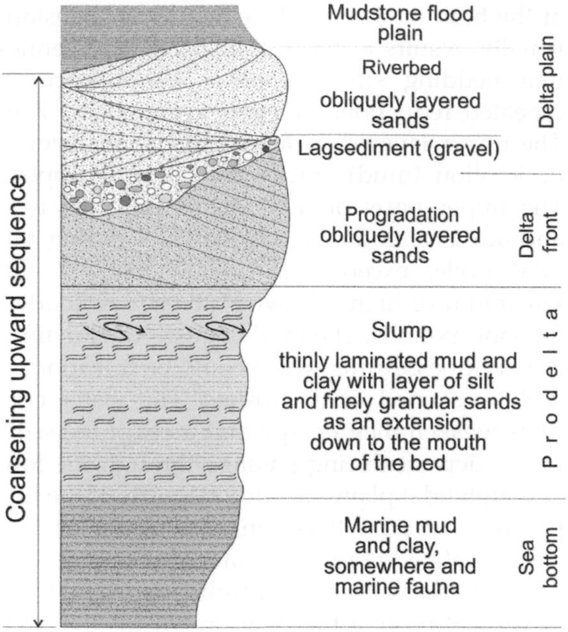

FIGURE 7.4 Vertical sequence of deposits in the sea delta, i.e. the complete delta sequence of coarsening upward.

The delta sequence of coarsening upward (Fig. 7.4) occurs due to progradation of the delta into the sea and usually starts with a dark mud from the edge part of the prodelta that contains sea fossils and participates on usual sediments of the seafloor. These are prodelta deposits that are deposited in the initial stage of formation of the delta, while delta front is still very far away. The depositions from lower part of delta front follow due to progradation of delta and accumulate thinly laminated pelite deposits (silt, clay, and mud). The deposits from the delta front are represented by fine-grained sands with wavy lamination or sandy sediments with slope stratification. The sequence ends, due to progradation of the delta and the extension of the riverbed on delta plain which constantly progresses toward the sea (Fig. 7.3(A)). The delta sequence can end with a thick series of inclined layered sandstones deposited in wide meandering riverbed (Fig. 7.4). Such delta sequence can be covered in mud and fine-grained sandy sediments of tidal plains if the delta plain progradates into sea establishing marine shallow with prevailing activities of tides and ebb (Fig. 7.3). If the relative rising of the sea level (by faster subsidence of the bottom portion than delta sediment accumulation) overfloods the delta system, then clayey or muddy marine sediments will accumulate on the delta sequence. These are the insulator rocks that allow the accumulation and retention of oil and gas within a kilometer of widespread collector rocks of sand and gravel sediments, deposited during progradation of the delta front. The accumulation will also occur in the riverbeds and channels formed during delta plain progradation across delta front.

7.2.3. Sand Bodies in Coastal Marine Environments (Beaches and Offshore)

In the inshore (or onshore close to shore) marine environment of deposition with clastic sedimentation, on the foreshore, shoreface, and offshore, large sand bodies in the form of tidal plains, sandbank, barrier islands, reefs and sand ridges, as well as submarine dunes or sand waves are precipitated.

The division of depositional environment on clastic shores is primarily based on the level of the tide and ebb, as well as the level of storm waves, and is defined as follows:

1. Backshore is the area above medium tide level which is under the influence of sea only in storm waves and high tides.
2. Foreshore is the flattened area between medium and low level of tide.
3. Shoreface is the slightly inclined zone of seabed between medium tide level and nice weather wave base.
4. Offshore transition is the zone between seabed and storm weather wave base.
5. Offshore is the area below nice weather wave base to the edge of fold on submarine slope.

Besides the ebb (movement of a tide back toward the sea) and tide currents, the activities of waves have significant influence on the processes of sedimentation in the lower clastic shores. The seashores and offshore constantly receive large amounts of sand and mud from low shore region. This sand-in-wave activity comes to foreshore and shoreface, at the same time the mud and part of the sand with return currents come to shoreface and offshore. The sand bodies of large dimensions may be potential collectors and reservoirs of oil and gas occurring in large tidal plains, shoreface, and offshore.

Tidal flats are areas on the low sea shores and low islands, which are flooded by sea during high tide and ebb, and are largely out of the main sea. The lower parts of the tidal plains are mostly flooded by sea, except for a very short time at the lowest ebb. These are the suitable places for accumulation of sands in the form of sand waves which on the surface contain asymmetric ripples and the size of grain from sea to shore reduces. The middle parts of the tidal plains are about half the time flooded by sea and half the time above sea level during the tidal cycles. The sands carried by tidal currents accumulate at the bottom. The mud carried by suspension, usually, results in flazer with wavy and lenticular bedding structure as a consequence of repeated reductions of water energy (Fig. 7.5). The upper layered surfaces in general develop desiccation (mud) cracks and current ripples. The upper parts of tidal plains are flat areas and are located above the sea level during the tidal cycle, except for a short time at the maximum of high tide when the sea is flooded for short periods. The mud deposits on it mainly as very fine detritus which could be transported with weakened tidal currents. The water currents weaken their energy and most of the sediments deposit during passage through the low and high tidal plains. Muddy deposits of the upper part of tidal flats can have more or less clearly defined horizontal lamination, numerous desiccation cracks, and bioturbation in the form of vertical rooting traces of organisms that live in the mud.

The most important characteristic of sediments deposited in siliciclastic tidal plains is the frequent vertical changes of sand–muddy sediments that are characterized by frequent changes of flazer, lenticular, and wavy

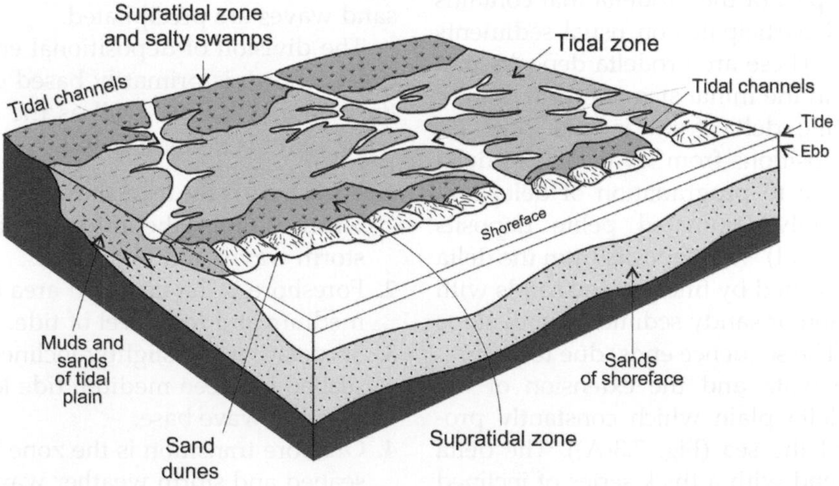

FIGURE 7.5 The morphology of the tidal plain with the system of tidal channels and sand dunes which form a barrier between the shoreface and tidal plains.

stratification (Fig. 7.5). The sand layers are deposited during the heavy flow of water at higher energy, and mud (pelite interbeds) is deposited during weak tidal currents. The underwater sand dunes (sand waves) may occur in some places on the tidal shallows by the influence of storm tides and waves. The sand dunes are mostly well developed with thickness between 1 and 2 m in wide tidal channels on the tidal plains. This could happen due to migration of sands at dominant currents, typically at strong return currents down the channels. The water drainage moves from the valley into the trough of channels at low tide.

Low sand ridges and reefs, underwater dunes, or sand waves (megaripples) occur in the lower part of the highly energetic foreshore with prevailing tidal currents where the difference between the level of low tide and high tide is greater than 1 m and on shoreface with existing tidal currents. Sand ridges or reefs are typically rugged or separated from each other by shallow tidal channels which are subsequently filled with sand (Fig. 7.6).

Long shore bars or elongated coastal sandbanks have an asymmetrical cross-section, and consist of slightly inclined (4–6°) laminae in the direction of sea. It occurs during the transition from foreshore to shoreface (Fig. 7.6). Sandbank slopes inclined toward the mainland are mostly composed of units of planar stratification inclined with an angle of 10°–30°.

"Barrier islands" and sandbanks occur along the coast with prevailing wave activity and microtidal conditions (tides <2 m) by accumulation of sand, mostly parallel to the coastline because of the activities of the waves and coastal currents. Barrier islands have the shape of long and narrow sand islands separated from each other with several permanent tidal backwaters or bays. In general, a series of disconnected sand islands or sandbanks are formed parallel to the shore (Fig. 7.6) along with narrower or wider rip channels (channels with feedback drainage of coastal currents) or tidal backwaters (backwaters through which water drains at low tide).

"Lagoon" is a shallow water body separated by barrier islands or reefs from a large body of water (sea). It may connect with open sea, be filled with tidal brackish backwaters, and have limited communication with the open sea. The muddy sediments are often deposited in the lagoon. If the barrier complex has a progradation tendency, then the coastline and/or barrier complex moves increasingly into the sea. The sandy

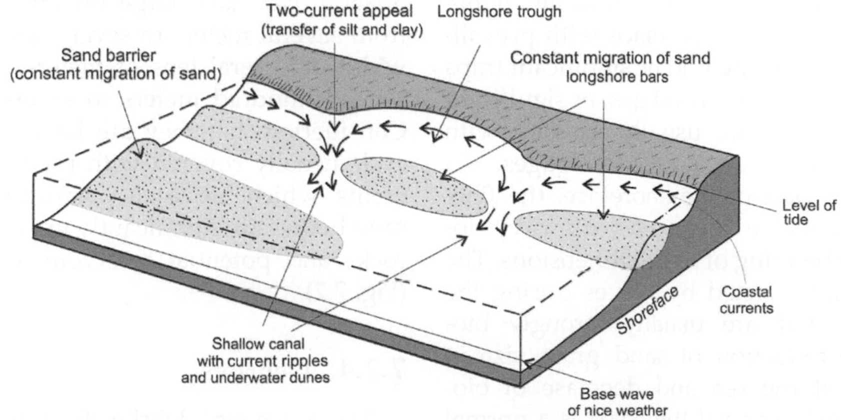

FIGURE 7.6 Sandbanks deposition on shoreface by transferring of sand with coastal currents and rip currents that occur on the confluence of two currents and operate between sandbanks in the rip channels.

sediments form a characteristic sequence of enlargement with vertical sorting of mud and sands with increasing grain size upward due to progradation of muddy deposits in offshore and/or the inner shelf precipitate barrier sands. The fine-grained sediments, from the calm water suspension, deposit in the lagoons that separate the barrier islands and the coast during the nice weather. This deposition is usually in the form of sludge and thinly laminated changes of clay and powder. The accumulation of such lagoon sludge is very slow. In the humid climates, and lagoons with warm water, mud usually contains a high proportion of organic matter, including the remains of plants, which arrive in rivers.

There may exist underwater dunes or sand waves (megaripples) oriented parallel to the coastline (Fig. 7.7). These are inclined layers of sandy sedimentary bodies that formed underwater in depths (>1 m) at high-energy shoreface, subjected to strong tidal currents (storm tides), and high-energy foreshore. Underwater dunes of larger dimensions (height between 75 and 150 cm and length between 10 and 20 m) are relatively stable sedimentary bodies that can be preserved from the resedimentation. The sediments are in the form of slanted layers of "clinostratified foresets" (whose slopes at most inclined, leeward side between 10° and 20°), often found in the sediments of the lower part of the highly energetic foreshore and shoreface with prevailing tidal currents. They could be significant traps for the accumulation of oil and gas, or significant hydrocarbon reservoirs, usually isolated with sediments of tidal plains or lagoon sludges.

In the lower part of the shoreface, the fine-grained sediments are deposited with wavy ripples and wavy bedding of small dimensions. The deposits are not affected by waves during the nice weather, and are usually strongly bioturbated. The reduction of sand grain size in the direction of the sea and decrease of bioturbidity intensity toward the coast is a normal feature. The waves affect the whole deeper part of submerged shores up to storm weather wave base during stormy weather. The currents of waves bring sediment to shore while moving to all shallow parts of the shores. Therefore, during the stormy weather at the bottom part of submerged shore there are sudden aggradations (rapid vertical accumulation of sediment), such that the currents pull from the shallow areas. The sand and mud deposited during the long period of nice weather is suddenly redeposited in the form of gradated layers at the same place or nearby in a short time, by the power of storm waves with oscillating current raised from the bottom of the lower part of shoreface. A special layer form of structure occurs by the combination of these energy conditions and the changes of two different flow regimes, one for nice and the other for the stormy weather. This sedimentary structure is most typical feature of "hummocky cross-bedding" or "hummocky cross-stratification" (HCS). It is made in fine-grained sand in association with wave ripples and fine-grained silty deposits as a consequence of the weakening storm waves and precipitation during the nice weather. The resulting layer is known as "tempestite", "tempestite sequence", or "storm layer" (Fig. 7.7).

Sand bodies in the form of underwater sand dunes, sand ridges and stripes, as well as barrier islands built on the foreshore, shoreface, and offshore can have large dimensions (thickness from several meters to several tens of meters, a width of several tens of meters, and length of several hundred meters to several kilometers). Considering that they are later in depositional cycle usually covered with mud and clay sediments which serve as insulator rocks, such sand bodies are significantly important reservoir rocks and potential reservoirs of oil and gas (Fig. 7.7).

7.2.4. Debrites

The cogenetic "debrite–turbidite" beds occur in a variety of ancient and modern zoned facies of mass flow systems with distinctive basic

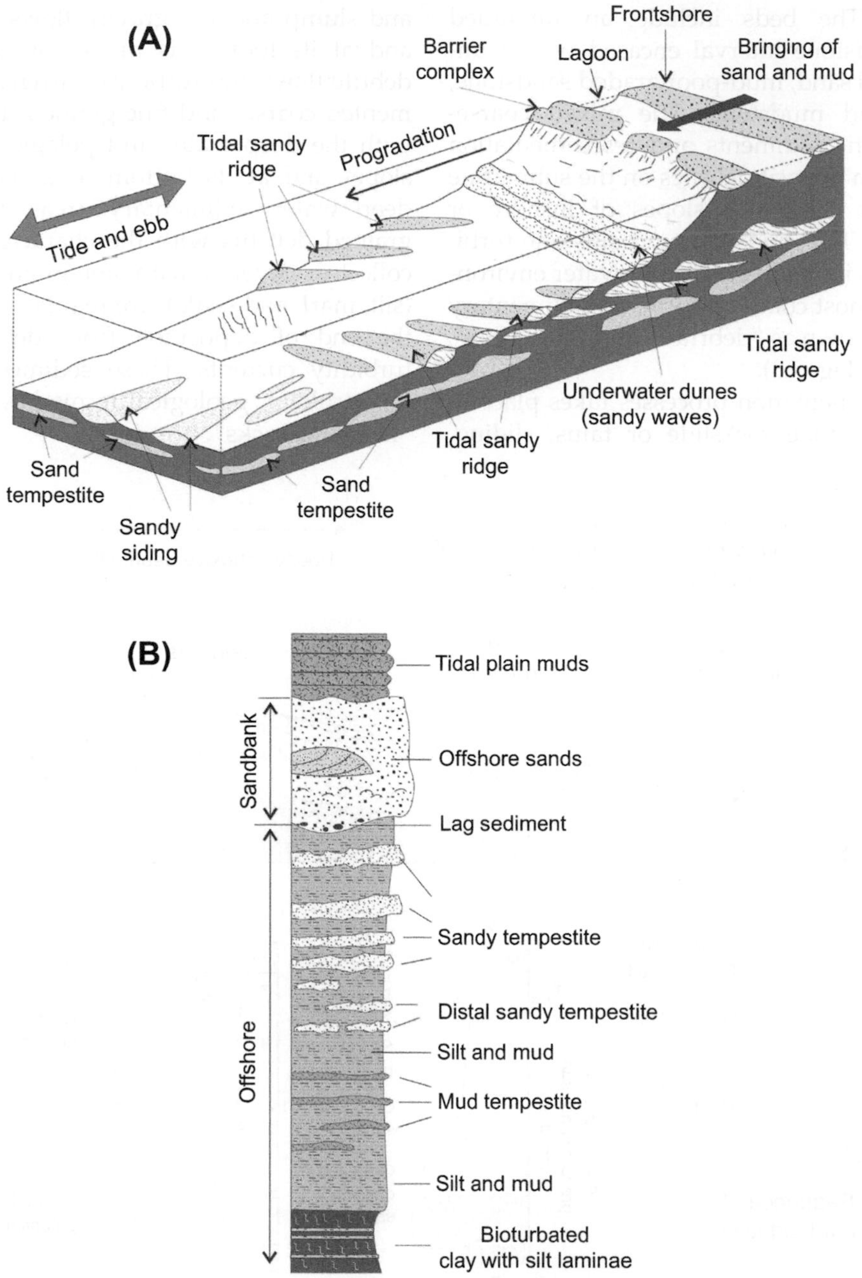

FIGURE 7.7 Conceptual diagram showing (A) offshore depositional systems with prevailing tidal action *(Source: Partially modified after Einsele[13])* due to the intensive input of sand–muddy sediments from the land and its precipitation with tidal currents. There are, according to arrangement from the coast toward the sea, deposit of large sand bodies: tidal sand ridges, underwater dunes, and sandy stripes and thinner interbeds of sand as sand "tempestite". (B) Complete enlargement of upward sequence caused by progradation of the sand barrier complex in offshore with the order of deposition of sand bodies (collector rock) within muddy sediments (insulator rocks).

characters. The beds include an ungraded muddy sandstone interval encased within unsorted gravel sand, mud-poor graded sandstone, siltstone, and mudstone. The mixed coarse-grained clastic sediments and resedimentation detritus often deposit debrites on the submarine and forereef calcareous slopes of inshore or shallow sea. The mudstone and siltstone turbidite deposits in relatively deeper water environments. The most common forms of sedimentary bodies are known as "debrites" and "turbidites" (Fig. 3.4 and Fig. 7.8).

The resedimentation processes takes place in the form of stone rockslide or talus, sliding, and slump and the gravity flows on the slope and at its foot. This mass flow is known as debrite flows and turbidity currents. The resedimented coarse- and fine-grained detritus mixes with the deep water and pelagic sediments at slopes and at its bottom or at the bottom of deep water sedimentary areas. Such coarse-grained detritus with the characteristics of the collector, covered with fine-grained sediments (silt, marl, and shale), are regularly formed after the end of deposition from debrite flow or turbidity currents. These sedimentary bodies, in petroleum geological terms have the role of "insulator rocks".

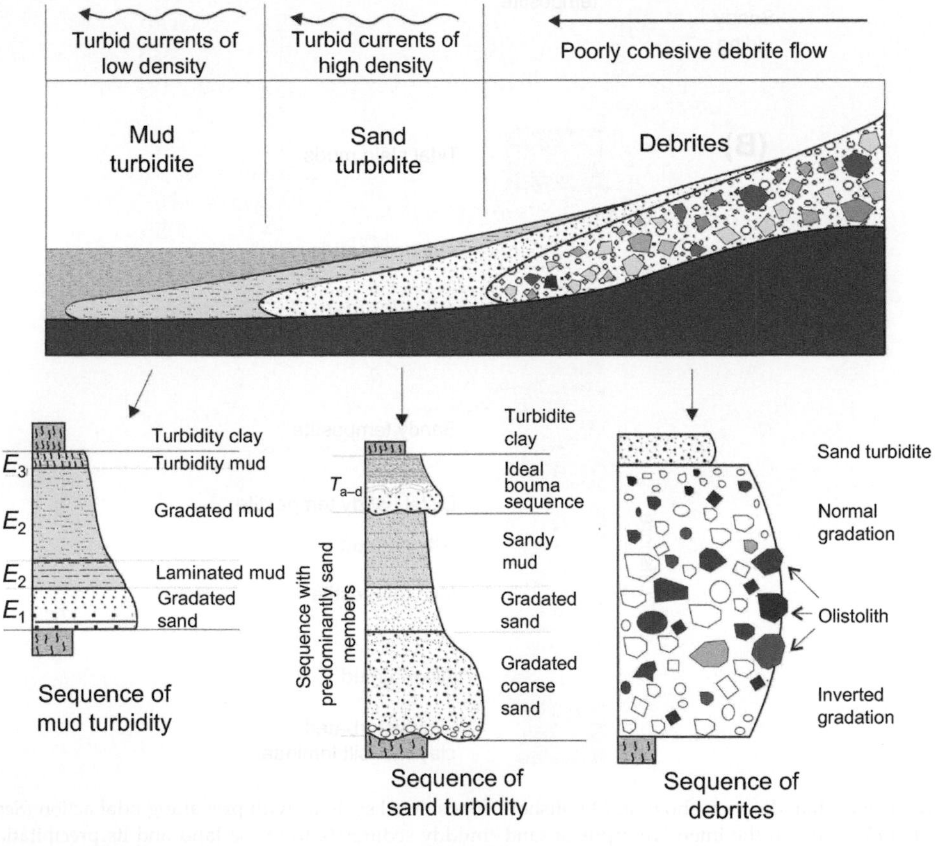

FIGURE 7.8 Idealized view of the horizontal and vertical distribution of debrite and turbidite facies on the slope and part of basin plain that occurred by sedimentation from gravitation currents with display of structure of debrites, sequence of sandy turbidites, and mud turbidites sequences. *Source: Modified after Einsele.*[13]

Debrites or deposits of debrite flows (detrite) consist of highly heterogeneous fine to boulder rocks. The constituents vary from those of the sludge which contains only a little sand detritus and clasts with dimensions of blocks ranging from a few centimeters to several tens and even hundreds of meters in diameter (olistolith). They occur at steep and vast submarine slopes or forereef slopes as a consequence of the caving or collapsing (Fig. 7.8). The main factors in foundation of debrites are tectonic processes of crushing and physical weathering of rocks at steep relief, especially tectonic grinding at faulting stress. Debrites, especially limestone debrites, are usually represented by megabreccia layers of sedimentary bodies in the shape of covers or sheets. It predominantly contains fragments and grains which have clast support and contain small amounts of fine-grained matrix (Fig. 7.8). Rockfall breccias are similar to the foundation of debrites, except with relatively short transport of clasts without formation of debrite flow. It typically just falls down the steep slope and accumulates in the immediate foot of such land. It also accrues, sometimes, underwater slopes, generally at the base of canyons or gullies at the land or seafloor. The debrite breccias also consist of materials from the poorly sorted coarse granular detritus and individual large blocks of olistolith (Section 5.5.2.2).

Debrites can cover areas from the several thousand square kilometers after moving hundreds of miles across long gentle slopes of only 1°–2°. On the steep underwater slopes and cliffs, their distribution is generally limited to slopes or foot of the slope or cliff. The thickness of debrites ranges from several tens of centimeters to several tens and even hundreds of meters. In some cases, for example, in a sufficiently long underwater slope, debrites could be directly covered by sediments or turbidites of muddy flow rich in powder-muddy matrix that contains only few individual larger clasts. In this way, debrites usually exceed in coarse, sandy, and often even mud turbidite (Fig. 7.8).

7.2.5. Turbidity Fans

The sediments deposited from turbidite currents are called turbidites (Section 5.2.2.1). Their basic lithological and sedimentological features are regularity of vertical sorting detritus of specific granulometric and textural–structural characteristics and form fan-shaped sedimentary bodies (Fig. 3.4, Figs 7.8 and 7.9), as a consequence of sedimentation way from turbidite flows.

The largest grains, which have been transferred by dragging, from the lower part of the turbidity current deposit initially and quickly (Section 5.2.2.1). Thereafter, all the smaller grains, which have been dragged from forehead of turbidity current and in suspension, travel longer and slowly deposit. There will be sediments with gradational bedding (graded coarse sand in sequence of sandy turbidites in Fig. 7.10). Once all the materials are deposited from forehead and immediately after forehead of turbidity current, which usually has a vast number of towing large granular gravel or sand detritus, the speed and turbulence of the current are reduced. The remaining small sandy grains transfer only by suspension. At this stage, mainly sand-sized detritus and occasionally silt deposit in the form of thin horizontal laminae of sand. The sand grains are no longer carried in suspension with further decrease of current speed and are transferred by dragging at the bottom. So it begins deposition of fine-grained sand detritus and some clay particles with the formation of wavy, irregular, or sloping bedding (Fig. 7.10).

The turbidity currents transfer only detritus of dimensions of clay and dust in suspension by dragging negligible amount of fine sand grains. The flow velocity will be low and slow and sediment deposited at this stage presents poorly parallel bedding. The current speed will be so slow at the end and it can transfer only the smallest particles of silt and clay in suspension with slow deposition. The small detritus from muddy

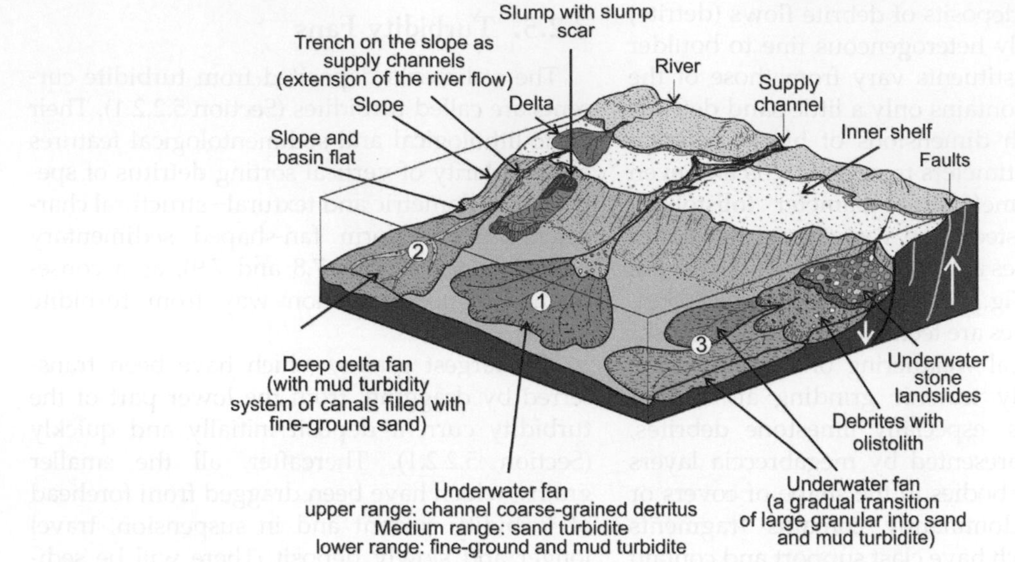

FIGURE 7.9 Main shapes of sedimentary bodies of "submarine fan" at the seafloor formed by the deposition of detritus from gravity flows. 1, supply channel; 2, trench at the slope as the extension of the river flow with the delta to offshore and inner shelf; and 3, slump scars and slumps at submarine slopes. *Source: Modified after Stow*[45] *and Einsele.*[13]

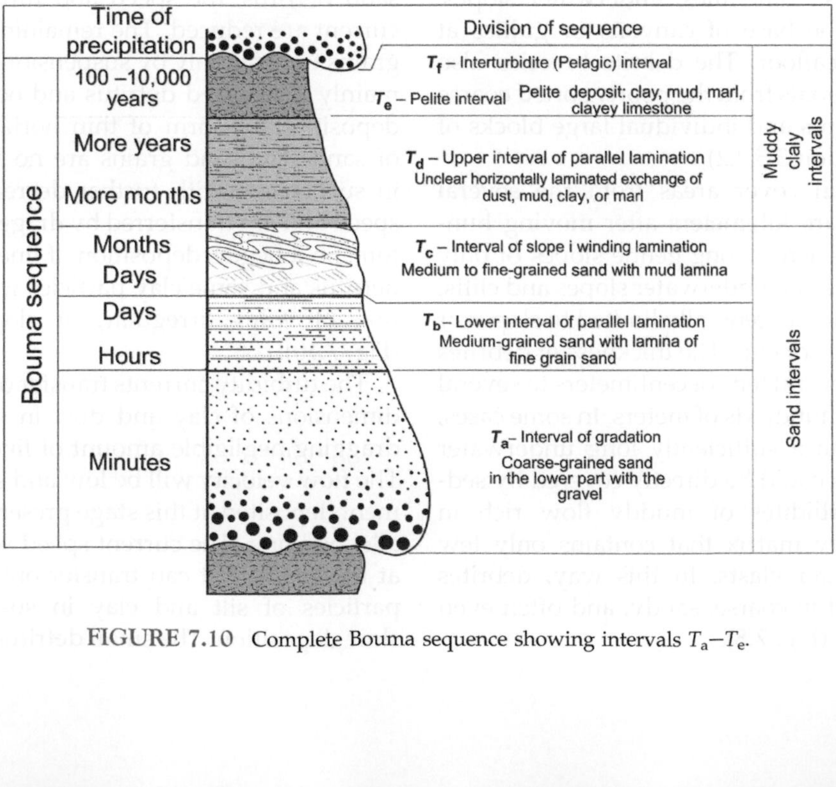

FIGURE 7.10 Complete Bouma sequence showing intervals T_a–T_e.

flow deposits regular basin material, mostly carbonates and skeletons of planktonic organisms. The deposition is called "Bouma sequence" (Fig. 7.10) to explain the origin and composition of such sequence deposited from turbidite flows.

Bouma sequence includes a complete depositional unit of medium grain turbidites formed by detritus from a turbidity current under optimal conditions of its development. It is characterized by specific vertical sequence of sediments with the specific granulometric composition and textural characteristics the bottom upward consisting of five turbidite intervals: T_a, T_b, T_c, T_d and T_e. The first three (T_a = grade interval, T_b = the lower interval of parallel lamination, and T_c = interval of wavy lamination) consist of pebbly sand or sandy sediment, the fourth (T_d = upper interval of parallel lamination) of muddy and/or marly deposits, and the fifth (T_e = pelite interval) of clayey limestone sediments.

The entire space occupied by material deposited from turbidity current is called turbidity fan, because the shape of sedimentary bodies looks similar to fan (Fig. 7.9). If the turbidity currents transfer material to the basin from the same source over time, i.e. within one larger turbidity fan, usually we speak about two areas: one near the source of material, the proximal region, and another away from sources of material, the distal region. The distal region mostly considers the flattened parts of the basin away from the slopes or peripheral parts of the turbidity fans. The proximal parts of turbidity fans contain Bouma sequences with the very thick gravel–sand intervals T_a, T_b, and T_c. The distal parts of fans contain very thin sand intervals or only contain pelite intervals—usually consisting of very thin interval T_c and slightly thicker intervals T_d and T_e.

The study of turbidites generates great practical value, especially when exploring reservoirs of oil and gas. This is because the upper proximal parts of turbidite fans, rich in sandy intervals T_a–T_c, are generally good potential collectors. On the other hand the lower distal parts in which clay intervals dominate T_d and T_e, are excellent insulator rocks, as illustrated by the schematic arrangement of debrites, sandy turbidites, and mud turbidites in Fig. 7.8.

7.3. MAIN FORMS OF COLLECTOR SEDIMENTARY BODIES IN CARBONATE ROCKS

Limestones and dolomites are important primary collectors of oil and/or gas among the carbonate rocks. These hydrocarbons are associated at primary porosity as a result of sedimentary processes. The greatest significant components are high-energy shallow water carbonates, peritidal carbonates, carbonate bodies of reef limestones of carbonate platforms, outside carbonate platforms, reef limestones and bioclastic reefs biocalcarenites, as well as the calcareous turbidites and debrites or allodapic limestones.

7.3.1. Carbonate Platforms

The most important forms of sedimentary bodies of limestones and dolomites as collector rocks of oil and gas are high-energy carbonate platform with limestone sandy bodies; peritidal carbonates; limestones of restricted shoals, lagoons, and inner shelf; as well as the ridge carbonate in particular, and on the slopes of platforms; carbonate debrites and turbidites or allodapic limestones.

Most of the carbonate sediments are made or formed in an environment with the long-term conditions almost exclusive of carbonate sedimentation which are generally known as the carbonate platform. The "carbonate platform" implies a vast area in which long-term conditions and environments of shallow marine carbonate sediment depositions are maintained, which results in the emergence of large thickness of these rocks. Shallow sea conditions are

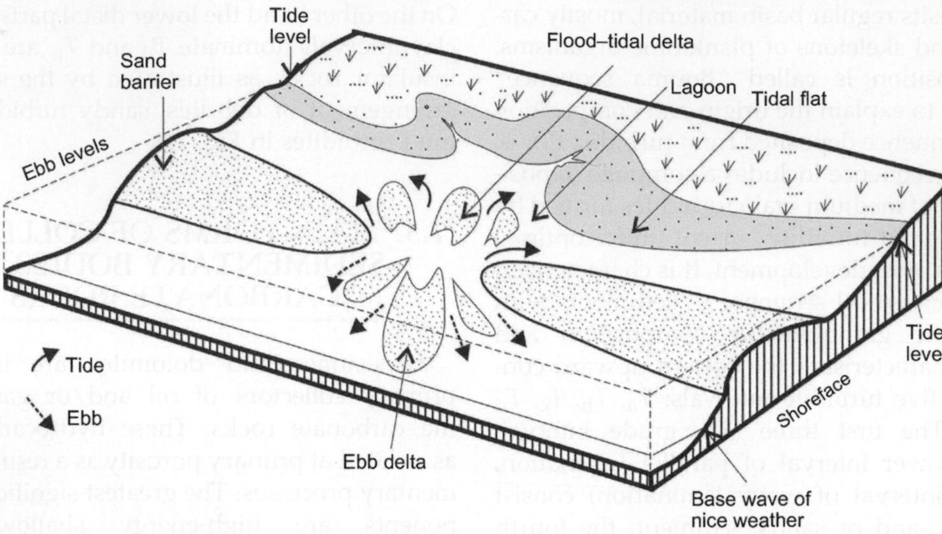

FIGURE 7.11 Conceptual model of carbonate depositional system of carbonate sand bodies on high-energy shore above the nice weather wave base which contains sand barrier islands, lagoon with sand deltas generated by activity of tides on flood-tidal delta, and deltas formed by activity of ebb on exit of channel (ebb delta) into shallow sea.

maintained with the deposition of several hundreds or thousands of meters thick shallow marine carbonate sediments on the carbonate platform during one or several geological periods, as for example, the case at Adriatic carbonate platform of older Jurassic to Paleogene period.

Shallow sea carbonate platform depositional system includes high-energy shallow water environments, peritidal environments, restricted shoals, lagoons, the environment of "inner shelf", as well as the complex environments of the ridge.

7.3.1.1. Carbonates of High-Energy Shallows

The formation of large carbonate sand detritus and systems of carbonate beach-barrier islands occur in the platform shallows in areas with middle- and high-energy waves, if the differences of tide and ebb are less than 3 m. There are lagoons, and behind these barrier islands, there are the tidal plains that connect to the open sea with tidal backwater or tidal channel (Figs 7.11 and 7.12).

The typical sediment of platform shallows, with high energy of water, is bioclastic detritus composed of skeletal grains and peloids, interspersed with a significant proportion of ooids (Fig. 5.51). The sediment is dominated by skeletal, peloid and/or ooid greystone. The grain size of detritus is usually increased from shoreface toward foreshore. However, concentration of coarse granular skeletal detritus is still visible at foreshore, as well as reduction of the share of fine-grained peloid detritus, especially of carbonate mud, due to their rinse with strong tidal currents, storm tides, and waves. The general increase in grain size in the shore direction, the increases in the level of sorting, and the degree of roundness of bioclastic detritus due to increased abrasion occur in parallel. The important features of limestones deposited at foreshore or in high-energy tidal zone and on tidal bars are cementation-type "shore rocks", cementation of grains with fibrous and acicular aragonite, and

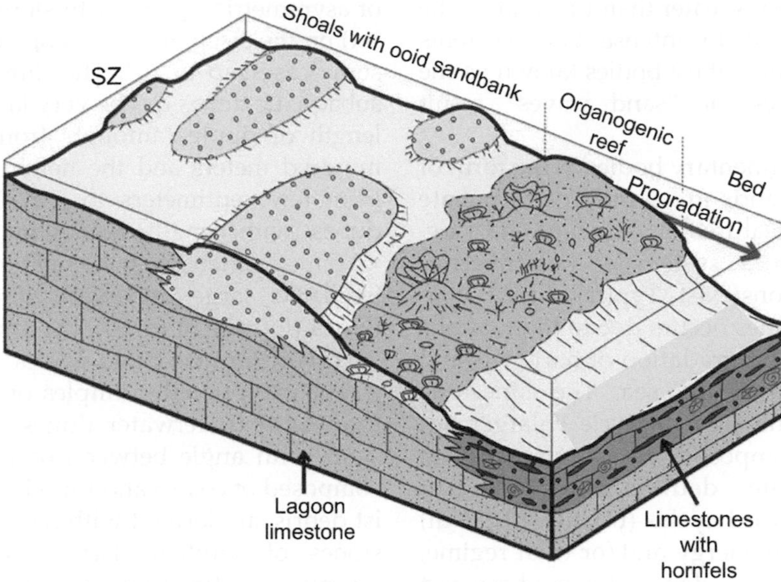

FIGURE 7.12 Interpretation of depositional environments in the younger Jurassic of Gorski-Kotar mountainous region (Croatia) with the ooid sandbanks in platform shallows and organogenic reefs at edge of the intraplatform trough; SZ, Shallow Zone. *Source: Ref. 52.*

micrite composed of high magnesium calcite cement. The characteristic depositional feature at the upper part of the foreshore is well-sorted carbonate sand with the planar layered surfaces. The rocks contain fine-grained well-sorted pellet, skeletal, ooid, or bioclastic greystones. The upper layered surfaces show asymmetrical wave ripples of small amplitude and frequent tidal channels filled with coarse sediments (greystone, rudstone, and coarse limestone).

These are the main environments of accumulation of ooids (ooid greystones) and bioclastic and/or skeletal carbonate sands with relatively small proportion of carbonate muds (bioclastic, skeletal, and peloid greystones). Ooid and skeletal detritus are derived from crushing and breaking of skeletons of mollusks, echinoderms, foraminifera, and corals.

Tidal deltas are carbonate sand bodies formed by the accumulation of carbonate sand at the mouths of tidal channels and tidal backwaters into the lagoon (Fig. 7.11). They have a clear cross-stratification of large, medium, and small size with slope of inclined layers mainly in the direction of the coast.

Deltas created by the current tide (ebb delta) are carbonate sand bodies at the exit of tidal channels and backwaters toward the sea. These deltas are formed by sand transfer with water currents at ebb, i.e. by strong withdrawal of water from the lagoon through tidal channel or tidal backwater at low tide (Fig. 7.11). This is characterized by clear cross-stratification with general inclination slanted layers in the direction of the sea.

At high-energy water shoreface, the carbonate platforms are commonly found at depths of 5–20 m in the area between low tide level and nice weather wave base. The high-energy shallow water sedimentary bodies can occur in the form of sand barrier (Fig. 7.11) and ooids and/or bioclastic sandbank (Fig. 7.12). At high-energy foreshore with existing tidal currents, where the difference between the level of low tide and high tide is greater than 1 m, and on the shoreface subjected to intense tidal currents (storm tides), sedimentary bodies known as the underwater dunes, or "sand waves" result ("megaripples").

Carbonate sedimentary bodies in the form of sandy barrier islands are typical of carbonate shoals with the features of shoreface and foreshore on which a system of beach-barrier islands and lagoons exist. Typical vertical continuity of limestones occurs due to fluctuations of sea level and progradation of barrier island in the direction of the sea. The limestone sequence is in the form of cycle enlargement upward and composed all the larger and coarser carbonate detritus. Thickness of enlargement upward cycles (usually 10–30 m) depends on wave energy and/or tidal regime, which determines the depth of wave base and the height of beach. The sequence of cycles from bottom up consists of:

1. mudstone with the features of storm sediments deposited in deeper part of subtidal or offshore;
2. peckstone and greystone with HCS, deposited on a submerged shore or shoreface;
3. coarse greystone and rudstone deposited on the foreshore; and
4. on the top of the cycle sometimes and coarse-grained greystones with cross-stratification of large dimensions.

Carbonate submarine dunes, except on the foreshore with incidence of high tidal currents, occur on the shoreface subjected to intense storm tidal currents and/or storm waves. They have the shape of asymmetric ripples with sloping layers (foreset) on the steep, leeward slope of ~10°, and in some cases up to 20°. The dimensions of such subaquatic dunes can be very large, and so their length of ripples "*L*". Carbonate submarine dunes, except on the foreshore with incidence of high tidal currents, occur on the shoreface subjected to intense storm tidal currents, and/or storm waves. They have the shape of asymmetric ripples with sloping layers (foreset) on the steep, leeward slope of ~10°, and in some cases up to 20°. The dimensions of such subaquatic dunes can be very large, so that their length of ripples amounts from 10 to several hundred meters and the height of ripples "*H*" from few centimeters to 5 m. The subaquatic dunes with length L <20 m and height H <1.5 m are often given the old term megaripples. Similarly, large and very large underwater dunes (L from 10 to >100 m and H from 1.5 to >5 m) are termed "sand waves".

The exceptional examples of slanted layered foresets of underwater dunes of large dimensions with angle between 6° and 10°, mainly composed of coarse and fine clear gradated rudist debris, are located within Cenomanian limestones of southern Istria and in Vinkuran quarry, Croatia. Cenomanian limestone contains very notable fossil faunas including ammonites that are scarcely known elsewhere in Britain.

7.3.1.2. Peritidal Carbonates

Peritidal carbonate sediments include carbonates deposited on the carbonate platforms in low-energy subtidal and supratidal environments, especially those deposited in the intertidal zone and tidal flat, i.e. peritidal environments. Peritidal environments include different low-energy environments on carbonate platforms in which the action of tide and ebb currents is felt (peritidal area around the high tide and low tide). The peritidal carbonates are classified into three main zones (Fig. 7.13):

1. *Subtidal zone* is the lowest part of the tidal plains or peritidal areas of low-energy lagoons or restricted shoals which is always located below the low tide level at a depth of <10–15 m at optimal ecological conditions for the development of green algae.
2. *Intertidal zone* and tidal plain is the part of the tidal flat located above the low tide and below the normal tide levels.

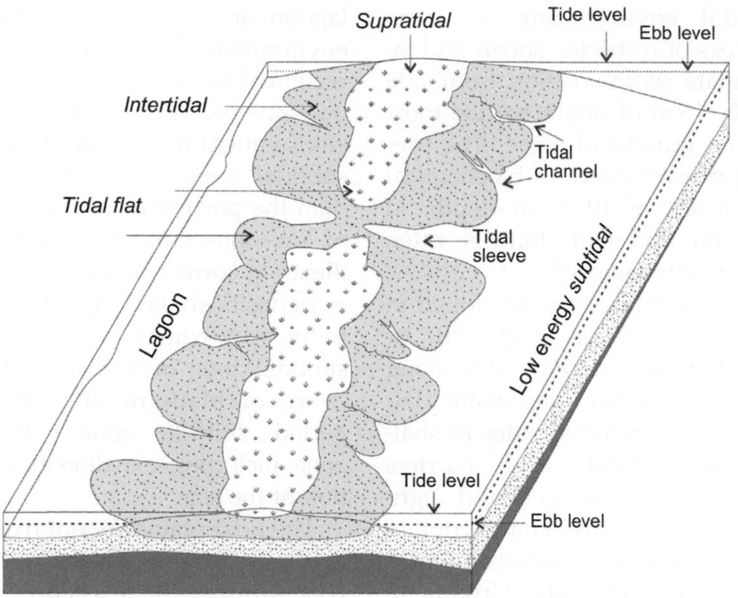

FIGURE 7.13 Conceptual diagram showing peritidal environment of deposition on carbonate platform and classification.

3. *Supratidal zone* is the highest part of the tidal plain or mild, flat shores with low relief. It is only occasionally flooded by sea and typically during the stormy waves, or the occasional high tides, or is only damped with seawater due to high waves splashes. The inshore Sabkha environment is normally considered to be in this zone.

Carbonate sediments deposited on supratidal environment are represented by changes of limestones and early diagenetic dolomites. The characteristic structures for these rocks are "fenestral" and "geopetal" with internal sediment, irregular laminations occasionally potentiated with thin bituminous films, desiccation cracks, and inserts of desiccation breccias and intraclastic rocks. These rocks occur in the accumulation of detritus flooded with high tides and/or by storm waves on supratidal zone, i.e. storm tidal sediments. The fossil remains are few in number or completely absent.

The intertidal zone, also known as the foreshore, seashore, or littoral zone, is the area existing above water at low tide and underwater at high tide or the area between high and low tide marks. Intertidal environments are very gently inclined or nearly flat areas along the coastline or along the coast of low islands between the levels of normal tides (Figs 7.11 and 7.13).

The supratidal zone, also known as the splash or spray or supralittoral zone, on the other hand, is the area above the spring high tide line on coastlines and estuaries that is regularly splashed, but not submerged by ocean water. Seawater penetrates these elevated areas only during storms with high tides. The supratidal zone slopes toward the sea-facing side on shallow subtidal zone and protected low-energy lagoon or bay (Fig. 5.70 and Fig. 7.13). An important feature of the tidal zone and tidal plains is their constant interchangeable flooding with seawater during tides and their emergence during ebb.

Shallow subtidal environments including shallow marine areas of restricted shoals and lagoons with carbonate sedimentation from tide level and up to 10–15 m of depth are the most suitable for optimal growth of algae. The predominant portion of carbonates of biochemical origin results in a sea of 10–15 m depth, in shallow subtidal environment, that is often referred to as "carbonate factory".

Tidal plains are intertidal areas developed in large surfaces along coastline (Fig. 5.70) or around islands and sandbanks (Fig. 7.13), along the protected side of barriers on carbonate platforms, as well as along protected sides in shallows and lagoons behind sand barriers (Fig. 7.11) and/or organogenic reefs and high-energy sandbanks. Tidal plains usually contain numerous tidal channels that are constantly connected with the lagoon or subtidal environment.

Progradation is one of the most important sedimentary processes in peritidal environments, especially tidal plains, tide, and supratidal zones. This occurs when the mass balance of sediment into the sedimentary deposition is such that the volume of incoming sediment is greater than the volume of the sedimentary platform that is lost through subsidence, sea-level rise, and/or erosion. In order to make progradation possible, it is essential that inflow and accumulation of sediment is always greater than the transfer away (erosion) of sediment from the area with the stagnation or low amounts of sea-level lift. The ratio of sinking (subsidence) of the carbonate platform is less than the portion of deposition. This, along with relatively lowering of sea level, leads to rapid filling of the lagoons or shallows and shifting of the coastline in the direction of the sea (Fig. 5.1). The progradation occurs also by gradual mild global or relative lowering of sea level due to tectonic and other allocyclic processes. The environment with greater depth of water, such as lagoons and shallow subtidals, gradually fills with sediments and passes in environments with shallow water such as intertidal or supratidal. This happens due to filling the lagoon or shallow subtidal emerged intertidal environment. The coastline also moves in the direction of the sea. The consequence of the progradation, i.e. migration of shallow marine environment in the direction of the sea or in the subtidal zone and larger portions of deposit than the portion of subsidence, is the formation of limestone deposits. The limestone sequences illustrate correct vertical sorting of certain types of deposits shallowing upward peritidal cycles. In the approximate portions of deposition in the subtidal zone of 30 m in 1000 years and subsidence of platform of 20 m during the same period, subtidal zone will become the tidal zone, including coastline movement in the direction of the sea.

The tidal plains are not fixed to one place, but gradually migrate on the tidal plain or subtidal zone during the existence and progradation. The low-energy tidal plains and tidal zones are typically shallow, mostly less than 3 m deep and up to 100 m wide. Tidal channels form on the tidal and intertidal plains (Fig. 7.13). The tidal channels are shallowing and narrowing at the lower end toward upper part of the tidal zone or tidal plains. The tides of seawater penetrates deeply into the tidal plain, and it can strongly erode its deposits, and also brings carbonate detritus from the lagoon or subtidal zone. The lateral migration of tidal channels in relation to progradation of tidal plains is very slow. They still cause a specific type of deposition, filling of tidal channels, and thus the vertical sorting of sediment in the form of sequences fining and shallowing upward. The bottom of the channel is normally filled with coarse-grained detritus in the form of lag sediment composed of intraclast semilithified peritidal sediments, as well as large bioclasts remaining after rinsing of sludges and sandy limestone. The majority of the channel fill muddy carbonate sand (pelletal–skeletal wackestone to peckstone) has been highly bioturbated. The typical laminated peritidal carbonates deposit over the tidal channel after the channels are completely filled.

The main feature of the tidal zone and tidal plains on carbonate platforms, in terms of low energy, represent thinly laminated calcareous muddy sediments with an abundance of cyanobacteria that form stromatolite, algal laminations, and desiccation cracks. The typical phenomenon of tidal limestones often stands for fenestral composition; that is emergence of a number of pores of irregular or elongated shape that are partially or completely filled with cement and internal sediment. This incurs as a result of rotting cyanobacteria meadows on intertidal sediments followed by lithified carbonate mud. It can also be caused by cracking or peeling of mud at its sudden drying and accumulation of gas bubbles in sediment.

Subtidal limestones in association with tidal and supratidal limestones, as peritidal carbonates, often appear as the initial members of the shallowing upward cycle. Peritidal carbonates are often due fenestral and "vuggy" porosity, good reservoir rocks.

7.3.1.3. Carbonates of Restricted Shoals, Lagoons, and Inner Shelf

Shoal or sandbar or sandbank and gravelbar is linear landform within or extending into a body of water (river, lagoon, lake, and sea), typically composed of sand, silt, or small pebbles. Environments of restricted shoals, lagoons, and calm water shoal are included in the inner part of the carbonate platform ("inner shelf"). It also includes low-energy shallow water environments of carbonate deposition that are separated from the open sea by some morphological barrier, such as reefs or sandbanks. These also include vast low-energy shallows on the inner part of the carbonate platform without direct influence of water mass of the open sea. It is characterized by elevated water temperature and little elevated salinity. Lagoons almost completely, and bays partially, are closed from shallow sea areas by barrier islands, reefs, and system of tidal plains. The connection to the open sea of lagoons and bays is extremely limited and poor. The water depth is not strictly determined as it depends on the morphology of the seafloor and shallow sea system. While the term shoal and lagoon normally implies shallows with low energy of water and more or less elevated salinity, the calm water shoal (or low-energy shoal) involves mainly shoal with low energy of water and normal salinity.

Restricted shoals and lagoons are characterized by an abundance of algae, benthic foraminifers, calcareous sponges, hydrozoa, bryozoa, worms, and gastropods, which are adapted to life in terms of photic or euphotic zone exposed to sufficient sunlight. The other components and conditions of limestones are abundant in low-energy carbonate mud, pellets, peloids, aggregate grains, and algal oncoids. These limestone deposits are mostly rich in carbonate mud, fecal pellets, oncoids, green algae, ostracods, benthic foraminifers, gastropods, and thin testaceous bivalves. Intensive oncoid envelopment of the skeleton, and bioerosion processes, capture, and paste of fine sediment on thin fibers and mucus of cyanobacteria, are especially characteristic of the restricted shoals and lagoons. Grains, either skeletal or nonskeletal, that are embedded in the carbonate mud, are poorly sorted, unrounded, have low degree of sphericity, and are generally without any particular orientation. They have no grain but have muddy support (wackestone, peckstone, to floatstone).

"Inner shelf" encompasses shallow sea environments with predominantly low energy in a very wide area of inner part of carbonate platform, from the coastline to a depth of 50–200 m (photic border zone) and/or outside part of shelf with or without a ridge complex. The limestone deposits of restricted shoals, lagoons, and generally protected part of the carbonate platform (inner shelf) are often very bioturbated (see Section 5.3.1.5) and contain abundant traces of rooting and digging, and burrow-making. They may all belong to one type or have several different types of organisms. Such deposits often become blotchy due to intense bioturbation, as for instance, the case of "blotchy limestone" of the Middle Liassic around the coasts of United Kingdom and Yugoslavia.

The final result of such intense bioturbation is complete homogenization of deposits without any clear individual traces of rooting, digging, and burrow-making.

In petroleum and geological terms, peritidal and lagoon carbonates can be extremely brilliant source rocks for oil and gas because of their high content of organic matter.

7.3.1.4. Carbonate Bodies of Reef and Perireef Limestones in Carbonate Platform

Reef complex includes environments that form large and resistant organogenic reef that has great impact on waves, topography, and relief as illustrated in Fig. 7.14 and is expressed as follows:

1. Reef slope is the relatively steeply sloping surface or the very steep side of reef.
2. Forereef environment is the high-energy area of the front, facing the open seaside of organogenic reef or reef barrier.
3. Reef front is the front part of organogenic reef which is exposed to waves, and its highest and most exposed part is reef crest.
4. Solid reef core is the central part of organogenic reef and reef flat, which is composed of skeletal reef grid with most resistant reef-making organisms.
5. Back-reef environment is the shallow sea area at the rear side of reef or reef barrier that is unlike forereef and reef front, protected from direct wave action.
6. Back-reef lagoon is the shallow sea, protected from waves with reef area at low energy of water, and occasionally with high salinity.

The organogenic reef formations occur along the edges of the carbonate shelf, platforms, or ramps where they form distinctive carbonate depositional system, in morphology, relief, and dimensions including its ecological, biological, climatic, chemical, physical, and hydrological features. The biological, physical, and chemical processes with very different mutual interactions participate in the formation of reef and reef complex. These processes can be grouped into four main types: (1) building or construction,

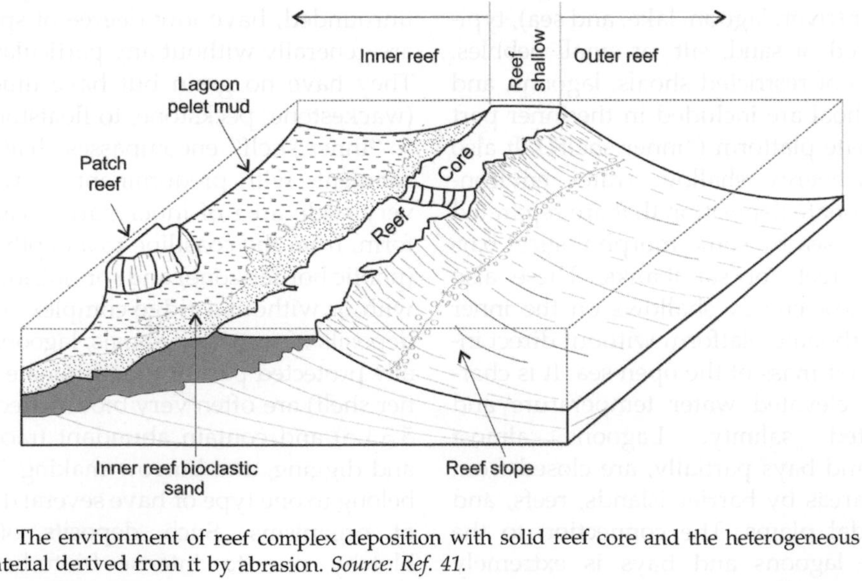

FIGURE 7.14 The environment of reef complex deposition with solid reef core and the heterogeneous and contiguous fragmentary material derived from it by abrasion. *Source: Ref. 41.*

(2) destructive or processes of reef degradation, (3) sedimentation, and (4) cementation.

Building or construction processes include biogenic processes of growth of reef-making organisms with carbonate skeletons. Primarily it is the accumulation and binding of sediments through the organisms that live in colonies. The life forms are eventually lithified during the positional growth along with their compete habitats. Abundance of holes occurs due to growth of irregular skeletal grids during the building process.

Destructive processes or processes of reef degradation cause damages and degradation of reef growth due to its physical destruction by the activity of waves and sea currents and bioerosion. Mutual relations between the skeletal grid intensity and speed of growth, physical destruction of reef with waves, and intensity of bioerosion are key factors in the development of reef. This maintains the balance between construction and destruction process at their creation. The physical degradation of reef is continuous and parallel process with their construction due to abrasion activities of the waves and ocean currents. It is particularly intense during the occasional storm and hurricane waves that can destroy large quantities of organogenic reef and colonies of organisms forming the skeletal reef grid. Bioerosion processes deploy drilling, scraping, breaking, and digging in life activity of different organisms. These are slow processes of degradation of reef that, however, become very important factor in their degradation passing through long time. It finally leads to the emergence of cavities, causing a high permeability and porosity of reef limestones. In the destructive processes, a huge amount of reef biodetritus of different grain sizes arises, which is deposited near or further away from the reef itself.

The formation of reef are the result of deposition and accumulation of large amounts of detritus during the sedimentary process. The volume of detritus is generated by abrasion and bioerosion of reef skeletal grid and generally organogenic reef creations. The detritus, that originates outside of reef environment or detritus drifted from neighboring environments, involves the sedimentation process to a lesser part. The interskeletal pores and caverns in the reef, and cavities formed by bioerosion, are very quickly filled with internal sediments. The sediments are carried by sea currents and stamped with hitting of the waves and tides in the cavity surface of the reef, especially on reef front or reef crest, in the form of multiple generations of internal deposits.

The cementation processes play a significant role in the formation and lithifaction of reef. The process is active in the early phase of reef emergence with the direct participation of seawater in the pores of the ridge structure. Cementation is most intense on the reef front and the reef crest. The saturated seawater, brought by wave activity, constantly inserts, retains, and evaporates in the pores and cavities. The fossil reef limestones are very significant reservoir rocks of liquid and gaseous hydrocarbons (oil and natural gas) due to their high porosity. The processes of their cementation are subject to numerous complex physical and chemical conditions. In general, we can say that in most of the fossil reef limestones the main types of cement are located as early diagenetic marine aragonite and magnesium calcite cements.

Reef front and reef core are made of nonlayered and massive organogenic limestones mostly built of the skeletons of reef-making organisms with well-preserved skeleton (reef) grid. At least 10–40% of skeleton remained lithified at the site of growth that includes various forms of bioherm, biostrome, or biolithite. The remaining part of most of the rock mass makes the larger or smaller skeletal debris of reef-making organisms, and the cavity filled with cement and/or with internal sediment the "stromatactis structure".

In forereef environments, coarse, poorly sorted, and unrounded skeletal debris of reef-making organisms and reef limestones precipitate. These components are created in the destruction of reef due to heavy abrasion by wave action, especially during storm and hurricane bad weather. Limestones, deposited in forereef environment, are characterized by poorly expressed slope stratification of large dimensions, with layers that are inclined toward the open sea, i.e. down the reef slope at an angle of 35°–60°. The incline slope depositional units—foreset—usually have a wedge shape. The accumulation of large amounts of reef biodetritus in forereef environment, especially right beside the reef slope, create sloping stratified sedimentary bodies with a wedge form of foreset, leaning down the slope. The continuous deposition of these sedimentary bodies place one over the others from reef in deeper sea. Reef slope gradually fills in and becomes a shallow sea over which it continues expansion or progradation, particularly in the phase of lowering of sea level as shown in Fig. 7.15. The main limestone type of forereef environment is reef breccias, bioclastic rudstones and floatstone made of biolithite limestone fragments (centimeter to millimeter), and skeletal grid and skeletal debris of reef-making organisms.

Back-reef environment, especially back-reef lagoons, are characterized by the key feature of low-energy water, appropriate for the deposition of only fine-grained carbonate detritus and life of organisms possessing significantly different ecological features than the organisms living in the reef core and reef front. The back-reef lagoons are especially favorable for the development of calcareous algae and seaweed. The limestone deposits of black-reef environment are mainly composed of fine-grained, well-rounded, and medium-sorted reef skeletal debris, shells of some foraminiferas, skeletons of green algae, oncoids, and peloids. The dominant types of sediment are bioclastic sands i.e. greystone to peckstone, and away from reef in the lagoon often wackestone and even mudstone (pellet mud) precipitate. The back-reef sediments move away from the reef core and gradually pass into the micrite-rich lagoon limestones. The circulation gets weaker and salinity rises higher as the energy of water lowers. The

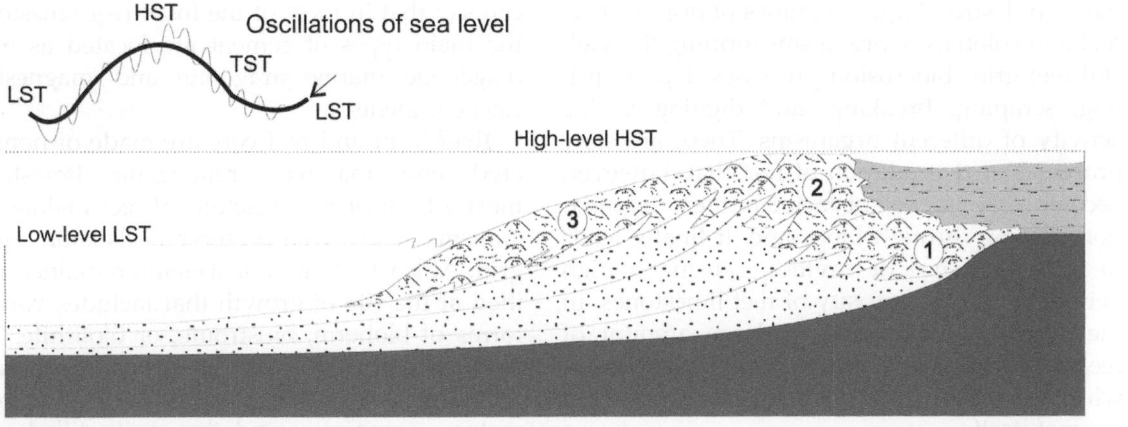

FIGURE 7.15 Development and growth of organogenic reef depending on the relative sea-level fluctuations: Development of the reef body as a complete depositional sequence occurs during a complete cycle of relative sea-level change: (1) low state of sea-level (LST) and high state of sea-level (HST), characteristic progradation of reef over forereef clinostratified sediment; (2) transgression (TST), characteristic aggradation growth shown in figure above; and (3) reduction of sea level, back to the low state (LST) in which highly prevalent progradation of reef body over forereef sediments and where reef-making organisms follow reduction of sea level and grow only at favorable depth.

sediments will be higher in proportion to fecal pellets, aggregate grains, skeletons of gastropods, ostracodes, and, certainly micrite and algal oncoid. The back-reef sediments will actually assume characteristics of sediments deposited in lagoons and restricted shoals.

Reef–perireef limestones of carbonate platforms have great significance as reservoir rocks for liquid and gaseous hydrocarbons (oil and gas) due to their great dimensions and high porosity. The economic potentiality of reservoir excels with the subsequent precipitation processes of impermeable deep sea or lagoon silty limestones acting as cover/insulator rocks.

7.3.2. Carbonate Debrites and Turbidites or Allodapic Limestones

The carbonate debrites consist of clasts of various dimensions and fine- to medium-grained marl or clay matrix in such mutual ratio that carbonate clasts include matrix support (clasts "swim" in fine-grained matrix). The carbonate clasts originate from the strong physical weathering and erosion of carbonate rocks outside the depositional area. It belongs to extraclasts, and partly from the erosion of some older sedimentary rocks within the same depositional area or intraclasts. The limestone debrites occur on steep and vast slopes of carbonate platforms or on forereef slopes as a consequence of collapse process or collapsing due to brecciation process of platform carbonates in tectonic deformations. This generates during the course of the formation of great systems of normal faults in the Earth's crust and tectonic crushing in overthrusting. The limestone debrites are usually represented by layers of megabreccias with sedimentary body in the shape of cover or sheet. It is mainly composed of carbonate clasts which have clast support and contain small amounts of fine-grained matrix, such as "stone dust" of finely disintegrated limestone and/or dolomite rock debris.

The limestone turbidites or allodapic limestones, deposited from turbidity currents, are accumulated in environments of deeper sea on slopes of same mechanisms and clastic sediments. The limestone turbidites deposited from turbidite currents of high density, i.e. calcirudite (coarse-grained limestone) or limestone breccias, are restricted to the edges of the basin with steep slopes and on forereef slopes. They are characterized by textural features, typical of coarse turbidite. The limestone turbidites deposited from low-density turbid currents encompass characteristics of Bouma turbidites with sequences of thickness between 50 and 30 cm. The submarine distribution channel generally exists on slope, and turbid fan-shaped sedimentary body will generate from these turbidity currents. The deposits are composed of Bouma sequences with interval T_{a-c} containing coarse limestone and sandstones with or without shallow marine reef and perireef fossil detritus, T_{d-e} intervals of mudstone and/or wackestone, and occasionally clay mudstone and wackestone and/or marl and marlstone. There may be significant differences between the limestone and siliciclastic turbidites. The complete Bouma sequences in limestone turbidites are rarer and have vertical zoning difference, much fewer lamination, and convolution due to relatively weak thixotropy of carbonate mud. That is why limestone turbidites deposited from turbid currents of low density have been given the name "allodapic limestone".

The allodapic limestone precipitates from turbid currents and forms a sequence characterized by vertical zoning of the three characteristic zones.

First zone consists of three parts:

1. Lower part consists of poor and then the well-sorted shallow marine fossil detritus and lithoclasts showing imbrication and gradation (common reversed gradation).
2. Middle part contains fine-grained carbonate detritus.
3. Upper part contains thin laminated carbonate mud.

Second zone consists of two parts:

1. Fine-grained limestone with a flat layer surface and horizontal lamination.
2. Fine-grained limestone with wavy lamination and sometimes with convolution.

Third zone contains:

Marls or clayey limestones that gradually take on more characteristics of pelagic sediments upward. Fossil detritus of shallow marine benthic organisms lack in this zone and are found in the open sea (pelagic) fossils.

The allodapic layers of limestone can be traced continuously on a large expanse of a few hundred meters to several kilometers. The thickness of the layers is typically from 1 cm to several meters. The allodapic limestones or calcareous turbidites are often significant as reservoir rocks for oil and gas. They are characterized by relatively high porosity compared to the densely packed micrite (pelagic) limestones within which the turbidity currents caused greater sedimentary bodies or turbidity fans. The allodapic limestone type of greystone, peckstone, and rudstone contain large amounts of skeletal and nonskeletal detritus transferred from the reef environment or sandbanks. This high primary intergranular texture will contain high secondary porosity too. This is caused by the diagenetic processes of dissolution. The surrounding basin micrite deposits, rich in organic matter, will be the potential source rocks from which the hydrocarbons migrate during diagenetic processes in allodapic limestones under anaerobic conditions of deposition at a greater depth.

7.3.3. Reef and Perireef Bioclastic Limestones Outside the Carbonate Platforms

Reef and perireef limestones, except on carbonate platforms (Section 7.3.1.4.), occur in shallows and inshore parts of sea depositional area that does not belong to the carbonate platform. In such cases, the reef limestones follow transgression on older rocks with the gradual advance of sea and sinking of earlier land.

The reef limestones outside of the carbonate platform are typically connected to the morphologically prominent relief of rocks of different ages that form underwater cliffs or islands in shallow sea with high energy of water and good aeration. It happens in the warm seas and provides ideal environmental conditions for the exuberant growth and development of reef-making organisms, especially corallinacea algae, bryozoans, corals, and some species of sessile foraminifers (nubecularia) and thick-shell bivalves, particularly oysters. The sedimentary siliciclastic detritus, mixed with lesser or greater amounts of fossil debris, originating from the reef-making organisms, dominates away from the shore and in deeper parts of the marine depositional area without morphologically prominent submarine cliffs and islands, as shown in Fig. 5.24 and Fig. 7.16. The reef limestones in such conditions generally form a few tens of meters thick bioherms associated with wide belt of perireef limestones (biocalcrudite and biocalcarenites) formed by accumulation of vast amounts of reef-making organisms' skeletal debris.

Bioherms contain organogenic skeletal grids and are made up of the fossilized skeleton in the place of growth. The basic reef grid or "reef skeleton" originates in the inshore submarine cliffs of crystalline limestone or dolomite composition. It consists of corallinacea algae skeleton and corals or corallinacea algae and bryozoa. The space within skeletal lattice are filled by bryozoans, sessile foraminifers, incrusted cyanobacteria, and bioclasts. It is composed of debris of different organisms created by destruction and abrasion of reef-making organisms. The large extent and thickness of perireef limestones mainly contains debris of reef-making organisms, bioherms with the rise of sea levels and

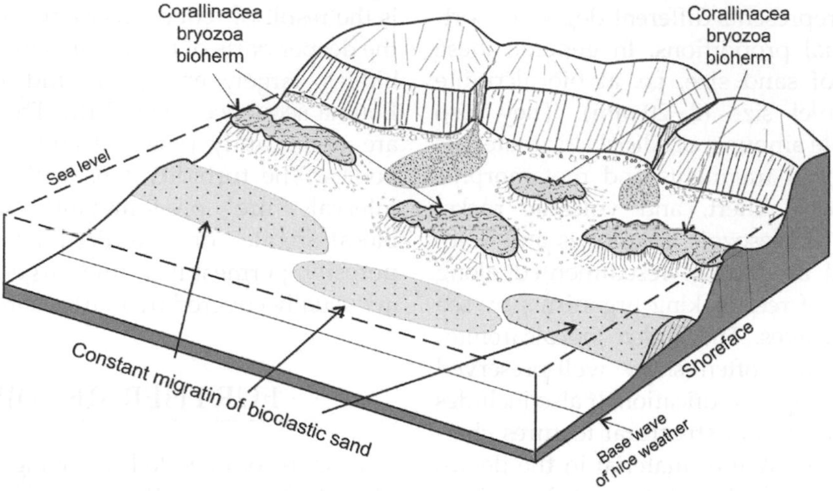

FIGURE 7.16 Conceptual diagram showing depositional environments of corallinacea reef–perireef limestones in the Baden of the large Pannonian Basin or Carpathian basin in East-Central Europe.

gradual flooding and sinking of the morphologically indented relief. It is made up of older Paleozoic and/or Mesozoic rocks during the Badenian (Middle Miocene) in the Pannonian basin covered by large expanses of coastal shallows. The greatest part of them, however, is ruined and destroyed by the activity of waves and is redeposited with progradation mechanisms closer to or away from the reef in the form of perireef bioclastic limestones.

The perireef bioclastic limestones contain large amounts of skeletal debris deposited in the perireef shallows with turbulent water. It indicates strong- and long-term destruction of organogenic reef creations with processes of bioerosion and wave abrasion, and the fact that the final volume and area of reef is significantly lower than their actual initial distribution (Fig. 7.16).

The greatest significance as reservoir rocks of oil and gas in the Pannonian Basin have Badenian, and sometimes Paleocene, reef and perireef limestones (bioherms, biocalcarenites, and biocalcrudites), since they are the most widespread transgressive reef–perireef carbonate sediments in the Tertiary sediments of the Pannonian Croatian territory. It forms large sedimentary bodies characterized by very high intergranular, interskeletal, and intraskeletal porosity and permeability. The Badenian reef limestones are typically belong to bioherms, dome morphological type of organogenic reef (Fig. 5.54(B)). The resulting growth and lithification of corallinacea algae colonies for several generations and bryozoa, together with thick testaceous shellfish (oysters), inhabits on shallow marine cliffs (Fig. 7.16). It consists mainly of corallinacea algae colonies, incrusted bryozoans and shellfish shells, and sometimes large benthic foraminifers. They may also contain a small proportion of siliciclastic ingredients, generally grains of sand of quartz and fragments of granitoid rocks, and sometimes in addition fragments or pebbles of quartzite, crystalline schist, limestones, and dolomites. These are clasts carried by sea currents and waves on organogenic reef and remain petrified and embedded in the ridge formations.

The Baden perireef limestones, known as corallinacea (lithothamnium) and corallinacea-bryozoan biocalcrudite and biocalcrudite limestones, consist of corallinacea algae bioclasts (lithothamnia), bryozoans, urchins, and shellfishes. It represents different degree of sorting and mutual proportions. In general, these are bioclasts of sand size, i.e. at biocalcrudite debris of gravel size (2–20 mm). They also contain variable amounts of siliciclastic materials (quartz, granitoid fragments and metamorphic rocks, sandstone, chert, and volcanic rocks) and/or clasts of Mesozoic limestones and dolomites, as well as detritus resedimented along with bioclasts of reef-making organisms by sea currents and waves. The Baden biocalcarenites and biocalcrudites often show well-preserved gradation and slope stratification. It also includes numerous textural and structural features characteristic of deposition of material in the deltas, shores, sandbanks, and tidal channels by activity of waves and tidal currents. A significant portion of reef–perireef detritus in the Pannonian basin is the result of redeposits by turbidity current in the deeper parts of the sea area in the form of turbidites, largely as T_a, T_b, and T_c intervals of Bouma sequences (Fig. 7.10). The turbidity fans are significantly potential and unique collector bodies. The turbidity fans with T_a, T_b, and T_c intervals are predominantly composed of biocalcarenite or biocalcrudite having high porosity, permeability, and larger thickness and are usually covered by marls as insulating rocks.

FURTHER READING

It is recommended referring Dott et al.,[11] Miall,[30] Reading,[36] Reeckman,[37] Tišljar,[47,48,50,51] and Tissot[53] on depositional environment favoring petroleum.

CHAPTER

8

Mineral Deposits: Host Rocks and Origin

OUTLINE

8.1. Definition 261
8.1.1. Mineral 261
8.1.2. Rock 262
8.1.3. Mineral and Rock Deposit 262

8.2. Classification of Minerals and Mineral Deposits 263
8.2.1. Mineral Classification System Based on Chemical Composition 263
8.2.2. Geographic Distribution 263
8.2.3. Depth of Occurrence 263
8.2.4. Mode of Occurrence 265
8.2.5. Nature of Mineralization 265
8.2.6. Structural Control 265

8.3. Host Rocks 265

8.4. Genetic Model 266
8.4.1. Magmatic 271
8.4.2. Sedimentary 272
8.4.3. Metamorphic 274
8.4.4. Volcanogenic Massive Sulfide and Volcanic-Hosted Massive Sulfide 275
8.4.5. Black Smokers Pipe Type 276
8.4.6. SEDEX/Stratiform 276
8.4.7. Mississippi Valley Type 277
8.4.8. Manto–Chimney - Replacement 277
8.4.9. Irish 278
8.4.10. Pennine 278
8.4.11. Alpine/Bleiberg 278
8.4.12. Skarn 278
8.4.13. Residual 279
8.4.14. Placer 279

Further Reading 279

Minerals are non-renewable wasting assets and once taken out of Mother Earth, can never be replenished by human ability. ***Haldar***

8.1. DEFINITION

Let us revise our understanding of some common terminology to explain and establish the mutual relationship between minerals, mineral deposits, the host rocks that bear specific mineral or group of minerals within it under certain physicochemical conditions, classification systems based on various characteristic features, and their origin.

8.1.1. Mineral

"Mineral" is a naturally occurring homogeneous substance, usually inorganic, and symbolized by unique atomic structure and chemical formula with minor deviation (addition or

http://dx.doi.org/10.1016/B978-0-12-408133-8.00008-0

subtraction) of certain basic compositional elements. It is habitually in solid form, other than a few like mercury, natural water, and fossil fuel (oil and gas). The minerals of solid form are stable at normal temperature and pressure.

Minerals can be described and identified in hand specimen by a number of physical properties that vary to a large extent. The most common and distinguishing characteristics include crystal structure (cubic, tetragonal, and hexagonal), color (colorless, white to all colors in rainbow), hardness (in Mohs scale between 1 (talc) and 10 (diamond)), luster (metallic, brilliance, glossy, glassy, and vitreous), streak (powder color on streak plate like black and cherry red), cleavage (basal, cubic and others), fracture (uneven and conchoidal) tenacity (brittle), specific gravity, and some quick tests like reaction to hydrochloric acid, magnetism, taste/smell, and radioactivity (refer Table 1.1). The identification of minerals can be confirmed by study of glass-mounted thin sections under microscope and chemical analysis in laboratory.

There are +4900 known mineral species and a majority of them have been approved by the International Mineralogical Association (IMA). The largest and most important class of rock-forming minerals (RFM), constituting +90% of the Earth's crust, are silicate minerals (quartz, feldspar, mica, pyroxene group, olivine, amphibole group, garnet, andalusite, zircon, kyanite, sillimanite, kaolin, etc.). The most abundant RFM are quartz (SiO_2), Fig. 8.1, orthoclase feldspar ($KAlSi_3O_8$), plagioclase feldspar ($CaNaAlSi_3O_8$), albite ($NaAlSi_3O_8$), as well as mica group such as muscovite ($H_2KAl_3(SiO_4)_3$) and biotite ($H_2K(MgFe)_3Al(SiO_4)_3$).

8.1.2. Rock

Minerals are different from rocks, on the face of it. "Rock" is an aggregate/assemblage of mineral(s) formed under natural process of igneous, sedimentary, and metamorphic origin. The rocks do not have a specific chemical composition.

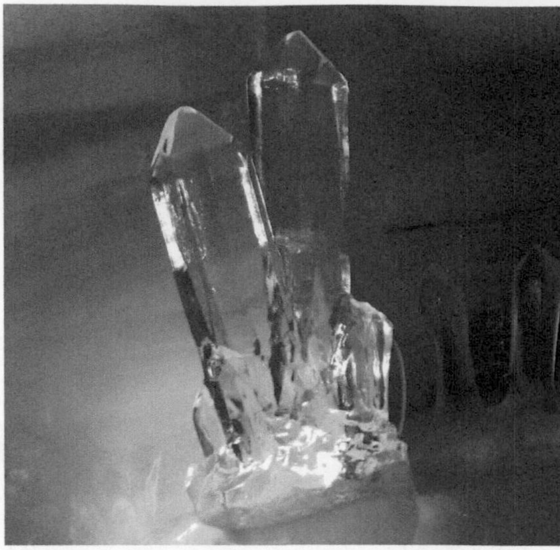

FIGURE 8.1 Model of stalagmite quartz crystal having natural form of hexagonal prisms closed by bipyramids at the Jungfrau museum (top of Europe) at 4158 m (13,642 ft), maiden summit of Bernese Alps Mountain, Switzerland.

The mineral constituents frequently vary widely and often transgress from one to another species. The rocks may contain large amount of organic matter as in fossiliferous limestone. Quartz is a mineral and may constitute monomineralic rock quartzite after metamorphic transformation. Similarly, calcite is a mineral and can form monomineralic limestone and marble after sedimentary and metamorphic changes, respectively. The most common rocks are basalt, granite, sandstone, limestone, quartzite, marble, granite gneiss, and mica schist. The characteristic features of common rocks are given in Table 1.2.

8.1.3. Mineral and Rock Deposit

"Mineral deposits" are aggregate of mineral and/or group of minerals in an unusually high concentration. The mineral deposits must have three-dimensional configuration that includes shape in plan and sectional view, continuity in strike and depth to represent volume, and size with average characteristics. The shape can be regular (iron ore, coal, and bauxite) to extremely irregular (gold and platinum) posing economic mining and extraction. The examples of mineral deposits are Broken Hill zinc–lead–silver deposit of Australia, Sudbury nickel–platinum–palladium deposits of Canada, Bushveld chromium–platinum–palladium deposits in South Africa, Jhamarkotra rock-phosphate deposit and Jharia coalfield in India, and Athabasca oil sands (crude oil) in northeastern Alberta, Canada.

The rock bodies can similarly be outlined by shape, size, and continuity and defined as deposit with commercial values. The rock deposits can serve as raw material sources for industry such as the limestone deposits of Egypt for preparation of lime, mortar, and cement. The rock deposits can directly be used after cutting, shaping, and polishing in road and building construction such as Makrana marble in Taj Mahal, India (Fig. 6.19), and limestone in "The Great Pyramid" at Giza, Egypt (Fig. 1.5(A)).

8.2. CLASSIFICATION OF MINERALS AND MINERAL DEPOSITS

The most common classification of minerals includes metallic, nonmetallic, and energy minerals. The metallic minerals are subclassified as (1) ferrous (iron, manganese, nickel, and cobalt), (2) nonferrous (copper, zinc, lead, tin, and bauxite), and (3) precious (gold, silver, and platinum). The nonmetallic minerals are composed of quartz, feldspar, mica, garnet, potash, sulfur, and salt. The energy minerals contain coal, petroleum, and natural gas.

Minerals and rocks can be classified in various formats based on the type and chemistry of minerals, geographic distribution, depth of occurrence, morphology, relation to host rocks, nature of mineralization, structural control, genetic model, economic gradation fitting with the overall characteristics, and perspective. A particular mineral or rock that will be formed from a certain combination of elements depends on the physical and chemical conditions under which the material forms. This, in turn, results in a wide range of colors, hardness, crystal forms, luster, and density that a particular mineral possesses (Tables 1.1 and 1.2).

8.2.1. Mineral Classification System Based on Chemical Composition

The most common mineral classification system is based and on the type and chemical composition of individual mineral. Once minerals have been grouped by chemical composition, they can be further separated into groups on the basis of internal structure. Native elements occur as metal (Au, Ag, Cu, and Sb), semimetals (boron, germanium, graphite, and silicon), and nonmetals (At, Br, Cl, F, H, I, N, O, P, and S). Minerals occur in various forms such as native elements to complex compounds of oxide, carbonate, silicate, sulfide, sulfate, sulfosalts, phosphate, etc. Mineral classification system can broadly be grouped into 10 categories or class following Gaines et al.[20] (Table 8.1).

8.2.2. Geographic Distribution

Mineral deposits can broadly be described based on geographic location and dimension. Mineral deposits often occur in group/cluster and are repeated over long distances along identical stratigraphic horizon and/or structural control like breccias zone and lineaments. Therefore, mineral-bearing environment can be described in part as individual small to large body/deposit (orebody) and in totality as spread over large area/distance (province) detailed in Table 8.2.

8.2.3. Depth of Occurrence

The mineral/mineral deposits occur in a variety of forms and formats. It may be exposed to

TABLE 8.1 Mineral Classification System by Chemical Composition

Class	Forms	Minerals
01	Native elements	Antimony (Sb), copper (Cu), gold (Au), silver (Ag), sulfur (S)
02	Sulfides and sulfosalts	Chalcopyrite ($CuFeS_2$), sphalerite (ZnS), galena (PbS), bournonite ($PbCuSbS_3$), tennantite ($Cu_{12}Sb_4S_{13}$), and tetrahedrite ($(Cu\ Fe)_{12}Sb_4S_{13}$)
03	Oxide and hydroxides	Quartz and amethyst (SiO_2), hematite (Fe_2O_3), cassiterite (SnO_2), boehmite (γ-AlO(OH)), and gibbsite, $Al(OH)_3$
04	Halides	Cryolite (Na_3AlF_6), fluorite (CaF_2), halite (NaCl), and sylvanite (KCl)
05	Carbonates	Calcite ($CaCO_3$), magnesite ($MgCO_3$), dolomite ($CaMg\ (CO_3)_2$), ankerite Ca (Fe, Mg, Mn) $(CO_3)_2$, smithsonite ($ZnCO_3$), cerussite ($PbCO_3$), rhodochrosite ($MnCO_3$)
06	Borates	Howlite ($Ca_2B_5SiO_9(OH)_5$) and kernite $Na_2B_4O_6(OH)_2 \cdot 3(H_2O)$
07	Sulfates	Barites ($BaSO_4\ 2H_2O$), Gypsum ($CaSO_4$), anglesite ($PbSO_4$)
08	Phosphates	Apatite ($Ca_5\ (PO_4)_3$ (F,Cl,OH)) and berlinite ($AlPO_4$).
09	Silicates	Andalusite-kyanite-sillimanite (Al_2SiO_5), beryl ($Be_3Al_2Si_6O_{18}$), amazonite ($KAlSi_3O_8$), garnet group: pyrope ($Mg_3Al_2\ (SiO_4)_3$), almandine ($Fe_3Al_2\ (SiO_4)_3$)
10	Organic minerals	Fossil-bearing limestone, coal, and oil shale

TABLE 8.2 Classification of Mineral Deposits Based on Geographical Distribution

Type	Description	Examples
Province or metallogenic province	Represents large area having essentially notable concentration of certain characteristic metal or several metal assemblages or a distinctive style of mineralization to be delineated and developed as economic deposits. The metallogenic province can be formed on various processes like plate tectonic activity, subduction, igneous intrusive, metal-rich epigenetic hydrothermal solution, and expulsion of pore water enriched in metals from sedimentary basin	Zn–Pb–Ag-bearing McArthur–Mt Isa inlier in Northern Territory, Australia (Fig. 8.2); gold province in Canadian shield; Pt–Pd–Ni–Cu–Au deposits in Sudbury basin, Canada; Bushveld igneous complex with Pt–Pd–Cr deposits, South Africa; Katanga and Zambian copper province; tungsten province of China; Zn–Pb–Ag deposits of Aravalli province, India
Region	Relatively smaller in size compared to province and controlled by stratigraphy and/or structure, for occurrence of specific mineral(s) at commercial quantity.	Kalgoorlie goldfield, Esperance region of Western Australia; Zn–Pb region of Mississippi Valley; copper region of Chile and Peru; diamond-bearing region of northern Minas Geraes, Brazil and at Kimberley, South Africa; Pacific and Central coal-bearing region of the United State; and rubies in high-grade metamorphic region of Kashmir region, India
District	Stands for one geographical area popularly known for the occurrence of particular mineral	Aeolian soils of Blayney District, NSW, Australia; Baguio mineral district in Philippines for copper deposits; new Mexico for uranium deposits; and East Singhbhum district for copper and Salem district for magnesite, India

TABLE 8.2 Classification of Mineral Deposits Based on Geographical Distribution (*cont'd*)

Type	Description	Examples
Belt	Represents narrow linear stretch of land having group of deposits of associated minerals	Colorado gold–molybdenum belt, USA; Khetri–Nimkathana copper belt, Rajpura–Dariba–Bethumni zinc–lead–silver belt, Rajasthan (Fig. 8.3) and Sukinda-Nausahi chromite belt, Orissa, India
Deposit	Stands for a single or a group of mineral occurrences of sufficient size and grade separated by natural narrow barren partings	Broken hill group of zinc–lead deposits, Australia; Red Dog zinc–lead deposit, Alaska; Zawar group of zinc–lead deposits, India; OK Tedi copper deposit, Papua new Guinea; Olympic Dam copper–gold–uranium–silver deposit, South Australia; Neves-Corvo polymetallic deposit, Portugal; and Stillwater group of platinum deposit, USA
Block	Well-defined area having mineral concentration wholly or partly of economic value	Broken hill main, Australia, and Bailadila deposit-14, Central Mochia, India

the surface and may or may not continue in greater depth. It may occur near/close to the surface like coal and lignite seams. Many of the mineral deposits are deep seated and hidden deep. The discovery, development, and mining will be easier for the former and complex for the latter type. A comparative statement is given in Table 8.3.

8.2.4. Mode of Occurrence

Minerals and rocks are found in varied forms in nature. The type or mode of occurrences are characteristic to certain groups of minerals and are classified as given in Table 8.4.

8.2.5. Nature of Mineralization

Mineral deposits can be classified based on the nature of their appearances in situ as described in Table 8.5.

8.2.6. Structural Control

Structural changes, tectonic movements, and surface weathering play a significant role over geological time as avenue for hydrothermal flow of mineralized fluids; accumulate and concentrate at suitable location; and remobilize and reorientate as postgenetic activity. The features related to mineralization control are deformation, weathering, joints, fractures, folds, faults, breccias, and zone of subduction. The structural features are given in Table 8.6.

8.3. HOST ROCKS

There are three types of rocks that host the mineralization, namely, igneous, sedimentary, and metamorphic (Table 8.7). Examples of igneous rocks are porphyry copper deposits in granite, platinum–palladium–chromium–nickel deposits in dunite, as well as peridotite, gabbro, norite and anorthosite, tantalite, columbite, and cassiterite in pegmatite. Ore deposits can exclusively be formed by sedimentation process like iron ore formation (Banded Iron Formation (BIF) / Banded Hematite Quartzite (BHQ)), zinc–lead deposits in dolomite, copper–gold in quartzite, and diamond in conglomerate and limestones. The deposits show bedded, stratabound, and often stratiform features having concordant relation with country rocks. Metamorphic rocks host important ore deposits

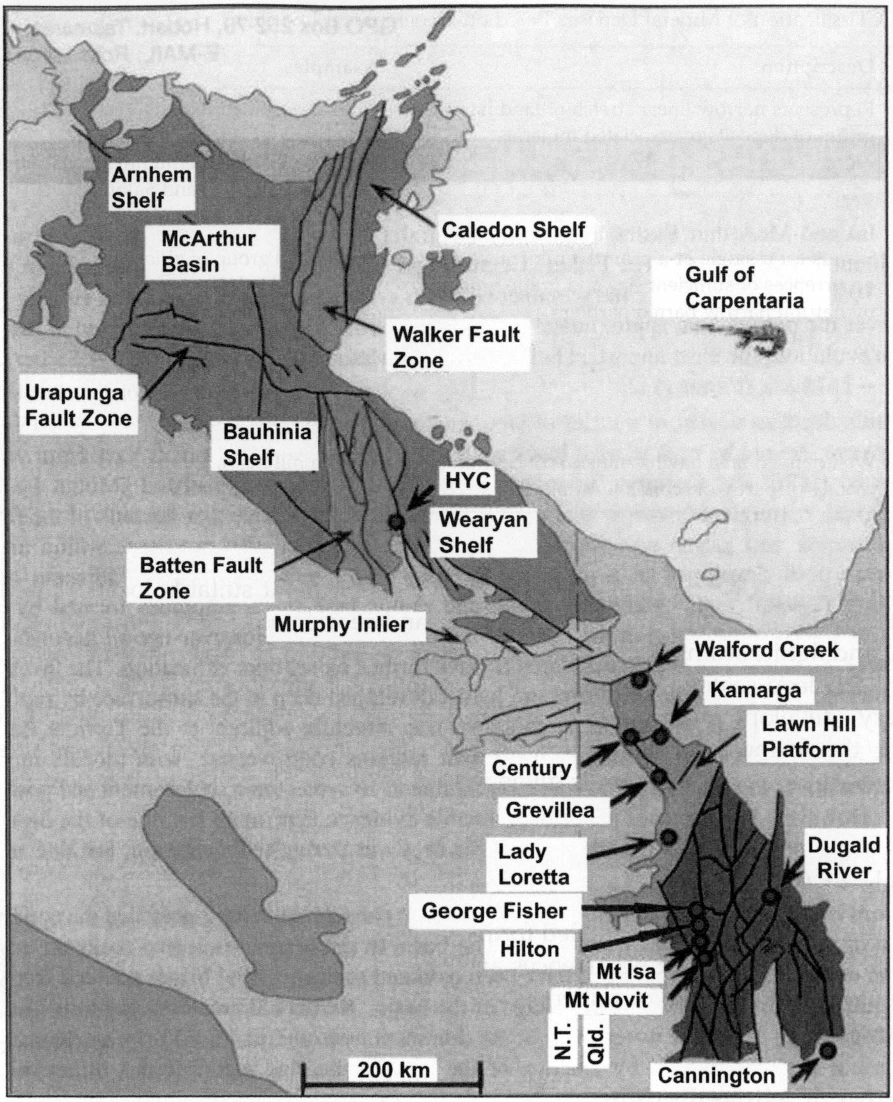

FIGURE 8.2 Distribution of significant stratiform and stratabound zinc–lead–silver deposits in Proterozoic province of North Australia. The deposits occur along NW-SE lineament. *Source: Ref. 23.*

generated as contact metamorphic aureoles. The ore deposits are garnet, wollastonite, andalusite, and graphite. The metamorphic equivalent of sedimentary and igneous rocks forms large deposits of marble, quartzite, and gneisses and is commonly used as building stones and construction materials.

8.4. GENETIC MODEL

The genetic model uses perceptions or formation of the ore genetic process based on direct and indirect evidences and knowledge of the host environments. It includes the overall specifications of how the geological forces act to

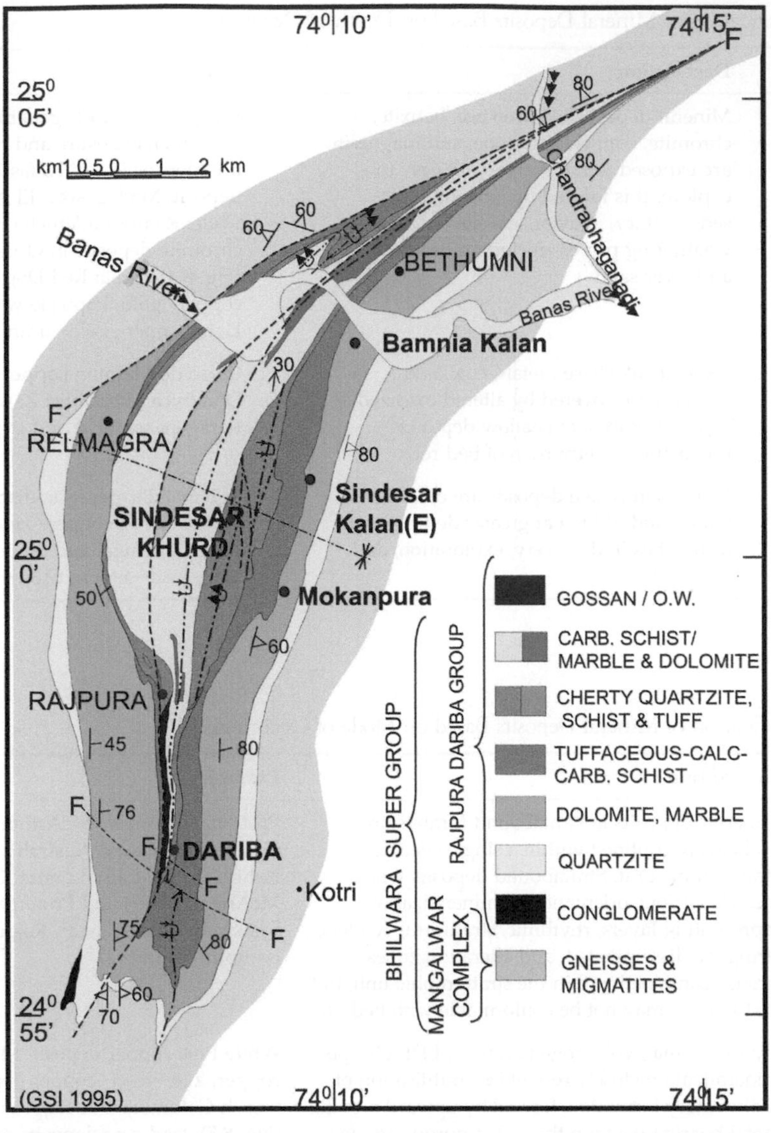

FIGURE 8.3 Position of significant zinc–lead–silver deposits in Rajpura–Dariba–Bethumni Belt, India. *Source: Ref. 24.*

influence the formation of the ore/orebody. These descriptive and interpretative features are used for enhancing the perception, and, based on the process of mineral formation, are classified as igneous, sedimentary, hydrothermal, metamorphic or syn-sedimentary exhalative (SEDEX), etc. The deposits are often deformed, metamorphosed, and remobilized as postdepositional phenomena—obscuring the primary structure. Many corroborating evidences are collected, assembled, and interpreted/postulated for a specific type of deposit before confirming a particular genetic process.

TABLE 8.3 Classification of Mineral Deposits Based on Depth of Occurrence

Type	Description	Example
Exposed to surface	Mineral deposits like iron ore, bauxite, chromite, copper, limestone, and magnesite are exposed to the surface and easy to explore. It is formed by igneous, sedimentary, alluvial, and eluvial weathering process including beach and river sand	Prospecting efforts get emphasis on looking for fresh rock exposure and newly derived boulders as is Adi Nefas Zn–Cu–Au–Ag deposit, Madagascar; El Abra Cu deposit, Chile; Rampura-Agucha Zn–Pb–Ag and chromite deposits in Orissa, Tamil Nadu (Fig. 8.4), India; Red Dog, Alaska; OK Tedi copper–gold, Papua new Guinea; and Olympic Dam copper–gold–uranium–silver
Shallow depth	Deposits like base metals, coal, and gypsum are covered by altered oxidized capping or exist at shallow depth or under thick overburden of bed rock	Cerro de Maimon copper–gold deposit at Dominican Republic; Zawar zinc–lead–silver; Raniganj coalfield and gypsum deposits, India
Deep-seated hidden deposit	Igneous intrusive deposits are often deep seated and hidden at greater depth posing difficulties in discovery, exploration, and extraction	Polymetallic deposits discovered in past are Neves-Corvo copper–zinc–tin at 330–1000 m depth, Portugal and Sindesar Khurd zinc–lead–silver at 130 m depth, India

TABLE 8.4 Classification of Mineral Deposits Based on Mode of Occurrences

Type	Description	Example
Stratabound	Minerals deposits in stratabound formats are exclusively confined within a single specific stratigraphic unit. Stratabound deposits will include various orientation of mineralization representing layers, rhythmic, stratiform, veinlets, stringers, disseminated, and alteration zones, strictly contained within the stratigraphic unit, but that may or may not be conformable with bedding	Proterozoic Mt Isa–McArthur Basin System of Northern Territory, Australia (Mt Isa, George Fisher, Hilton, Lady Loretta, Century, and McArthur River) and Proterozoic Middle Aravalli System in India (Zawar, Rajpura–Dariba and Rampura–Agucha)
Stratiform	Hydrothermal, volcanogenic, and "SEDEX"-type mineralization closely resembles stratification of sedimentary formation, formed by upward moving metal-bearing solution through a porous aquifer, and deposits ore minerals in the overlying pile of sedimentary strata of shale and carbonates. It may contain significant amount of fine pyrite and organic matter	White Pine copper deposit, Michigan, Zambia; copper, zinc–lead–copper deposits of Sullivan in British Columbia and Rajpura–Dariba in India (Fig. 8.5); lead zinc deposits of Broken hills in new South Wales, Mt Isa in Queensland; and McArthur river in Northern Territory, Australia
Bedded, layered, and rhythmic	Bedded deposits are formed generally by deposition and consolidation of sediments. Layered and rhythmic features are developed during the differential crystallization and segregation of mafic and ultramafic magma in a huge chamber over a prolong time	Banded Hematite quartzite (BHQ), limestone, lignite, and coal (Fig. 8.6). Bushveld platinum–chromite deposits, South Africa; Sittampundi Cr–Pt–Pd (Fig. 8.7); Sukinda Cr–Ni and Nausahi Cr–Pt–Pd, India

TABLE 8.4 Classification of Mineral Deposits Based on Mode of Occurrences (*cont'd*)

Type	Description	Example
Porphyritic	Igneous host rock consisting of large-grained crystal of quartz, feldspar, and amphiboles scattered in a fine-grained groundmass. Porphyritic refers to the texture of the rocks	Chuquicamata (690 Mt @ 2.58% Cu), Escondida and El Salvador, Chile; Toquepala, Peru; Lavender pit, Arizona; and Malanjkhand, India
Pipe/chimney	These deposits are relatively narrow in horizontal dimension and extensively large in vertical direction. Pipes may be formed by infillings of mineralized breccias in volcanic pipes, e.g. copper-bearing breccia pipes of Messina, South Africa	Diamond-bearing kimberlite/lamproite pipes at Kimberly, South Africa and Panna deposit, India

TABLE 8.5 Classification of Mineral Deposits Based on Nature of Mineralization

Type	Description	Example
Disseminated	Disseminated types of mineralization are formed by crystallization of deep-seated magma. The early formed in situ valuable metallic and nonmetallic minerals are sparsely disseminated or scattered as fine grains throughout or part of the host rock	Diamond in kimberlite pipes in South Africa, porphyry copper deposits at El Salvador, Chile; porphyry tungsten–molybdenum deposit at Yukon; Malanjkhand copper and Sargipalli lead-copper deposit, India
Massive	Massive deposits are characterized by substantial share of ore-forming minerals with very little gangue +60% sulfides in VMS-, VHMS-, or SEDEX-type base metals, chromite, and magnetite deposits	Zinc–lead–silver deposit of Red Dog, Northwest Alaska; Neves-Corvo, Portugal; and chromite deposits of Tamil Nadu, India (Fig. 8.4)
Veins and stringers	Veins, fissure, and lodes are tabular deposit formed by deposition of ore and gangue minerals in open spaces within fault, shear, and fracture zones	Polymetallic deposits of Silvania; Silver Cup, Lucky Jim, Highland Lass Bell in British Columbia; Zawar zinc–lead–silver (Fig. 8.8); Kolihan copper deposit; chromite–magnesite deposit at Sindhuvally, India
Sheet and stringers	Stringers are large numbers of thin, tiny, and closely spaced mineralized veins originating from the main orebody and often described as "stringer zone"	Zawar group of zinc–lead deposit, India (Fig. 8.9)
Stock work	Style is characterized by a large mass of rock impregnated by dense interlacing network of variously oriented irregular metal-bearing hydrothermal veins and grouped veinlets. Common occurrences are platinum-bearing sulfides, copper, gold, molybdenum, tin, tungsten, beryllium, uranium, mercury, and other metal ore	Nevada, copper- and tin-rich stock work at Neves-Corvo mine, Portugal; zinc–lead–silver mineralization at Sindesar Khurd (Fig. 8.10); and platinum–palladium–chromite mines at Boula-Nausali India

(*Continued*)

TABLE 8.5 Classification of Mineral Deposits Based on Nature of Mineralization (*cont'd*)

Type	Description	Example
Ladder veins	Ladder veins are regularly spaced, short, and transverse nearly fractures confined wall to wall within dikes or compact rock mass for considerable distance. Fractures are formed by contraction joints and filled with auriferous quartz or valuable minerals forming economic deposit	Examples of commercial ladder vein-type deposits are Morning Star gold mine in Victoria; molybdenite veins in New South Wales, Australia; and copper ladder veins in Norway

TABLE 8.6 Classification of Mineral Deposits Based on Structural Control

Undeformed	Most of the residual/weathered (laterite), alluvial (placer) and heavy mineral beach sand deposits are undeformed types	Bauxite deposit; beach sand deposits containing ilmenite, titanium, thorium, tungsten zircon of India, Indonesia, and Australia; and large platinum placer deposit at Ural Mountain, Russia
Joints and fractures	Joints and fractures are formed synchronous to primary formation or aftereffect caused by regional stress and break of rocks along which little or no movement occurred. Mineralization often concentrates along these regular and irregular planes	Magnesite deposits in Salem district town, Tamil Nadu, India, and Lennard Shelf zinc–lead deposit, Western Australia
Fold	Directed compression of the crust, resulting in a semiplastic deformation, creates folding of strata ("fold"). The fold closure, limb inflex zone, and axial planes are suitable for mineral localization. Mineral deposits are often folded during or after formation	Rajpura–Dariba zinc–lead–copper deposit (Fig. 8.11), Agnigundala lead–copper deposit, Sukinda chromite belt, India
Fault	Joints and fractures (Fig. 8.11) along which noticeable movements have occurred are called "fault", with displacement from centimeters to kilometers, thus creating challenges for exploration. Fault zones are favorable avenue and localization of mineralized solution for movement and concentration	Manto Verde Cu in Chile and many of the coal deposits are faulted
Shear zone	Shear is extreme rock deformation generating fractures, intense foliation, and microfolding due to compressive stress displaying wide zone of shearing in crushed rock mass with width varying between few centimeters and several kilometers. The interconnected openings of shear zone serve as an excellent channelways for mineral-bearing solutions and subsequent formation of deposits	Shear zones in orogenic belts host ore deposits. Shingbhum shear zone hosts copper–uranium mineralization. Chromite–magnesite veins developed in shear zone at Sindhuvally, Karnataka, India, (Fig. 8.12).

TABLE 8.6 Classification of Mineral Deposits Based on Structural Control (*cont'd*)

Breccia	Breccia is clastic sedimentary rocks of large sharp-angled fragments embedded in fine-grained matrix of finer particles or mineral cement, generated by folding, faulting, magmatic intrusions, and similar forces (tectonic breccias). Tectonic breccia zones represent crush, rubble, crackle, and shatter rock mass. Breccia/conglomerate differs in shape of larger particles due to transportation mechanism. Igneous/flow/pyroclastic breccias are rocks of angular fragments of preexisting igneous rocks of pyroclastic debris ejected by volcanic blast or pyroclastic flow	Intrusive gabbroic magma with sharp-angled fragments of Cr embedded in fine-grained gabbroic matrix containing PGE (Figs 8.13 and 8.14) within preexisting layered ultramafics at Nausahi, India. Zinc–copper–gold deposits of Saudi Arabia are hosted in volcaniclastic breccia. Fossil downs Zn-rich ore, Lennard shelf deposit, Western Australia, is closely related to major N–S fault, brecciated cavity filled in limestone reefs
Subduction zone	Subduction is a process of two converging tectonic plate movement where one slides under the other. The formation is associated with multidimensional tectonic activities like shallow and deep-center earthquakes, melting of mantle, volcanism, volcanic arc, plutonic ophiolite suites, platinum–chromium-bearing peridotite–dunite–gabbro–norite, movement of metal-bearing hydrothermal solution, and metamorphic dewatering of crust	The porphyry copper–gold belt extends north from central Chile into Peru and is associated with subduction of Pacific Ocean floor beneath the South American plate. The main Chilean porphyry copper belt hosts some of the largest open-cut copper mines in the world

TABLE 8.7 Mineral Deposits and Their Relation to Host Rocks

Type	Description	Example
Identical with host	Mineral deposits like granite, limestone, marble, quartzite, and slate are indistinguishable from the host rock	Limestone deposits of Egypt and Saudi Arabia (Fig. 8.15)
Different from host	Gold-bearing quartz veins act as an exclusive host for Au and are different from the surrounding rocks	Kolar gold deposit, Karnataka, India
Gradational contact	Forms around the vein systems with characteristic disseminated mineral distribution	Bulldog Mountain fine-grained sphalerite, Colorado, and Sargipalli lead–copper, India

8.4.1. Magmatic

"Magmatic" deposits are genetically linked with the evolution of magma that emplaced into the continental or ocean crust as plutonic (intrusive) or volcanic (extrusive) phenomenon. The mineralization is located within the rock types derived from the differential crystallization of the parent magma. The most significant magmatic deposits are related to felsic (granite and rhyolite), mafic (gabbro and norite), and ultramafic (peridotite and dunite) rocks formed

FIGURE 8.4 Massive chromite orebody exposed to surface near Karungalpatti village at Sittampundi belt, Namakkal District, Tamil Nadu, India (Mr Alan Mulligan of Inverse Activity Ltd, Perth (left) and Dr Tom Evans, Lonmin Plc, UK (right), during Reconnaissance field work).

FIGURE 8.6 Alternate bands of coal (shining-black) and shale (brownish-gray), Belatan mine, Jharia coalfield, India. *Source: Ref. 24.*

from the crystallization of felsic, mafic, and ultramafic magma (Fig. 8.16). Ore minerals are formed by the separation of metal sulfides and oxides in molten form within an igneous melt before crystallization. The deposit types include copper, chromite, nickel–copper, and platinum group of elements. There are several large magmatic deposits: they are Cr–PGE (Platinum Group of Elements) deposits at Bushveld Igneous Complex, South Africa; Ni–Cu–PGE deposits at The Great Dykes, Zimbabwe; Ni–PGE–Cr deposits at Sudbury, Canada;

FIGURE 8.5 Stratiform sphalerite (honey brown) ore forming mineralization in calc-silicate (bluish grey) host rock at Rajpura Dariba deposit, India. *Source: Ref. 24.*

Ni–Cu–PGE deposits at Stillwater Igneous Complex, Montana, USA; and Mo–Cu porphyry deposits at Malanjkhand, Cr–Ni ± PGE deposits at Sukinda–Nausahi belt (Orissa), and Byrapur, Layered Igneous Complex (Karnataka), India.

8.4.2. Sedimentary

"Sedimentary" type of deposits are formed by the process of deposition and consolidation of loose materials under aqueous condition. The sedimentary deposits are concordant and may be an integral part of stratigraphic sequence. It is formed due to seasonal concentration of heavy minerals like hematite on the seafloor. The structures consist of repeated thin layers of iron oxides, hematite, or magnetite, alternating with bands of iron-poor shale and chert. The large reserves and production are shared by China, Northwestern Australia, Brazil, India, Russia, Ukraine, South Africa, United States, Canada, Iran, Sweden, Kazakhstan, and Venezuela. Similarly, limestone deposits are formed by chemical sedimentation of calcium–magnesium carbonate ± fossils on the seafloor. The fossil-bearing sandstones, carbonates, and conglomerates may present large reservoir of petroleum and gas.

FIGURE 8.7 Layers of chromite (black) and Pt–Pd-bearing gabbro (white), Sittampundi Igneous Complex, Tamil Nadu, India. *Source: Ref. 23.*

Phosphate-bearing stromatolitic limestone may contribute large resources of phosphate fertilizer such as at Jhamarkotra, India (Fig. 8.17). Coal and lignite are formed under sedimentary depositional condition. The largest coal reserves are from the United States, Russia, China, Australia, India, Germany, Ukraine, Columbia, Canada, Indonesia, and Brazil. The largest reservoirs/production of petroleum are from Venezuela, Saudi Arabia, Canada, Iran, Iraq, Kuwait, United Arab Emirates, Russia, Kazakhstan, Libya, Nigeria, Qatar, China, United States, Angola, Algeria, Egypt, and Brazil.

Evaporite deposits form through the evaporation of saline water in lakes and sea, in regions of low rainfall and high temperature. The common evaporite deposits are salts (halite and sylvite), gypsum, borax, and nitrates. The original character of most evaporite deposits is destroyed by replacement through circulating fluids.

FIGURE 8.8 Well-crystalline veins and stringers of sphalerite (brown) and galena (steel gray) hosted in fine-grained massive proterozoic dolostone at Zawar group of deposits, India.

FIGURE 8.9 Sheeted veins and fine stringers of sphalerite and galena (dark) in dolostone host rock at Zawar deposit, India.

FIGURE 8.10 Stock work style of mineralization, characteristic of large rock mass impregnated by dense interlacing network of variously oriented irregular metal-bearing hydrothermal sulfide (sphalerite, galena, and pyrite) veins and grouped veinlets hosted by carbonaceous calc-silicate rock at Sindesar Khurd orebody, India. *Source: Ref. 24.*

8.4.3. Metamorphic

"Metamorphic" type of deposits are transformed alteration product of preexisting igneous or sedimentary materials. The reconstruction occurs under increasing pressure and temperature caused by igneous intrusive body or by tectonic events. Metamorphic mineral deposits are formed due to regional prograde or retrograde metamorphic process and hosted by metamorphic rocks. Minerals like garnet, kyanite, sillimanite, wollastonite, graphite, and andalusite are end products of metamorphic process. The large resources of garnet are from India (Fig. 8.18), China, Australia, and United States. The large resources of kyanite are from Brazil, India, Kenya, Mozambique, Nepal, Russia, Serbia, Switzerland, Tibet, and North Carolina and Georgia (USA). The major resources and production of wollastonite are from China,

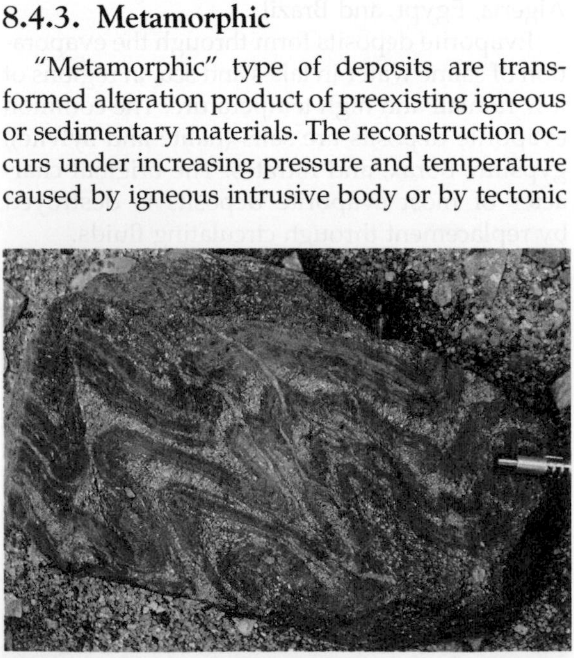

FIGURE 8.11 Stratiform pyrite–zinc–lead mineralization folded and microfaulted with mineral concentration at crests presenting saddle reef structure, Rajpura–Dariba deposit, India. *Source: Ref. 24.*

FIGURE 8.12 Layered chromite (black) and magnesite (white) veins developed in shear zones, Sindhuvally, Karnataka, India. *Source: Ref. 24.*

FIGURE 8.13 Field photographs of layered ultramafic igneous complex in the left (peridotite, gabbro, and norite) showing tectonic breccias zone (center and right) at Boula-Nausahi open-pit mine, Orissa, India.

FIGURE 8.14 Irregular fragmented chromite (black with white rims) in matrix of Pt–Pd-bearing gabbro from the tectonic breccia zone, Boula-Nausahi underground mine, Orissa, India. *Source: Ref. 24.*

FIGURE 8.16 Layered igneous complex forming from intrusive ultramafic magma hosts economic chromite (dark steel gray) deposits at Byrapur, Karnataka, India.

India, United States, Mexico, and Finland. The graphite resources are shared by China, India, Brazil, North Korea, Canada, and Sri Lanka. Copper deposits of Kennecott, Alaska and White Pine, Michigan, are formed by low-grade metamorphism of organic-rich sediments resting over mafic or ultramafic rocks. The low copper content of underlying source rocks liberate during a leaching process caused by passing of low temperature hydrothermal fluids. The fluids migrate upward along the fractures and faults and precipitate high-grade copper in the rocks containing organic matter.

8.4.4. Volcanogenic Massive Sulfide and Volcanic-Hosted Massive Sulfide

"Volcanogenic massive sulfide" (VMS) and "volcanic-hosted massive sulfide" (VHMS) type of ore deposits contribute significant resources of Cu–Zn–Pb sulfide ± Au and Ag, formed as a result of volcanic-associated hydrothermal events under submarine environments at or near the seafloor. It forms in close time and space association between submarine volcanism, hydrothermal circulation, and exhalation of sulfides, independent of sedimentary process. The

FIGURE 8.15 The hills of horizontal bedded limestone (30–50 m high) on either side of the Riyadh-Jeddah-Mecca National Highway, Saudi Arabia, gives a pleasant journey in the desert countryside. These limestone orebodies are identical to the host rock and excellent raw material for cement and lime-related industries.

FIGURE 8.17 Phosphate-bearing limestone showing unique algal stromatolite columns with an abundance of carbonate matrix. The columnar structure upward is the natural feature to receive sunlight and is protruded due to differential surface weathering. The Jhamarkotra rock phosphate deposit in India is a unique fossil assemblage in Proterozoic (~188 Ma) dolomitic limestone in the world.

deposits are predominantly stratabound (volcanic derived or volcanosedimentary rocks) and often stratiform in nature. The ore formation system is synonymous to black smoker type of deposit. Kidd Creek, Timmins, Canada, is the largest VMS deposit in the world. Kidd is also the deepest (+1000 m) base metal mine. The other notable VMS/VHMS deposits are Iberian Pyrite Belt of Spain and Portugal, Wolverine Zn–Cu–Pb–Ag–Au deposit, Canada, and Khnaiguiyah Zn–Pb–Cu, Saudi Arabia.

8.4.5. Black Smokers Pipe Type

"Black smokers" pipe-type deposits are formed on the tectonically and volcanically active modern ocean floor by superheated hydrothermal water ejected from below the crust. The water with high concentrations of dissolved metal sulfides (Cu, Zn, and Pb) from the crust precipitates to form black chimneylike massive sulfide ore deposits around each vent and fissure when it comes in contact with cold ocean water

FIGURE 8.18 Inferior quality garnet in mica schist, Rajasthan, India.

over time. The formation of black smokers by sulfurous plumes is synonymous with VMS or VHMS deposits of Kidd Creek, Canada, formed 2.4 billion years ago on ancient seafloor.

8.4.6. SEDEX/Stratiform

SEDEX type of ore deposits are formed due to concurrent release of ore-bearing hydrothermal fluids into aqueous reservoir mainly ocean, resulting in the precipitation of stratiform zinc–lead sulfide ore in a marine basin environment. The stratification may be obscured due to postdepositional deformation and remobilization. The source of metals and mineralizing solutions are deep-seated superheated formational brines migrated through intracratonic rift basin faults which come in contact with sedimentation process. In contrast, the sulfide deposits are more intimately associated with intrusive or metamorphic process or trapped within a rock matrix and not exhalative. The formation occurred mainly during the mid-Proterozoic period. SEDEX deposits are the most important source of zinc, lead, barite, and copper with associated by-products of silver, gold, bismuth, and tungsten. The examples are zinc–lead–silver deposits of Red Dog, northwest Alaska, McArthur River, Mt Isa, HYC, Australia, Sullivan, British Columbia, Rampura-Agucha, Rajpura–Dariba (Fig. 8.19), India, and Zambian copper belt.

FIGURE 8.19 Massive sphalerite (iron-brown) and galena (shining) in carbonaceous calc-silicate host rock sedimentary exhalative deposition at Sindesar Khurd SEDEX type in India. *Source: Ref. 24.*

8.4.7. Mississippi Valley Type

The "Mississippi valley-type" (MVT) deposits are epigenetic, stratabound, often rhythmically banded (stratiform) ores with replacement of primary sedimentary features predominantly carbonate (limestone, marl, dolomite, and rarely sandstone) host rocks. The mineralization is hosted in open space filling, collapse breccias, faults, and hydrothermal cavities. The deposits, formed by diagenetic recrystallization of carbonates (limestone/dolostone), create low-temperature hydrothermal solution that migrates to suitable stratigraphic traps like fold hinge and faults at the continental margin and intracratonic basin setting. The ore-forming minerals are predominantly sphalerite, galena, and barite. Calcite is the most common gangue mineral. Low pyrite content supports clean concentrate with high metal recovery of +95%. Some deposits are surrounded by pyrite/marcasite halo. Prospects can be defined by regional stream sediment and soil and gossan sample anomaly supported by aeromagnetic and gravity survey. There are numerous Zn–Pb–Ag sulfide deposits along the Mississippi river in the United States; Pine Point, Canada; San Vicente, Central Peru; Silesia, Southern Poland; Polaris, British Columbia; as well as Lennard Shelf (Fig. 8.20) and Admiral Bay, Western Australia.

8.4.8. Manto–Chimney - Replacement

"Manto–chimney - replacement" type of deposits are hosted by limestone and dolostone. The Manto orebodies are stratabound and even have stratiform irregular sheetlike to rod-shaped peneconcordant to transgressive mode of occurrence, usually horizontal to near horizontal in attitude. Manto deposits represent platformal to rift settings in epi/intra continental regions. These type of deposits may stack

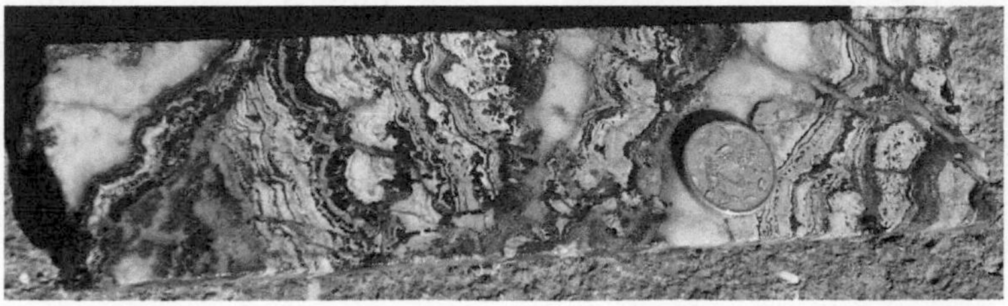

FIGURE 8.20 Sphalerite (yellow) and galena (black) mineralization in calcite (white) bands indicating different fluid-phase events, Lennard Shelf MVT deposit, Western Australia. *Source: Ref. 24.*

vertically one above the other and be connected by pipes or chimneys of dispersed mineralization. The major ore minerals are sphalerite, galena, sulfosalts silver, pyrite, pyrrhotite, and, locally, chalcopyrite ± Au. The orebodies are often affected by deformation and metamorphism. The mineralization is genetically replacement type by hydrothermal solutions (Fig. 8.21) at ~200 °C with/without development of skarns. The metal sources are from plutonic granite/granodiorite and introduction of hydrothermal fluids penecontemporaneous with sedimentation during Devonian to Jurassic period. The examples are Santa Eulalia silver–zinc–lead–copper–tin deposit at Chihuahua, Mexico, and similar deposit at Leadville, Colorado, USA.

8.4.9. Irish

"Irish" type of carbonate (limestone and dolostone) hosted sulfide deposits are stratiform and

FIGURE 8.21 Massive crystalline galena (shining steel gray) with remnant of partially replaced quartz vein/fragment (white).

stratabound, often dislocated by normal faults, and occur as riftogenic basin margin with the presence of basic volcanic and pluton rocks. Dolomitization and silicification along with silica-rich Fr oxide zoning are common. The major metallic minerals in order of abundances are sphalerite, galena, and chalcopyrite with minor amount of barium, silver, and cadmium. The usual age of formation is Carboniferous. Navan is the largest (preproduction ore reserves of 58 Mt @ 8.33% Zn, 2.05% Pb, and 244 t Ag) of the Irish Zn–Pb deposits and contains some of the most important evidence for epigenetic mineralization and is hosted in Lower Carboniferous platform carbonates of the early Courceyan Navan Group (~351 ± 4 Ma).

8.4.10. Pennine

The "pennine" type of deposits are typically hosted by limestone, and locally sandstone and dolostone. The deposits occur mainly as subvertical veins along margins of second-order basins within large rift basin platforms formed during Carboniferous to Triassic period. The major ore minerals are sphalerite, galena, and fluorite. The fracture-controlled fluorite-dominant zinc–lead deposits of Kentucky and Illinois are some of the examples.

8.4.11. Alpine/Bleiberg

The "alpine" deposits are hosted by dolostone, limestone, and marble occurring as concordant sheets, lenses, and discordant veins on rifted platform formed during Middle to Upper Triassic period. Dolomitization and silicification of host rock is very common. The major metallic minerals are sphalerite, galena, and pyrite ± sulfosalts, fluorite, barite, and anhydrite.

8.4.12. Skarn

The "skarn"-type deposits are formed in a process similar to that of porphyry orebodies. Skarn deposits are developed due to replacement, alteration, and contact metasomatism of the surrounding country rocks by ore-bearing hydrothermal solution adjacent to mafic, ultramafic, felsic, or granitic-intrusive body. It is most often developed at the contact of intrusive plutons and carbonate country rocks. The latter are converted to marbles, calc-silicate hornfels by contact metamorphic effects. The mineralization can occur in mafic volcanic and ultramafic flows or other intrusive rocks. There are many significant world-class economic skarn deposits: Pine Creek tungsten, California; Twin Buttes copper, Arizona, and Bingham Canyon copper, Utah, United States, OK Tedi gold–copper, Papua New Guinea; Avebury nickel, Tasmania; and Tosam Tin–Copper, India (reconnaissance stage).

8.4.13. Residual

The "residual" deposits are formed by chemical weathering process like leaching which removes gangue minerals from protore and enriches valuable metals in situ or at nearby location. The most important example is formation of bauxite under tropical climate where abundance of high temperature and high rainfall during chemical weathering of granitic rocks produces highly leached cover rich in aluminum. Examples are bauxite deposit of Weipa, Gove Peninsula, Darling Range and Mitchel Plateau in Australia; Awaso and Kibi, Ghana; East Coast, India; and Eyre Peninsula Kaolin deposit Australia. Basic and ultrabasic rocks tend to form laterites rich in iron and nickel, respectively. Nickel-bearing laterites, which may or may not be associated with platinum group of elements, are mined at New Caledonia, Norseman-Wiluna greenstone belt of Western Australia and Central Africa, Ni-bearing limonite overburden at Sukinda, India. The other residual-type deposits are auriferous laterites in greenstone belts (Western Australia); Ni–Co and Cr in laterites on top of peridotites, New Caledonia and Western Australia, respectively; and Ti in soils on top of alkali igneous rocks (Parana Basin, Brazil).

8.4.14. Placer

The "placer" deposits are formed by surface weathering and ocean, river, or wind action resulting in concentration of some valuable strongly resistant minerals of economic quantities. The placer can be an accumulation of valuable minerals formed by gravity separation during sedimentary processes. The type of placer deposits are, namely, alluvial (transported by a river), colluvial (transported by gravity action), eluvial (material still at or near its point of formation), beach placers (coarse sand deposited along the edge of large water bodies), and paleoplacers (ancient buried and converted rock from an original loose mass of sediment). The most common placer deposits are those of gold, platinum group minerals, gemstones, pyrite, magnetite, cassiterite, wolframite, rutile, monazite, and zircon. The California gold rush in 1849 began when someone discovered rich placer deposits of gold in streams draining the Sierra Nevada Mountains. The marine placer deposits of rutile, monazite, ilmenite, and zircon are currently being exploited along the coast of eastern Australia, India, and Indonesia.

FURTHER READING

Pirsson[33] and Dana[9] gave a comprehensive account of almost all minerals and description of rocks, their composition. Dana's New Mineralogy by Genies et al.[21] described a new outlook to the system of mineralogy. Host rock environments, structures and classification has been described in detail by Evans[21] and Haldar.[24]

CHAPTER

9

Resource Assessment and Economic Parameters

OUTLINE

9.1. Definition 282

9.2. Parameters 282
- 9.2.1. *Cutoff* 282
- 9.2.2. *Minimum Width* 283
- 9.2.3. *Ore* 283
- 9.2.4. *Ore Deposit* 283

9.3. Estimation Procedure 284
- 9.3.1. *Small and Medium Size* 285
- 9.3.2. *Large and Deep Seated* 285
  - 9.3.2.1. Cross-Section 285
  - 9.3.2.2. Long Vertical Section 287
  - 9.3.2.3. Level Plan 287
- 9.3.3. *Statistical Method* 288
- 9.3.4. *Geostatistical Method* 288
- 9.3.5. *Petroleum (Oil and Gas)* 290
  - 9.3.5.1. Analogy Base 290
  - 9.3.5.2. Volumetric Estimate 291
  - 9.3.5.3. Performance Analysis 292

9.4. Resource Classification 292
- 9.4.1. *Metallic/Nonmetallic Minerals* 293
  - 9.4.1.1. Conventional/Traditional Classification System 293
  - 9.4.1.2. USGS/USBM Classification Scheme 294
  - 9.4.1.3. UNFC Scheme 295
  - 9.4.1.4. JORC Classification Code 296
- 9.4.2. *Mineral Oil and Gas* 296

9.5. Mineral Economics 298
- 9.5.1. *Stages of Investment* 298
- 9.5.2. *Investment Analysis* 299
- 9.5.3. *Order of Magnitude Study/Scoping Study* 300
- 9.5.4. *Prefeasibility Study* 302
- 9.5.5. *Feasibility Study* 302

9.6. Overview 302

Further Reading 304

Introduction to Mineralogy and Petrology
http://dx.doi.org/10.1016/B978-0-12-408133-8.00009-2

9.1. DEFINITION

The basic concepts of mineral resource, reserve, and associated economic parameters must be understood to establish the mutual relationship or common link between them. Mineral resources and mineral (ore) reserves are defined by the quantity (tonnage) and average quality (grade/grades of elements) of in situ concentration of valuable mineral/minerals including gangue constituents. The mineral resources pertain to well-defined three-dimensional (3D) mineralized envelopes without emphasis on economic return on investment due to inadequate exploration data. The mineral/ore reserves on the other hand consider "cutoff" based on economic boundaries with adequate exploration information. The outer boundaries of orebody are drawn judicially controlled by economic (profit over investment) criteria between valuable minerals and waste rocks or between several grades of minerals of all possible bodies within the overall framework of mineralized horizon. The evaluation is based on the information generated during various stages of exploration from inception to date. Data are collected from all types of sampling program, validated with due diligence, and captured in the main database. In situ geological resources and grades are generally higher than recoverable reserves and lower than the average grade respectively. The resource and reserve are estimated in the same way with the only significant distinction of attaching economic return with the latter. The estimation of resource/reserve for petroleum is carried based on a similar concept.

9.2. PARAMETERS

The estimation of resource/reserve primarily depends on physical (shape and size), chemical (average content of elements), and economic (profit/loss on investment) criteria of the deposit under evaluation.

9.2.1. Cutoff

"Cutoff" is the most significant relative economic factor for computation of resource and reserve from exploration data. It is an artificial boundary demarcating between low-grade mineralization or barren rock and technoeconomically viable ore that can be exploited at a profit. The cutoff boundaries change with the complexity of mineral distribution, method of mining, rate of production, metallurgical recovery, cost of production, royalty, taxes, and, finally, the commodity price in international market. Change of any of these criteria individually or in combination gives rise to different cutoff and average grade of the deposit. Cutoff is apparently "static" on short term and "dynamic" on long term. Cutoff never changes on short-term basis. Market trend is continuously monitored over long-term perspective and situation may compel to change the cutoff or close the mining operation.

The concept works well in case of deposits with disseminated grade gradually changing from outer limits to core of the mineralization. In heterogeneous vein-type deposits with rich mineral/metal at the contacts, the cutoff has little application in defining the ore limits. In large-scale mechanized mining operations, the internal waste partings are unavoidable. The minimum acceptable average grade, defined by combination of alternate layers of ore and waste, is the basic criterion of decision making. In such situation, an even "run-of-mine" (ROM) grade is obtained by scheduling ore from a number of operating stopes with variable grades. A combination of ore veins and waste partings with marginal cost analysis will define the shape of orebody. The ore veins at the margins along with the internal waste must satisfy the cost of production by itself; otherwise the marginal vein should be excluded while mine planning. This is known as variable or dynamic cutoff concept.

9.2.2. Minimum Width

The ultimate use of reserves and grades are related to mine the orebody economically. Mining of ore, by open pit and underground methods, requires minimum width of the orebody for technical reasons. Narrow width of orebody restricts the vertical limit of open pit mining due to increase of ore to waste ratio with depth. A minimum of 3 m is suitable for semimechanized ore extraction in underground mining. However, greater the width of the orebody, larger will be the volume of ore production and higher the mechanization and ore man shift leading to low cost of production. Therefore, cutoff base mineralized zone computation is performed keeping in view the minimum width.

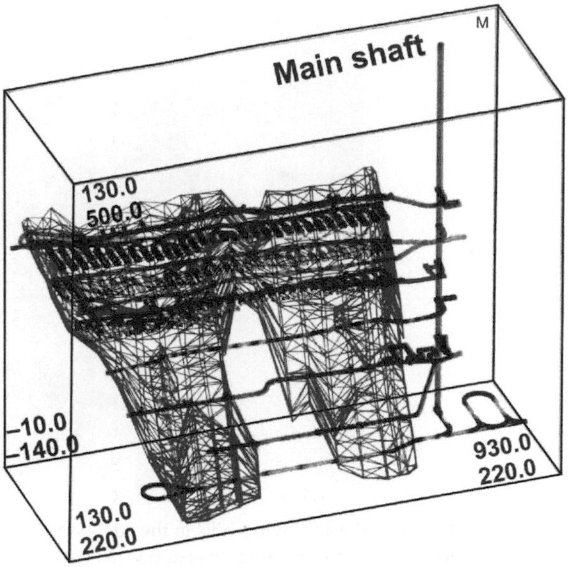

FIGURE 9.1 3D orebody wireframe model based on 50 × 50-m drill interval of main lode (South) at Rajpura-Dariba mine, India, processed by DATAMINE software in 1991. *Source: Ref. 23.*

9.2.3. Ore

"Ore" is defined as a solid naturally occurring mineral aggregate of economic interest from which one or more valuable constituents may be recovered by treatment. Therefore, ore and orebody include metallic deposits, noble metals, industrial minerals, rocks, bulk or aggregate materials, gravel, sand, gemstones, natural water, polymetallic nodules, and mineral fuel from land and ocean bed. All ores are minerals or its aggregate, but the reverse is not true.

9.2.4. Ore Deposit

An ore deposit is a natural concentration of one or more minerals within the host rock. It has a definite shape on economic criteria with finite quantity (tons) and average quality (grade). The shape varies according to the complex nature of the deposit such as layered, disseminated, veins, folded, and deformed. It may be exposed to the surface or hidden below stony barren hills, agricultural soil, sand, river, and forest.

Some of the important ore deposits are Broken Hill, Mount Isa, McArthur, HYC, Century, Lady Loretta, Lennard Shelf zinc–lead, Munni Munni platinum and Olympic Dam copper–uranium–gold deposits, Australia; Neves-Corvo copper–zinc–lead–tin deposit, Portugal; Sullivan zinc–lead deposit, British Columbia; Sudbury nickel–copper–platinum, Lac Des Iles palladium deposits, Canada; Pering zinc, Bushveld chromite–platinum deposits, South Africa; The Great Dyke platinum–nickel–copper deposit, Zimbabwe; Red Dog zinc–lead deposit, Alaska; Paguanta zinc–copper–silver deposit, South America; Stillwater platinum deposit, America; Bou Jabeur zinc–lead–fluorite–barite deposit, Tunisia; Hambok copper–zinc and Bisha copper–gold deposits, Eritrea; Noril'sk and Kola platinum deposits, Russia; as well as Rampura-Agucha, Rajpura-Dariba (Fig. 9.1) and Zawar Group of zinc–lead, Singhbhum copper, Bailadila iron ore, Sukinda chromium, Nausahi chromium–platinum, Kolar gold, Jaisalmer limestone, Jhamarkotra rock phosphate, Makrana marble, and

FIGURE 9.2 The flat low-lying hills in the east bank of River Nile are fully exploited by miners with virtually no or little mechanization, mainly for production of good-quality limes and cement situated close to Cairo city. The picture is taken while on a train journey between Cairo and Aswan, Egypt.

Salem magnesite deposits, India. There is no choice of preferential geographical location of orebody—it can be at a remote place or below a thickly populated city. It has to be accepted as it is and where it is. Moreover, ore deposits, being an exploitable nonrenewable asset, have to be used judicially at present and be left sensibly for the future (Fig. 9.2).

9.3. ESTIMATION PROCEDURE

The conventional and general procedure of mineral/ore reserves (tonnage, "T" or "t") and average grade (percent or part per million or grams per ton metal/nonmetal content) can be computed by geometrical methods.

The mineral resource and ore reserve potential of mineral deposit are estimated principally by one straightforward formula with minor variation. The unit of measurement is ton.

$$t = V \times \text{Sp.Gr.}$$

$$V = A \times \text{influence of third dimension}$$

$$\text{Total } T = \sum_{i=1}^{n} (t_1 + t_2 + t_3 \ldots t_n)$$

where

t or T = measured quantity in tons,
V = volume in cubic meter (m^3), and
A = area in square meter (m^2) derived by measurement from plans or sections of the geologically defined mineralized area of the deposit.
"Influence" of third dimension is the thickness of horizontal deposit like coal seam, bauxite, placer deposits, or drill section interval for base metal deposits.
Sp. Gr. = specific gravity, bulk density, and tonnage factor; although not truly synonymous, these are used in computation of tons by including likely volume of the void and pore spaces. Measurement of number of undisturbed drill cores or bulk samples is the most reliable means of establishing a tonnage factor.

The volume/tons can also be estimated by making 3D wireframe model of the deposit (Fig. 9.1).

The database takes into consideration of all samples and geological aspects collected during exploration program and uses in-house/commercial software.

The average grade of the deposit is computed by the standard formula:

$$\text{Grade}\,(g) = \sum (t_1 \times g_1 + t_2 \times g_2 \ldots + t_n \times g_n) / \times \sum_{i=1}^{n} (t_1 + t_2 \ldots + t_n)$$

where

t = tons of ore in subblock and
g = grade of sample

The average grade for linear samples (channel, drill hole), area (plan or section subblock), and volume can be computed by replacing "t" by "l", "a", and "v", respectively.

The resource and reserve estimation can be planned based on the type of deposits by conventional procedure. The quality estimated can be upgraded by applying statistical and geostatistical applications.

9.3.1. Small and Medium Size

Small- and medium-size deposits like coal/lignite seam, bauxite, and laterite are virtually flat and exist on or near the surface. The triangular, square, rectangle, and polygonal methods are point estimates by declustering of cells around the samples.

Triangles are formed by joining three adjacent positive intersections defining a block. The horizontal area of each block is measured and multiplied by the thickness of the mineralization to get the volume. The reserve is obtained by multiplying the volume with bulk Sp. Gr. of ore. The grade is computed by averaging the three corner values of the triangle. In the same way, squares, rectangles, and polygons can be created either by joining each positive drill hole or by perpendicular bisectrix around each borehole (Fig. 9.3).

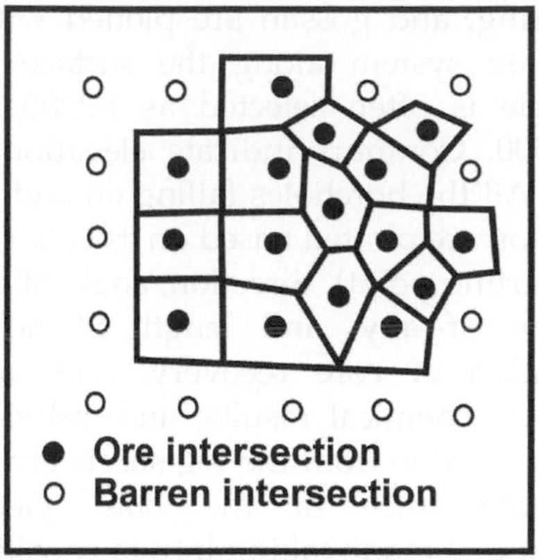

FIGURE 9.3 Conceptual diagram showing reserve estimation by square, rectangle, and polygonal method keeping sample at the center of the square or polygon. *Source: Ref. 23.*

9.3.2. Large and Deep Seated

Large and medium deposits are represented by zinc, lead, copper, chromite, and many other types of mineralization. These deposits exhibit large volume and longer continuity along strike and dip. The deposits are explored by regular grid pattern drill holes that enable creating series of cross-sections, composite longitudinal vertical section, and regular interval level plan between the top and bottom drill holes. The reserve/resources are computed with these cross-sections, long section, and level plans. The reserves and grade estimated by these three methods must show close agreement in tonnage and grade. If any major difference occurs between any two or all three procedures, it must carefully be checked and recomputed to arrive at near-identical results.

9.3.2.1. Cross-Section

Geological cross-section is a vertical image of the plane across the geological continuity of the area. The extent of section is limited by the available surface geological data and borehole information. The total surface features such as rock contacts, structures, mineralized signatures, weathering, and gossan are plotted with local coordinate system along the surface profile. The scale is often selected as 1:2000, 1:1000, and 1:500. Contours indicate elevation of the profile. All the boreholes falling on and around the section are plotted based on its collar coordinate (starting point), direction, angle of drilling, deviation, if any, and length of hole. The information of core recovery, rock contacts, structures, chemical results, and individual or composite value from the log-sheets are plotted along the trace of the hole. Geological correlation is made taking into to consideration the knowledge of the area and experience of the geologists. The orebody can be extended up to the surface if it is directly exposed, such as depicted by fresh mineralization or indirectly by signature like the presence of oxidation/gossan of sulfide deposits. Otherwise, the orebody will be treated as concealed type and shape will be drawn by drill information. The orebody configuration can be very simple consisting of a single vein or it can be multiple and give a complex type by splitting and coalescing with each other.

The total mineralized area is divided into several subblocks around each borehole intersection by halfway influence principle (Fig. 9.4). The halfway demarcation is made by joining midpoints of hanging and footwall mineralization contacts between two adjacent boreholes. The area of each subblock is measured by geometrical

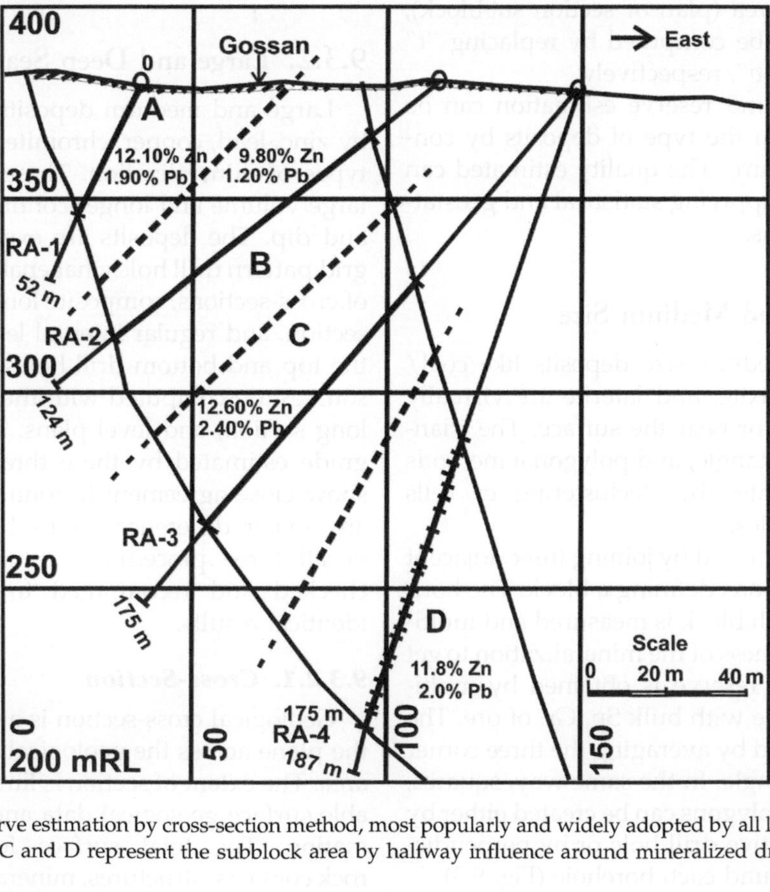

FIGURE 9.4 Reserve estimation by cross-section method, most popularly and widely adopted by all levels of professionals since decades. A, B, C and D represent the subblock area by halfway influence around mineralized drill core intersections. *Source: Ref. 23.*

formulas for rectangular, square, and triangular orebody. A planimeter or an overlay of transparent graph sheet, or AutoCAD software, can be used for measuring the area of irregular orebody. Planimeter is a drafting instrument used to measure the area of a graphically represented planar region by tracing the perimeter of the figure. The volume of the subblock is computed by multiplying the third dimension, i.e. half of drilling interval on either side. The extremities of the orebody at both the end sections can be logically extended to any distance less than or equal to half of the drill interval. Halfway influence on either side, for volume computation between sections, may introduce significant errors in tonnage and grade if similar configuration does not exist in the adjacent sections. It is recommended to draw longitudinal vertical section and level plan simultaneously to depict a reasonable 3D perspective.

9.3.2.2. Long Vertical Section

Longitudinal vertical section/projection is the creation of a vertical image along the elongated direction of ore geometry. The trace of the surface profile and subsurface position of mineralized information as gathered by drill holes and underground workings are plotted in the vertical plane. The negative information of drill holes is considered to delimit the mineralization from barren rocks. The total mineralized envelope on the longitudinal vertical section is divided into subblocks around the positive intersection with the principle of halfway influence (Fig. 9.5). The tonnage and average grade of individual subblock and total ore deposit are computed similar to the cross-section method.

9.3.2.3. Level Plan

Level plan is the horizontal plan image of any subsurface datum plane. It is very similar to

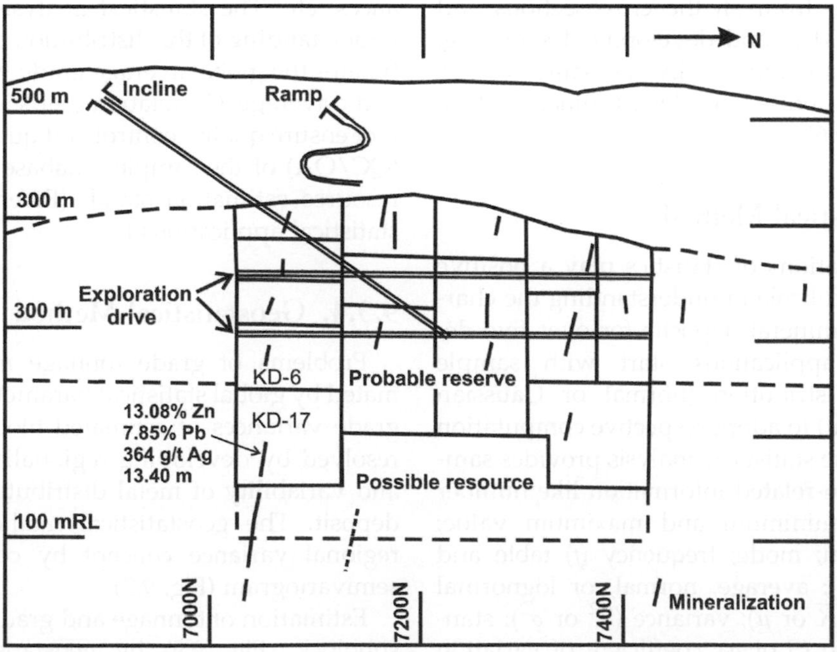

FIGURE 9.5 Estimation of reserve and grade by longitudinal vertical section—an alternative process to validate the estimate by other techniques. *Source: Ref. 23.*

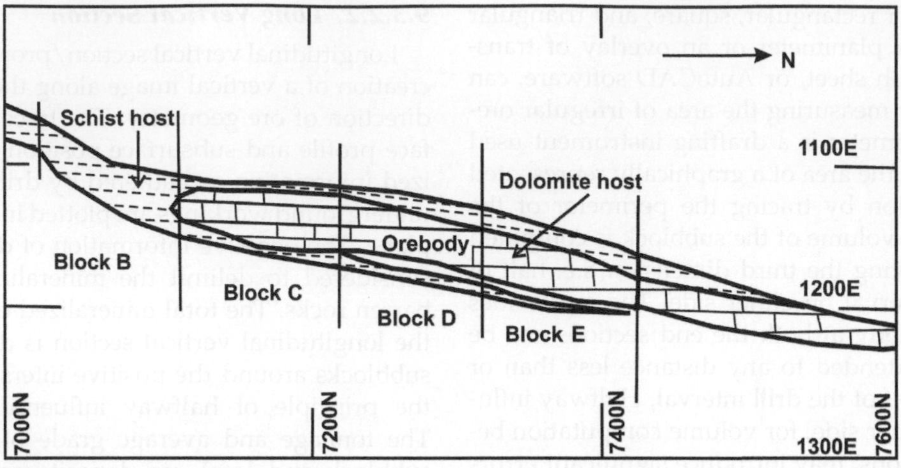

FIGURE 9.6 Conceptual diagram showing estimation of reserve and grade by level plan method—an alternative technique to cross-check estimates. *Source: Ref. 23.*

surface geological map to a large extent. Plan view of a particular level is created taking measurements from all the cross-sections and underground drill and development sampling. The reserve is computed by the same way as discussed for cross and longitudinal vertical section (Fig. 9.6).

9.3.3. Statistical Method

The applications of statistics play a positive and meaningful role in understanding the characteristics of mineral deposits for over five decades. The applications start with sample probability distribution (normal or Gaussian and lognormal) to adopt respective computation procedure. The statistical analysis provides sample population-related information like number of samples; minimum and maximum value; range; median; mode; frequency (*f*) table and plot (Fig. 9.7); average, normal, or lognormal mean grade ($\overline{X}$ or μ); variance ($\mathbf{S}^2$ or σ^2); standard deviation (*S* or σ); coefficient of variation (CV); confidence limit (CL); t-test; F-test; chi-square (χ^2) test skewness; kurtosis (K); covariance (COV) and correlation coefficient (*r*); scatter plot; regression and analysis of variances; etc. The statistical analysis enriches the understanding of the distribution pattern and estimates the global average grade with global CL of the average. Correlation coefficient and scatter plot ensure quality control and quality assurance (QC/QA) of the sample database on which the resource estimation stand. (Refer Haldar[23] for statistical applications.)

9.3.4. Geostatistical Method

Problems of grade tonnage mismatch estimated by global statistical parameters and wider grade variances of estimated blocks have been resolved by developing regionalized stationary and variability of metal distribution within the deposit. The geostatistical method works on regional variance concept by construction of semivariogram (Fig. 9.8).

Estimation of tonnage and grade for a mining subblock (Fig. 9.9) by geostatistical method (kriging) is complex and preferably needs well-tested in-house or commercial software with

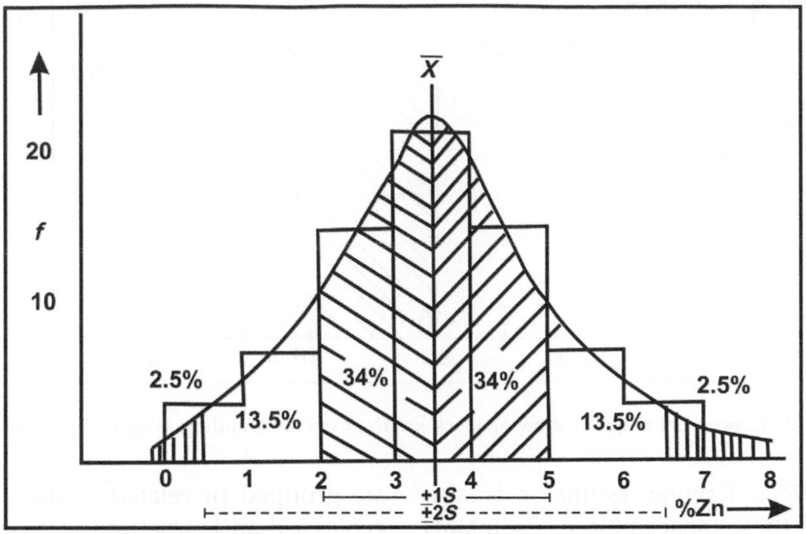

FIGURE 9.7 Relative frequency plot showing the percentage area covered by 1 and 2 standard deviations on either side of central value or mean grade of 1-m sample population and it represents a standard normal probability distribution. *Source: Ref. 23.*

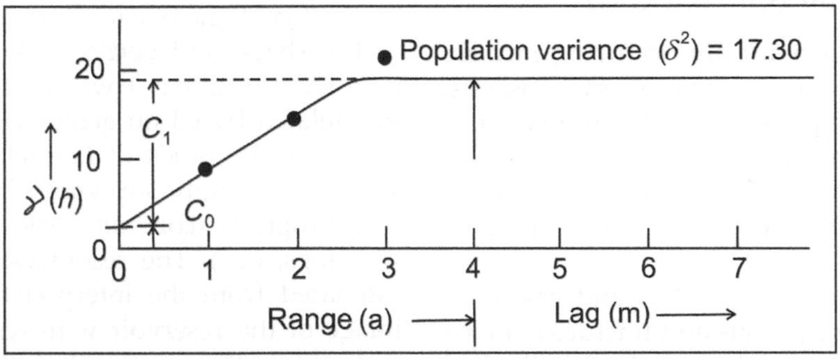

FIGURE 9.8 Standard drawing of standard semivariogram along/across the drill-hole samples or along the different directions of the mineral deposit. *Source: Ref. 23.*

knowledge of data processing. Principles of the estimation procedure are as follows:

1. It should be a linear function of the sample value x_i with block value of

$$W = \sum_{i=1}^{n} b_1 x_1 + b_2 x_2 + \ldots b_n x_n$$

where b_i is the weight given to sample w_i.

2. It should be unbiased. The expected value (μk) should be equal to the true block value (μW).

$$E\{(\mu k - \mu W)\} = 0$$

3. The mean squared error of estimation of μW should be a minimum.

$$E\left\{(\mu k - \mu W)^2\right\} = \text{a minimum}$$

The kriging estimator (μk) satisfies these conditions of linear function, unbiased estimation, and minimum variance. The corresponding error of estimation from sample (w_s) to block (W) is the

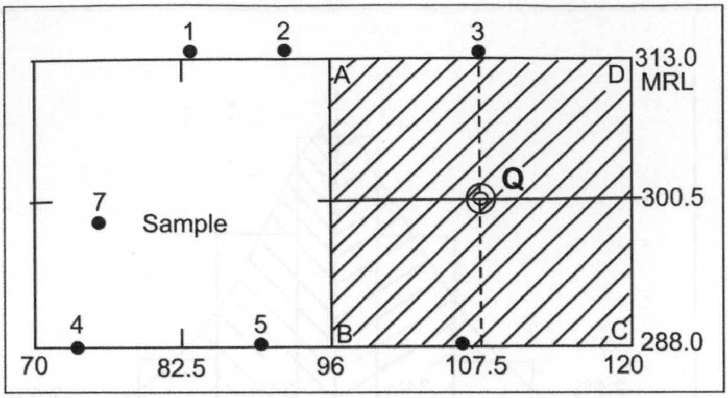

FIGURE 9.9 Conceptual diagram showing block kriging from surrounding samples 1 to 7. *Source: Ref. 23.*

kriging error (δ^2_e). Kriging estimator is also known as BLUE (best linear unbiased estimator) (Refer Haldar[23] for geostatistical applications).

9.3.5. Petroleum (Oil and Gas)

Petroleum is a naturally occurring mixture consisting of hydrocarbons in the gaseous, liquid, or solid phase. Petroleum may also contain nonhydrocarbons such as carbon dioxide, hydrogen sulfide, nitrogen, and sulfur. The chance of nonhydrocarbon content $\geq$50% is rare.

"Petroleum reservoir" or "oil and gas reservoir" is subsurface porous and fractured unique rock formation containing pool of individual and/or separate natural accumulation of moveable petroleum and gas, that is restricted by impermeable trap rock or water barriers and characterized by a single-pressure system. The oil and gas-bearing strata broadly includes sandstone, calcarenite, limestone, dolomite, argillite, and often fossil bearing with high porosity and permeability. "Traps" are of various types, such as, "stratigraphic", "structural", "hydrodynamic", and "seal". Traps are invariably underlying and overlying impermeable beds, dome, folds, faults, and structural unconformity.

"Field" is defined as an area consisting of a single and/or multiple reservoirs. The reservoirs are grouped or related to the same individual geological and structural feature, and/or stratigraphic condition. There may be two or more reservoirs in a field that are separated vertically by intervening impermeable rock, and laterally by local geologic barriers, or both.

The shape and vertical/lateral continuity of petroleum (oil/gas $\pm$ water) reservoir can be interrelated based on geological–geochemical–geophysical studies of favorable stratigraphy/structure strongly supported by intensive drilling program (core analysis, lithofacies data, well logs, etc.). The resources/reserves can be estimated from the interpreted 3D conceptual image of the reservoir with respect to oil, gas, and water. Three general methods can be discussed as (1) analogy base, (2) volumetric estimate, and (3) performance analysis.

9.3.5.1. Analogy Base

The characteristics of reservoir under consideration can be compared (well to well) with similar features of a producing reservoir for arriving at a possible resource base and average oil or gas recovery. The features include drill well spacing, lithofacies, rock and fluid properties, reservoir depth, pressure, temperature, pay thickness, and drive mechanism. The analog-based approach is the least accurate and little reliable method of petroleum reserve

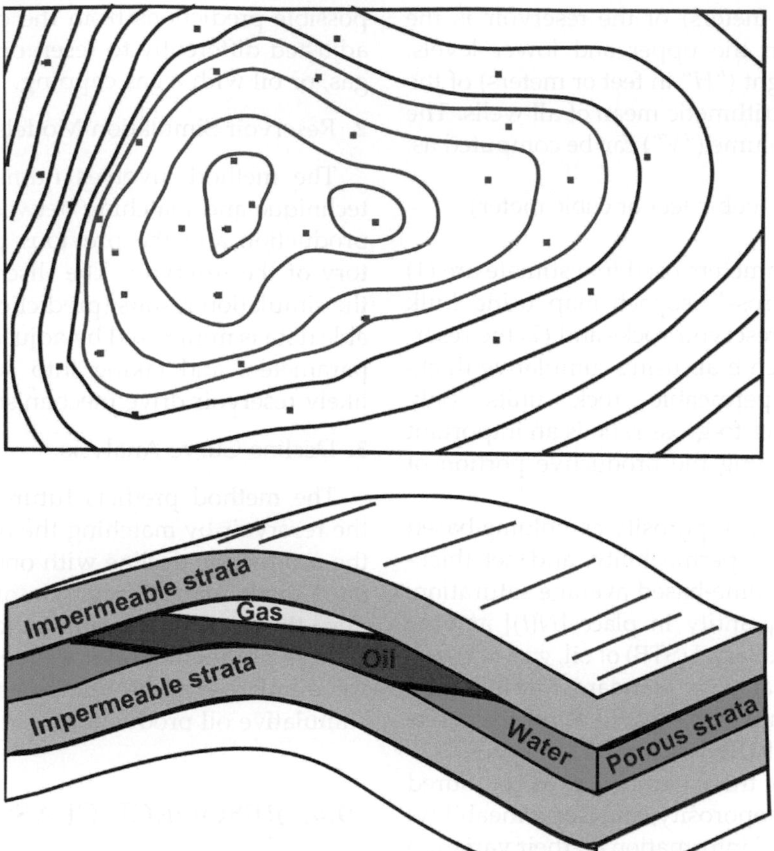

FIGURE 9.10 Conceptual diagram showing oil, gas, and water accumulation in an anticline structure as interpreted through geological–geochemical–geophysical studies and supported by drill holes (bottom). The reserves for each component are estimated by isopach method (top). •/• are location of positive/negative drill wells.

estimation. The approach simply gives an idea for further data collection, analysis, and interpretation and to adopt reliable estimation method.

9.3.5.2. Volumetric Estimate

The volumetric method determines the physical size of the reservoir from net thickness, the pore volume from void spaces, and the permeability and fluid content within the rock matrix. This provides an estimate of the hydrocarbons in place, from which ultimate recovery can be estimated by using an appropriate recovery factor (RF). Each of the factors used in the calculation have inherent uncertainties that, when combined, cause significant uncertainties and deviations in the reserves estimate.

The principle of volumetric approach is based on the information derived from the wells and supported by seismic survey into the interwell space by interpolation technique. The reservoir volume can be estimated by plotting the elevation of upper and lower levels of oil, gas, or water boundaries and making isopach map by contouring (Fig. 9.10). The average area (*"A"* in acres or square meters) of the reservoir is the average between the upper and lower levels. The average height ("H" in feet or meters) of the reservoir is the arithmetic mean of all wells. The total reservoir volume ("V") can be computed as:

$$V = A \times H \text{ (cubic feet or cubic meter)}$$

The main parameters used for estimate are (1) the reservoir "gross" isopach map using bulk thickness of the reservoir rocks and (2) the reservoir "net" isopach map using cumulative thickness of the permeable rock units only (Fig. 9.10). The net-to-gross ratio is an important parameter indicating the productive portion of the reservoir.

The reservoir rock porosity as volume-based weighted average, permeability, and net thickness product; volume-based average saturation; as well as net quantity in place [$N(t)$] at time "t" in Stock Tank Barrel (STB) of oil, gas, and water can be estimated by standard formula. The estimated ultimate recovery (EUR) of the reservoir is given by EUR $= N(t) \times$ RF.

The plotting of these parameters as contoured maps (isopach, isoporosity, and isopermeability) portray the critical information on their variation and distribution within the reservoir space and evaluate the reservoir pore volume and its fractions saturated with oil and gas (hydrocarbon volume). The numerical value of hydrocarbon resources/reserve estimate represents an outcome of "integrated" map analysis.

9.3.5.3. *Performance Analysis*

The performance analysis method aims at achieving the best reservoir performance prediction and works on the following:

1. Analysis based on Material Balance Equation

The method is based on the data obtained from previous reservoir performance and PVT (pressure-volume-temperature) analysis and involves some assumptions for the reservoir driving mechanism to minimize the range of possible predictions from the dataset. It is thus adjusted differently to reservoirs containing oil, gas, or oil with a gas capping.

2. Reservoir Simulation Models

The method involves numerical simulation technique and matching between the simulated production and the previous performance history of the reservoir. The discrepancy between the simulation results (prediction) and the available data is minimized by adjusting the reservoir parameters and taking into account the most likely reservoir drive mechanism.

3. Decline Curve Analysis

The method predicts future performance of the reservoir by matching the observed trend of the production decline with one or several standard mathematical methods of the production rate–time curves. The production decline curves include production rate vs time, production rate vs cumulative oil production, and water cut vs cumulative oil production.

9.4. RESOURCE CLASSIFICATION

The mineral resources and ore reserves are estimation of tonnage and grade of the deposit as outlined three-dimensionally with variation in sampling density, interpretations, and assumptions of continuity, shape, and grade. It stands as approximate and not certain until the entire ore is taken out by mining. Various sampling techniques are conducted at different density or interval with associated uncertainties. One part of the deposit may have been so thoroughly sampled that we can be fairly accurate of the orebody interpretation with respect to tonnages and grade. In another part of the same deposit, sampling may not be intensely detailed, but we have enough geologic information to be reasonably secure in making a statement of the estimate of tonnage and grade. The third situation may be based on few scattered samples on the fringes of the orebody. But we have some reasonable knowledge from extended parts of the orebody supported by geologic evidences and our understanding of similar deposits elsewhere to say that a certain amount of ore with certain grade may exist. Increase of sampling density in the lower confidence region will certainly enhance the status as mining proceeds.

Mineral resource and ore reserve classification system and reporting code have been evolved over the years by different countries exclusively on the basis of geological confidence, convenience to use, and investment need in mineral sector. Conventional or traditional classification system was in use during the twentieth century. New development took place from third and fourth quarters of the same century satisfying statutes, regulations, economic functions, industry best practices, competitiveness, acceptability, and internationally. There are several classification schemes and reporting codes worldwide such as US Geological Survey (USGS)/US Bureau of Mines (USBM) reserve classification scheme, United States; United Nations Framework Classification (UNFC) system; Joint Ore Reserves Committee (JORC) code, Australia and New Zealand; Canadian Institute of Mining and Metallurgy (CIM) classification; South African Code for the Reporting of Mineral Resources and Mineral Reserves; and The Reporting Code, United Kingdom.

The basic material and information for mineral resources and mineral reserves classification scheme and reporting code must be prepared by or under the supervision of "qualified persons (QPs)". A QP is a reputed professional with graduate or postgraduate degree in geosciences or mining engineering with sufficient experience (+5 years) in mineral exploration, mineral project assessment, mine development, and mine operation. The QP may preferably be in good standing or be affiliated with national and international professional associations or institutions. He is well informed with technical reports including QA, QC, data verification, discrepancy and limitations, estimation procedure, quantity, grade, level of confidence, categorization, and economic status (order-of-magnitude, prefeasibility, and feasibility study) of the deposit. He is in a position to make the statements and vouches for the accuracy and completeness of the contained technical report. This is a matter of professional integrity and carries legal risk.

The classification system is discussed for two categories: (1) metallic/nonmetallic minerals and (2) mineral oil and gas.

9.4.1. Metallic/Nonmetallic Minerals

9.4.1.1. Conventional/Traditional Classification System

The degree of assurance in the estimates of tonnage and grade can qualitatively be classified by using convenient terminology. In order of increasing exploration input creating high confidence level and technoeconomic viability, the categorization has been grouped as "economic reserves" and "subeconomic conditional resources". The economic ore reserves and subeconomic resources are further subdivided as Developed, Proved, Probable, and Possible (Fig. 9.11). The classification system helps the investor in decision making for project formulation. These terms are supported by experience, time tested, and well accepted over the years. The terminology is comparable with equivalent international nomenclature used by the USGS or Russian systems such as Measured, Indicated, and Inferred.

"Developed" reserves are exposed by trenches or trial pit on the surface for open pit or bounded on all sides by levels above and below, and raises and winzes on the sides of the block for underground mines. Close space definition or delineation drilling has been completed. The block is ready for "stope" preparation, blast hole drilling, blasting, and ore draw. The risk of error in tonnage and grade is minimal. The confidence of estimate is ~90%.

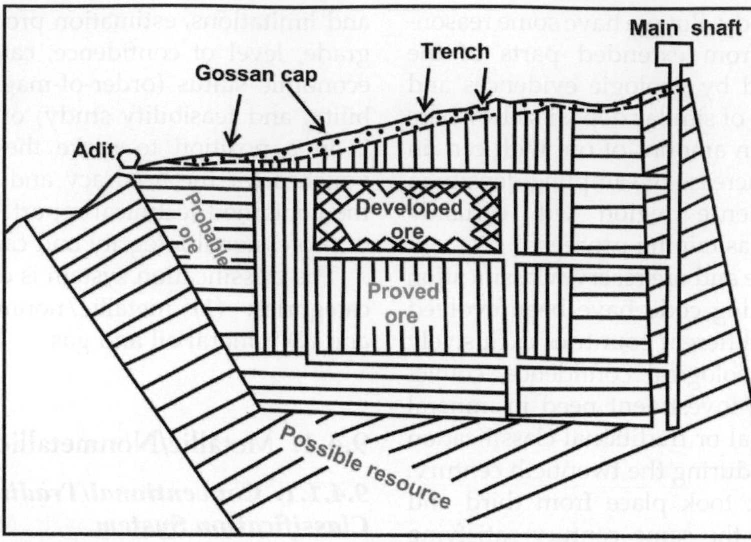

FIGURE 9.11 Conventional reserve classification systems showing various categories of reserve and resources based on enriched geological experience—a simple option for small players in mining industry. *Source: Ref. 23.*

"Proved" or "Measured" reserves are estimated based on samples from outcrops, trenches, development levels, and diamond drilling. The drilling interval would be +200 m for simple sedimentary bedded deposits (coal seam and iron ore) with expected continuity. The deposit is either exposed by trenches or trial pit for open pit and development of one or two levels for underground drilling. Further stope delineation drilling will continue to upgrade the reserve to the developed category. The confidence of estimate is ~80%.

"Probable" or "Indicated" reserve estimate is essentially based on widespaced surface and underground drilling. The opening of the deposits by trial pit or underground levels is not mandatory to arrive at this category. The confidence of estimate is ~70%.

"Possible" or "Inferred" resources are based on few scattered sample information in the strike and dip extension of the mineral deposit. There would be sufficient evidences of mineralized environment within the broad geological framework having confidence of ~50%. The possible resource will act as sustainable replacement of mined out ore.

> "Stoping" is the removal of the broken ore from an underground mine leaving behind an open space known as a "stope". Stope is a 3D configuration of in situ ore material designed for mining as an independent subblock in underground mining. Stopes are excavated near perpendicular to the level into the orebody. The excavated stopes are often backfilled with tailings, development waste, sand, and rocks from nearby area. The fill material is mixed with cement at various proportions to increase the strength of solidification. There are various stoping methods such as shrinkage, cut and fill, sublevel, vertical retreat mining, etc.

9.4.1.2. USGS/USBM Classification Scheme

The USGS collected nationwide information of geological resources and developed a

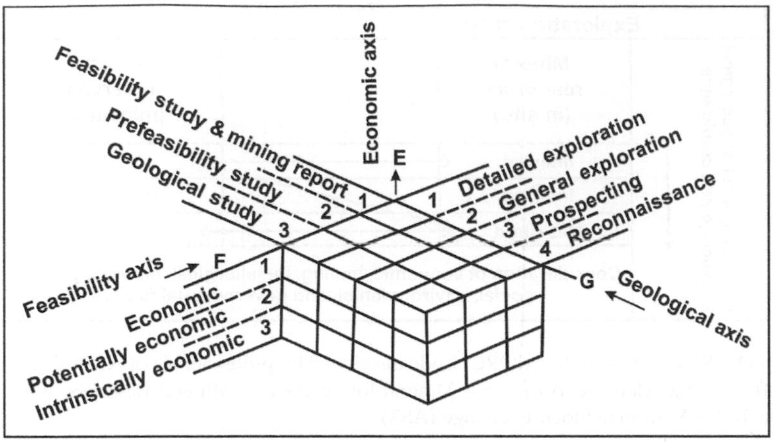

FIGURE 9.12 Resource and reserve scheme by United Nations Framework Classification (UNFC) system adopted by many countries including Government of India. *Source: Ref. 23.*

classification system (1980) in two combined axis (1) increasing degree of geological assurances and (2) increasing economic feasibility with division of identified resources and unidentified resources. The identified resources include Measured (Proved) and Indicated (Probable) reserves and Possible resources grouped under Economic and Subeconomic subclass. The Unidentified Resources include Hypothetical (Prospective) and Speculative (Prognostic) types. The resource classification scheme gives emphasis to Identified Sub-Economic resources for future target. It also initiated the concept of probability of existence of undiscovered esources simply on hypothetical and speculative ground. The USGS/USBM mineral resource classification system conveys a common classification and nomenclature, more workable in practice and more useful in long-term public and commercial planning.

9.4.1.3. UNFC Scheme

The UNFC system is a recent development in reserve categorization (E/2004/37-E/ECE/1416, February 2004). The scheme is formulated giving equal emphasis on all three criteria of exploration, investment, and profitability of mineral deposits. The format provides (1) the stage of geological exploration and assessment, (2) the stage of feasibility appraisal, and (3) the degree of economic viability. The model is represented by multiple cubes (4 × 3 × 3 blocks) with geological (G) axis, feasibility (F) axis, and economic (E) axis. The three decision-making measures for resource estimation are further specified with descending order as:

Geological axis (G) →

1. Detailed exploration
2. General exploration
3. Prospecting
4. Reconnaissance

Feasibility axis (F) →

1. Feasibility study and mining report
2. Prefeasibility study
3. Geological study

Economic axis (E) →

1. Economic
2. Potentially economic
3. Intrinsically economic

The scheme is presented in 3D perspective (Fig. 9.12) with simplified numerical codification

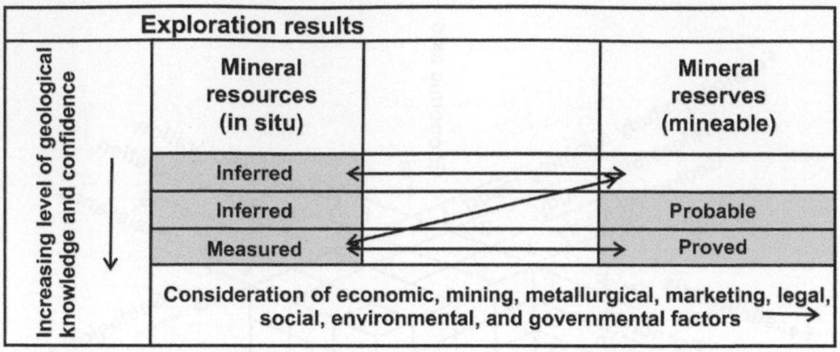

FIGURE 9.13 Joint Ore Reserve Committee (JORC) Code developed by professionals of Australian Institute of Mining and Metallurgy (AusIMM) showing relationship between Mineral Resources and Mineral Reserves. JORC-compliant organizations are registered with the Australian Stock Exchange (ASX).

facilitating digital processing of information. The scheme is internationally understandable, communicable, and acceptable across national boundaries under economic globalization that makes easy for the investor to take correct decision.

9.4.1.4. JORC Classification Code

The Minerals Council of Australia, The Australian Institute of Mining and Metallurgy, and The Australian Institute of Geoscientists established the Australian Joint Ore Reserve Committee or JORC for public reporting of Exploration Results, Mineral Resources, and Ore Reserves. The scheme was formulated on the basic principles of transparency, materiality, and competency. The other organizations represented on JORC are the Australian Stock Exchange (ASX) and Securities Institute of Australia, and registered with the New Zealand Stock Exchange (NZX) listing rules. All exploration and mining companies listed in ASX and NZX are required to comply with JORC Code and regulate the publication of mineral exploration reports on the ASX. Since 1971, the codes are being effectively updated for comparable reporting standards introduced internationally. The JORC Code applies essentially to all solid mineral commodities including diamond and other gemstones, energy resources, industrial minerals, and coal. The general relation between Exploration Results, Mineral Resources, and Ore Reserves classifies tonnage and grade estimates. The format reflects the increasing levels of geological knowledge and rising confidence. It takes due consideration of mining, metallurgical, technical, economic, marketing, legal, social, environmental, and governmental factors. The scheme imparts a checklist for authenticity at each level. Mineral Resources are concentration or occurrence of mineral prospects that eventually may become sources for economic extraction. It is placed in the Inferred category. Mineral Reserve on the other hand is the economically mineable part of Measured and/or Indicated ore. It includes dilution and allowances on account of ore losses, likely to occur when the material is mined. The relationship between mineral resources and mineral reserves is presented in Fig. 9.13.

9.4.2. Mineral Oil and Gas

The fundamental principles of petroleum (crude oil and gas) resource subdivision were established by McKelvey1 in 1972 on the same concept developed for metallic/nonmetallic minerals. The classification system has undergone changes and the simplest model is shown in Fig. 9.14.

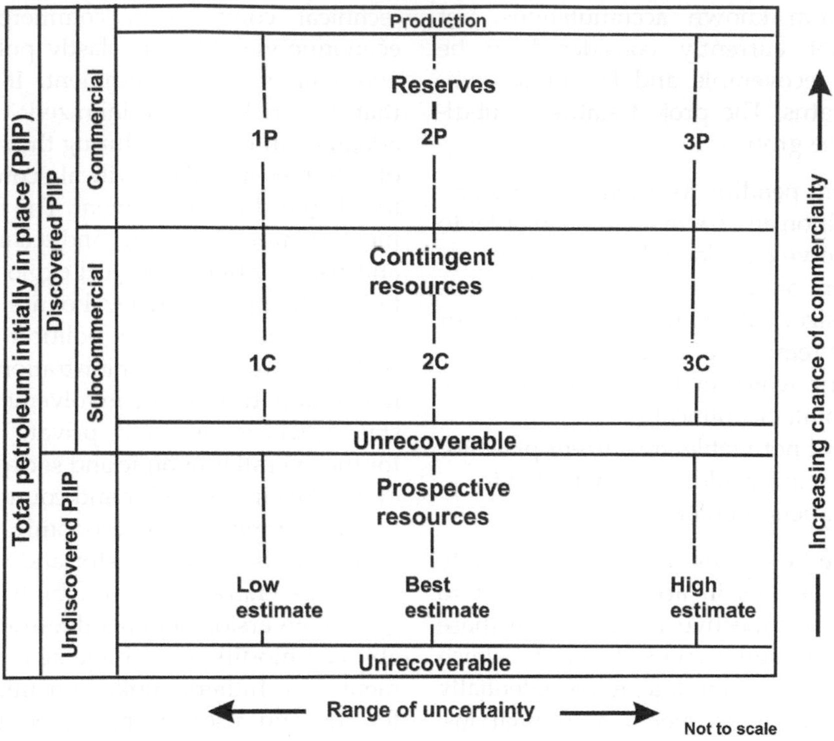

FIGURE 9.14 A graphical presentation of resources classification system adopted internally for petroleum and gas. The system defines the major recoverable reserves at successive phases of production, development, and planning. It also distinguishes Contingent, Prospective, and Unrecoverable resources. *Source: Modified after Society of Petroleum Engineers.*[44]

The reserves are a subset of resource base. The entire resource base is generally accepted to be all estimated quantities of petroleum contained in the subsurface, as well as quantities already produced. Resource is the volume estimates derived for an accumulation, and reserves is only to be quoted for a known accumulation.

"Reserves" are those quantities of petroleum which are anticipated to be commercially recovered from known accumulations from a given date forward and must satisfy the four criteria: discovered, recoverable, commercial, and remaining. Reserves are classified under three categories of Proved (1P), Probable (2P), and Possible (3P) with decreasing range of geological uncertainties and increasing chance of commerciality (Fig. 9.14). Reserves are commercially viable. The project status is subdivided into three categories.

1. Currently on production and marketing.
2. Under development and all essential approvals have been obtained.
3. Planned for development as it satisfies all the criteria for reserves, and there is a firm intent to develop, but detailed development planning and/or necessary approvals/contracts have yet to be finalized.

"Contingent Resources" are those discovered and potentially recoverable quantities that are currently not considered to satisfy the criteria for commerciality. Contingent Resources are those quantities of (crude oil and gas) which are estimated, on a given date, to be potentially recoverable from known accumulations, but which are not currently considered to be commercially recoverable and fall under sub-commercial status. The project status is subdivided into three groups:

1. Development pending as it requires further data acquisition and/or evaluation in order to confirm reserve criteria and commerciality.
2. Development on hold: the reservoir describes significant size, but awaiting development of a market or removal of other constraints to development, which may be technical, environmental, or political.
3. Development not viable: no current plans to develop or acquire additional data due to limited production potential.

"Prospective Resources" are those potentially recoverable quantities in accumulations yet to be discovered. Prospective resources are those quantities of petroleum (crude oil and gas) which are estimated, on a given date, to be potentially recoverable from undiscovered accumulations. This category stands as undiscovered and uneconomic as on date. The project status of prospective resources is subdivided into three types.

1. Prospect: potential accumulation is sufficiently well-defined to represent a viable drilling target.
2. Lead: potential accumulation is currently poorly defined and requires more data acquisition and/or evaluation in order to be classified as a prospect.
3. Play: recognized prospective trend of potential prospects and requires more data acquisition and/or evaluation to define specific leads or prospects.

9.5. MINERAL ECONOMICS

Mineral exploration is an opportunity-based investment venture with high degree of risk associated at each stage of activity. The risks are governed by geological uncertainties, technical competency, commercial necessity, economic viability, and lastly political stability and will of the Government. It is significant that the risks are minimized by generating adequate information during the various phases of activities and their critical economic analysis to safeguard the investment. The first two activities are largely scientific and technical in nature and come under geological exploration, estimation with appropriate level of accuracy, mining, and beneficiation. The political stability and will of the regional administration are socioeconomic in nature by the involvement of federal—state governments and private entrepreneurs for the overall economic and social development of the area in particular and country as a whole.

The commercial and economic aspects are not in the hands of the investor and mainly rely on market scenario. The feasibility analysis can guide conversion of mineral resources to marketable commodity with adequate return on investment. The hitherto unknown mineral resource for the end users is processed through three well-defined stages, namely, exploration, development, and production. The investment decision for each stage spins around interrelated components of "Resource, Risk, and Revenue".

9.5.1. Stages of Investment

Three well-defined stages of exploration, development, and production can convert a once unknown mineral resource into a profitable commodity. The sequential approach of reconnaissance, prospecting, and detailed exploration can establish an economic deposit. The development phase provides creation of infrastructure facilities, mining methods, entry system to the mine, and designs of mineral processing roots. There will be net outflow of cash or "negative cash flow" during exploration and development and the amount is capitalized. In the production stage, the operating cost is met through the revenue generated ending with "positive cash flow" of different magnitude into the project.

9.5.2. Investment Analysis

The investment opportunity of a mineral project is evaluated and compared with the cost at different stages of exploration, development, and production with the expected revenue to be earned during the first 10–15 years of mine production. If the benefits are higher than the associated cost, then the opportunity is worth considering. The method of investment analysis begins with estimation of the resource to be spent on exploration, development, production, royalty, taxes, and other activities vs the revenue expected to be received from the sale of end product (Fig. 9.15). The Fig. 9.15(B) clearly indicates that the cash flow during exploration and

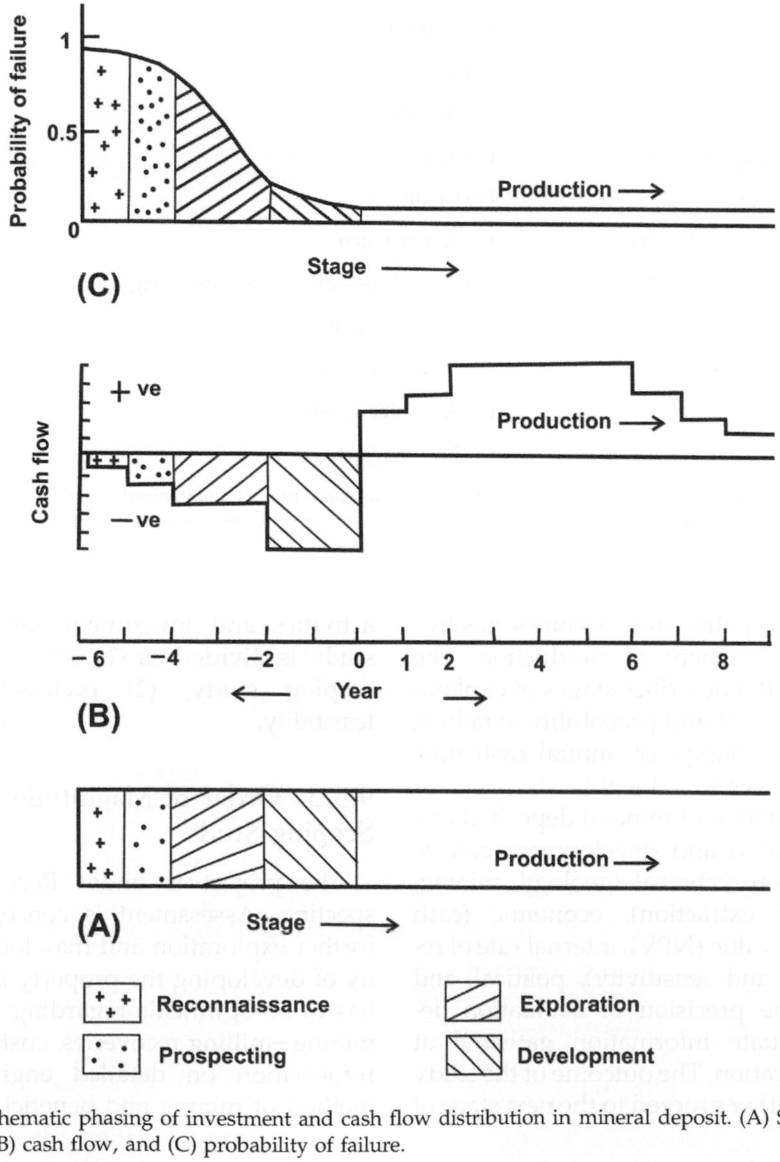

FIGURE 9.15 Schematic phasing of investment and cash flow distribution in mineral deposit. (A) Stages of exploration and development, (B) cash flow, and (C) probability of failure.

TABLE 9.1 Annual Cash Flow Concept in US Dollars

Item Head	Action
Gross sales revenue	(−) Transportation, smelting/refining and downstream ore processing charges
	(−) Royalties
	(−) Less operating costs
Net operating revenue	(−) Noncash items
	(−) Depreciation
	(−) Depletion
	(−) Amortization
Net taxable income	(−) Taxes
	(+) Credit
Net income after tax	(+) Noncash items
Net operating cash flow	(−) Capital costs (initial and sustaining)
	(−) Working capital
	(−) Exploration costs
	(−) Acquisition costs
	(−) Land payments
Net cash flow	Investment decision, financial statement

development is negative and becomes positive with the commencement of production. The Figs 9.15(A) and (C) describes stages of exploration and development and probability of failure, respectively. The concept of annual cash flow computation is given in Table 9.1.

Economic valuation of mineral deposit at any stage of exploration and development can be assessed based on technical (geology, mining, processing, and extraction), economic (cash flow, net present value (NPV), internal rate of return (IRR), risk, and sensitivity), political, and social needs. The precision of evaluation depends on adequate information gathered at that point of operation. The outcome of the study will indicate to either proceed to the next stage of activities and investments or withdraw. The study is divided as (1) order of magnitude or scoping study, (2) prefeasibility, and (3) feasibility.

9.5.3. Order of Magnitude Study/Scoping Study

The project is under Reconnaissance/Prospecting. Assessment is conceptual to decide further exploration and may look for a possibility of developing the property into a mine. One has to be optimistic regarding reserves grades, mining–milling recoveries, costs, and revenues. Information on detailed engineering design, method of mining and beneficiation, as well as operating and capital costs are borrowed from experience, reports, case studies, and published literature on similar type of deposits. This type of economic review being conducted during exploration tenure forms the groundwork and acts as an excellent guide to improve the area of information base. The main purpose is to create the ability of the investor for a "go-or-non-go decision". Case study of a base metal deposit during exploration has been discussed later. A conceptual scoping study on hypothetical data input is given in Table 9.2.

TABLE 9.2 Standard Data Collection Format for Order of Magnitude/Scoping Study

	Parameters	Unit	Total*
	Ore reserves	Mt	9
	Grade (Zn + Pb) + Ag	%	13.28
	Mine capacity	tpa	300,000
	Mining recovery	%	80
	Mine life	Year	24
	Mine dilution	%	15
	Concentrate grade	%	52
	Operating cost/ton	$	40
	Capital cost (CAPEX in Million)	$	50
	Treatment charges/t concentrate	$	180
	Metal price	$	1100
	Particulars		Million $
1	Gross in situ value		
	Zinc equivalent metal		1341.01
2	a. Mine & milling loss		377.92
	b. Smelting & refining charges		399.38
	c. Concentrate handling & transport cost		36.66
	d. (a + b + c)		813.96
3	Revenue at mine head (1−2)		527.05
4	a. Operating cost		189.85
	b. Capital sustaining cost		14
	c. Total (a + b)		203.85
5	Gross income (3−4)		323.20
6	Depreciation allowance		50.00

(*Continued*)

TABLE 9.2 Standard Data Collection Format for Order of Magnitude/Scoping Study (*cont'd*)

	Parameters	Unit	Total*
7	Taxable income (5–6)		273.20
8	Tax @46%		125.67
9	Net income (7–8)		147.53
10	Cash flow		
	a. Before tax (5)		323.20
	b. After tax (6 + 9)		197.53
11	Capital costs		50.00
12	Exploration cost		1.30
13	Net present value (NPV)		
	a. Before tax		107.17
	After tax		33.64
14	Remarks on investment		Viable

* *Figures are approximates.*

9.5.4. Prefeasibility Study

Prefeasibility study is a more detailed approach to more definite and factual information with well-defined ore geometry, reserve, and grade of higher confidence of ~80% accuracy, availability of infrastructure, proposed mining plans including scale of production, operating cost and equipments (not detailed engineering), bench-scale mineral process route and recovery, as well as economic analysis including sensitivity tests, environmental impact, and legal aspects. Experimental mining, pilot ore dressing plant using bulk samples, and other relevant detailed information may be required as a follow-up. Prefeasibility provides a more realistic picture of project viability. The project is either under mining lease or ready to apply.

9.5.5. Feasibility Study

Feasibility study is the final phase of target evaluation based on sound basic data with much greater detail analysis of the property toward development of mine and plant leading to production. All previous estimates are modified and finalized with the availability of every detail on geology, engineering, and economics. The majority of the ore reserves and grade is in the Proved and Partly Developed category. Detail engineering on mining and beneficiation plant is made. Capital and operating costs are set. Cash flow analysis with NPV, IRR, and sensitivity to different assumptions regarding revenues, costs, discount rates, and inflation are realistic and more authentic. Environment impacts with possible mitigation measures, and government formalities are expected to be cleared. Economic viability of the project is assured. Feasibility report acts as a "bankable" document for sources funding from potential financial institutions, equity, and joint venture.

9.6. OVERVIEW

The mineral reserves and resources are recovered by surface (Fig. 9.16) and underground (Fig. 9.17) mining depending on various

FIGURE 9.16 View of Jhamarkotra rock phosphate mine, India, with a team of visiting geoscientists. The mine is planned to be 7 km long, 700 m wide, and 280 m ultimate open pit limit @ 2 Mt ore and 16 Mt overburden waste per annum capacity (December, 2008).

FIGURE 9.17 Underground mine view of a main cross-cut starting from the central main shaft to the orebody at Boula-Nausahi chromite deposit, main production haulage (December, 2009).

physical, technical and economic criteria. Crude oil and gas is recovered by pumping out of the Earth's crust. Minerals, oil, and gas as produced from the crust are often in the form of low grade or complex intermixed with valuable components and waste materials. These intermediate products are processed (beneficiation, Fig. 9.18) to make a concentrate of higher values. Finally, the concentrate is smelted (Fig. 9.19) and/or refined to the highest purity metal or

FIGURE 9.18 A typical bank of froth flotation cells in operating circuit to benefit sulfide ore and generate high-grade concentrate.

FIGURE 9.19 Panoramic view of new hydrometallurgical smelter of Hindustan Zinc Limited at Dariba, Rajasthan. The smelter has annual production capacity of 210,000 t zinc and 100,000 t lead metal, and 160 MW captive power. *Source: HINDZINC Annual Report 2012–2013.*

nonmetal that can be input for the manufacturers or directly used by the consumer in the society. The prime uses of metallic and nonmetallic minerals and fuels are discussed in Table 1.2.

FURTHER READING

Haldar[24] gave detail procedure of resource and reserve estimation and reviewed the classification system including mineral economics. Popoff[34] elaborated principles and conventional methods of geological reserve classification. The classification are continuously updated by USGS and USBM,[58,59] UNFC system[57] and The Australian Joint Ore Reserves Committee (JORC)[65] for public reporting. Society of Petroleum Engineers provided *guidelines for the evaluation of petroleum reserves and resources.*[44]

CHAPTER

10

Hazards of Minerals—Rocks and Sustainable Development

OUTLINE

10.1. Definition 306

10.2. Natural Hazards 306

10.2.1. Earthquake 306

10.2.2. Volcano and Volcanism 307

10.2.3. Glacier and Avalanche 309

10.2.4. Lightning 309

10.2.5. Forest Fire 310

10.3. Hazards of Minerals 310

10.3.1. Apatite 311

10.3.2. Arsenic 311

10.3.3. Asbestos 311

10.3.4. Bauxite 311

10.3.5. Cinnabar 311

10.3.6. Clay 311

10.3.7. Coal 311

10.3.8. Corundum 312

10.3.9. Feldspar 312

10.3.10. Fluorite 312

10.3.11. Galena and Cerussite 312

10.3.12. Graphite 312

10.3.13. Gypsum 312

10.3.14. Mica 312

10.3.15. Pyrite 312

10.3.16. Radon Gas 313

10.3.17. Silica 313

10.3.18. Talc 313

10.3.19. Wollastonite 313

10.4. Hazards of Rocks 313

10.4.1. Granite 313

10.4.2. Limestone 313

10.4.3. Sandstone 313

10.4.4. Slate 314

10.4.5. Rock-Fall 314

10.4.6. Balancing Rocks 314

10.4.7. Rock Fault 315

10.5. Hazards of Exploration 315

10.6. Hazards of Mining 316

10.6.1. Baseline Monitoring 316

10.6.2. Surface Land 316

10.6.3. Mine Waste 317

10.6.4. Mine Subsidence 317

10.6.5. Mine Fire 318

10.6.6. Airborne Contaminations 319

10.6.7. Noise 319

10.6.8. Vibration 320

10.6.9. Water Resources 320

10.7. Hazards of Mineral Beneficiation 321

10.8. Hazards of Smelting 321

Introduction to Mineralogy and Petrology
http://dx.doi.org/10.1016/B978-0-12-408133-8.00010-9

10.9. Hazards of Refining 322
10.10. Sustainable Mineral Development 322
10.10.1. Mineral/Mining Sustainability 322
Further Reading 323

10.1. DEFINITION

The hazards are the sources of potential damage, harm or adverse effects on human, animals and plants. The hazards from minerals and rocks represent the health danger of men and women at workplace or at home, animals in general and loss of agriculture due to toxic effects. The damage on human and animal health is the direct effect of minerals and rocks either due to chronic exposure at workplace or due to consumption in any form through carrying agencies.

10.2. NATURAL HAZARDS

Many of the hazards, caused by the nature or naturally occurring events, are directly or indirectly related to the rocks and minerals (geological phenomenon) such as earthquakes, volcanoes, glaciers, lightening and forest fire. Many of the natural hazards are interrelated, for example, earthquakes can cause tsunami.

10.2.1. Earthquake

An "earthquake" is caused by a sudden release of stored energy within the Earth's crust that creates radiating seismic waves. The "focus" is the point where an earthquake originates (source) due to underground induced seismicity, explosions, volcanic eruptions, plate movement along, plate margin, and fault movement. The "epicenter" is the point on the Earth's surface that is directly above the focus. The magnitude of the earthquake is measured by "seismometer" that determines and monitors the ground vibration, focus and epicenter. The effects of earthquakes are most horrifying among all other natural calamities causing extensive damages to life, properties and natural geomorphologic changes (Fig. 10.1).

The intensity of the earthquake, assessed by the seismometer, is recorded as "seismographs" at various observation stations located at strategic points all over the World. "Richter magnitude scale" assigns a number between 0 and 10 (Table 10.1) to quantify the energy released during an earthquake.

The earthquake events can be predicted indicating the likely range of magnitude to occur in

FIGURE 10.1 "Sangetser Lake (Jo-Naga-Tseir)", 42 km north of Tawang, Arunachal, India. The vast lake emerged during 1971 earthquake causing massive land subsidence of 6–7 ft and subsequently water filled by flash flood in 1973. The depth of sinking can be measured from the dead trees still erected over the blue water. The lake attracts the observation of nature's beauty, serenity and sanctity and gifts a peaceful rest.

TABLE 10.1 Division of Richter Magnitude Scale and Expected Impact of Earthquake

Magnitude	Description	Average Effect	Example (magnitude)
<2.0	Micro	Felt by sensitive people and birds and animals.	Tremors are often felt in earthquake-prone region
2.0–2.9	Minor	Felt by few and no damage to buildings.	
3.0–3.9	Minor	Minor tremor but rarely any damage.	Israel May 23, 2008 (3.3)
4.0–4.9	Light	Noticeable shaking and rattling noise with minimal damage.	Neunkirchen, Germany, October 20, 2004 (4.5)
5.0–5.9	Moderate	Damage to weak buildings.	China, November 26, 2006 (5.2–5.7)
6.0–6.9	Strong	Damage to building without earthquake-resistant structure and low casualties.	Tasmania Sea: January 26, 1992, (6.9); Long Beach, CA: 1933, (6.3)
7.0–7.9	Major	Moderate to extensive damage to building and fair casualties.	Philippines, 1976 (7.9); Gujarat India, January 26, 2001 (7.7)
8.0–8.9	Great	Devastating and major damage to buildings and other structure with extensive casualties.	Ecuador, January 31, 2006 (8.8); Assam–Tibet, August 16, 1950 (8.6)
9.0–9.9	Great	Completely devastating, severe damage to all or most buildings, change in ground topography and enormous death toll.	Chile May 22, 1960 (9.5) and March 28, 1964 (9.2); Sumatra, December 26, 2004 (9.1)
>10.0	Great	Completely devastating.	Not recorded

Source: USGS and others.

specific region and time window. This is possible by scientific studies of the seismographs at the monitoring stations of seismometer.

The significant precautions to minimize the loss and damage of life and properties are to strictly follow earthquake-resistant structural design of buildings particularly in earthquake-prone region and move to open space away from the residence and workplace with slightest tremors felt.

The major earthquake-prone countries include Afghanistan, Argentina, Armenia, Australia, Burma, Chile, China, Colombia, Egypt, Ethiopia, India, Indonesia, Iran, Italy, Japan, Kazakhstan, Mexico, New Zealand, Pakistan, Russia, Turkey and United States.

10.2.2. Volcano and Volcanism

A "volcano" is an opening or rupture on the surface of the Earth's (planet) crust or ocean floor through which hot magma, ash and gases erupt from the deep seated magma chamber. The "volcanism" is the phenomenon of eruption of molten magma, ash and gases to the Earth's crust or ocean floor. The most suitable location and cause for volcanoes are tectonically active diverging and converging plate movements, as well as stretching and thinning of the Earth's crust. The examples can be cited by the mid-oceanic ridges such as the "Mid-Atlantic Ridge" and the "Pacific Ring of Fires" caused by divergent (pulling apart) and convergent (coming together) tectonic plates, respectively.

The status of a volcano can be active or extinct. An *active* volcano is the one that has at least erupted once during the last 10,000 years. An active volcano may be erupting currently or dormant and presume to erupt any time in geological future. An *extinct volcano* has not had an eruption for at least last 10,000 years and is not expected to erupt again in a comparable timescale of the future. "Mount Etna", Italy is one of the most active volcanoes in the World and is in an almost constant state of eruption activity. Mount Etna extends an extensive fertile volcanic soil support system to agriculture (vineyards and orchards) spread across the lower slopes of the mountain. The other most active volcanoes are Mauna Loa, Hawaii; Mount Nyamuragira, Congo; Kilauea volcano, Hawaii (USA); Santa Maria, Guatemala; Peak of the Furnace (Piton de la Fournaise), Eastern Reunion Island in the Indian Ocean; Stromboli, Italy; Mount Yasur, Vanuatu in the South Pacific; Lascar Volcano, Chile; Sangay Volcano, Ecuador; Mt St Helens, USA; Barren Island (Fig. 10.2), Andaman, Indian Ocean, India; and Popocatépetl Volcano, located at 55 km from Mexico City (Fig. 10.3). The examples of dormant volcanoes are Mauna Kea, Hawaii; Mount Edziza, Canada; Mount Fuji, Japan; Mount Rainier, USA; and Western Victoria, Australia. The examples of extinct volcanoes are Mount Ashitaka, Japan; Hohentwiel, Germany; and Mount Buninyong, Australia.

The volcanic activities such as lava, mud and pyroclastic flows and related events (earthquakes) can be threats to life, properties and infrastructures. The various forms of effusive lava flows may cause different hazards. Pahoehoe or basaltic lava is smooth, undulating and ropy. The "Aa" is stony hard, blocky and burning lava with rough and rubble surface. Lava flows normally follow the topography, sinking into depressions and valleys flowing down the volcano. The flows will bury roads, farmlands, crops, housing colonies, vehicles and other forms of livestock and properties standing on the way. Lava flows are dangerous. But people get enough time to evacuate out of affected areas due to slow movement of lava flow. People can also mitigate this hazard by not moving to valley or depressed areas around a volcano.

Pyroclastic material is mixture of assorted debris (dust, ash, cinders, bombs and blocks) generated by volcanic eruptions. The different kinds of pyroclastic materials pose different hazards. The dust and ash cover the automobiles and houses, rendering them unfit to drive and stay. The pyroclasts add extra weight to roofs causing the house to collapse. Inhaling of ash and dust over extended period causes long-term respiratory and lung damage. Cinders are flaming pieces of ejected volcanic material that may set fire to houses and wooded areas. Bombs

FIGURE 10.2 An active volcano in Barren Island, Indian Ocean, erupting thick column of fiery smokes during 1991. The small island is 3 km wide and contains a 1.5 km wide crater. *Source: Late Dr D. Haldar.*

FIGURE 10.3 The land base Popocatépetl active volcano, located at 55 km from Mexico City, triggered an eruption plume or cloud of ash and smoke rising at least 3 km above the crater, on May 19, 2013. *Source: Times of India, Kolkata Edition, May 20, 2013.*

and blocks run the risk of hitting various objects and people within range of the volcano.

The pyroclastic materials, mix with water from a stream or river, change the watercourse into a fast-moving mudflow. The nature of flow is fast-moving extremely hot mass of air and debris that charges down the sides of a volcano during an explosive eruption. The thicker and/or more fast-moving mudflow is more potential to destroy life and properties in its path. The mudflows damage and wash away buildings, wildlife and cars and can prove difficult to escape. The debris flows and mudflows that travel into river or stream cause flash flood and pollute the water, making it unsafe to drink.

The volcanic-associated earthquakes produce topographical deformation and/or destruction of life, buildings and other properties.

10.2.3. Glacier and Avalanche

A "glacier" is a large persistent massive body of ice that forms where the accumulation of snow exceeds its melting and sublimation over many centuries (Fig. 10.4). The glaciers slowly deform and flow due to stresses induced by their weight creating crevasses, blocks and columns of ice and other characteristic features. The glaciers scrape underlying rocks to form accumulation of unconsolidated debris of rocks and soil. This glacial change of landforms is known as "moraine". Glaciers form exclusively on land.

FIGURE 10.4 View of the Aletsch Glacier, a large flat valley of snow and ice lying just south of the summit of Jungfrau, which is one of the main peaks of Bernese Alps, Switzerland. Photo from: Top of Europe at 11,782 ft or 3571 m, September, 2009.

An "avalanche" is a rapid flow of snow down the slope, typically triggers in the starting zone by gravity or mechanical failure and accelerates fast during its journey.

Glaciers and avalanches are dynamic bodies of snow and ice, and change the speed of flow with time and topography. It may washout life and properties that stand on its way of movement. Stepping into crevasses and collapsing of snow bridge are common dangers to life. The lakes of glacial and avalanche often outburst causing flash flood and disaster to the people in the mountain ranges. The movement of glacial–avalanche is unpredictable and difficult to plan in advance.

The example of glacial–avalanche disaster, triggered by massive earthquake, could be cited from Mount Huascarán in Peru, reporting casualty between 6000 and 15,000 people. The formation of moraine-dam lake is common in the high mountain ranges like Alps, Andes, Himalaya, and Rockies and elsewhere due to retreat of glacial tongues. Moraine dam often becomes weak under high pressure from swelling water due to retreat of glacier tongue, crumble creating flash flood in the region.

10.2.4. Lightning

"Lightning" is a massive electrostatic discharge or sudden flows of electricity between two objects, say, by contact of electrically charged regions within clouds catalyzed by fine mineral dust, or between cloud and the Earth's surface. The charged regions within the atmosphere momentarily equalize themselves through a lightning flash (Fig. 10.5) or a *strike* if it hits an object on the ground. There are three

FIGURE 10.5 A lightning flash during a heavy thunderstorm over the city of Kolkata, India. *Source: Ananda Bazar Patrika, daily news paper, Kolkata, early May 2013.*

FIGURE 10.6 Forest fire at the highway connecting between Broome air port and Lennard Shelf base metal deposit, NW Australia. The day temperature was 47 °C on November 9, 2010.

types of contact: from a cloud to itself, between clouds, and between a cloud and the ground objects. The lightning phenomena can be seen by flash and heard by accompanying thunder. If one hears the sound, then he is safe from striking because the speed of sound is much slow than light.

Lightning transmits tremendous heat, high-volt electrical energy, magnetic forces of great magnitude and high-energy radiation to the striking objects (life and properties). Lightning is awfully dangerous to human life and kills about 10,000 people around the world every year and injures about 100,000 people. Lightning burns and destroys objects (properties and tall trees) on the ground. Lightning has the ability to create forest fire and large-scale power outage to damage communication and electrical system.

10.2.5. Forest Fire

A "forest fire" or "wildfire" or "bushfire" is an uncontrolled fire in an area densely enriched with abundant growth of inflammable vegetation that occurs in the countryside (Fig. 10.6). The wilderness of plant growth is due to quality of top soil, heavy rainfall and plants with high resistance. The forest fire is of extensively larger size and propagates out in high speed from its original source. It is potentially strong to change direction unexpectedly. The fire is capable to jump gaps like roads, rivers and fire-breaks. The ignition can be initiated by lightening, spark from rock-falls, spontaneous combustion, volcanic eruption, coal seam fire, extreme heat in environment and human negligence. The fire can be prevented by isolating the spread by trenches in the ground, spray of sand, water and chemicals. The hazards include loss of human and animal life and forest wealth. The smoke, ash and dust damage the lungs and respiratory system.

10.3. HAZARDS OF MINERALS

Minerals and metals are one of the essential components for the growth of human society. Needs of survival taught the prehistoric Paleolithic men the uses of stones as tools even before 20,000 years ago. There are about 5000 mineral species existing in the Earth's crust and about 4650 of these is approved by the International Mineralogical Association. The silicate minerals represent over 90% of the Earth's crust and balance includes sulfides, carbonates, etc. The discovery of minerals, estimation, mining and uses become many folds with the advent of civilization and is continuing till today.

A major proportion of minerals are immensely significant to human uses and good numbers of them are associated with inherent risk on human health and living. The risk factors of minerals/ mineral deposits are least in in-situ position. But hazards aggravate on exposure by mining and coming in contact with air and water or any other reasons. This worsens in case of radioactive minerals. Some of the hazards due to minerals are discussed.

10.3.1. Apatite

Apatite is natural calcium mineral ± fluorine and occurs as phosphate rock and mainly occurs in Canada, United States, Europe and Russia. Apatite is used as a source of phosphorus and phosphoric acid and fertilizers. Skin contact, inhalation or ingestion may cause irritation of skin, eyes, nose, throat or gastric system.

10.3.2. Arsenic

Arsenic occurs as native forms as well as in various proportions in the minerals arsenopyrite, realgar, cobaltite, enargite, and tennantite. Arsenic is a toxic element and is harmful even to low concentration in domestic and drinking water, occupational exposure and food, if exposed over long periods. Health hazards suspected to be caused by arsenic include hypertension, bronchitis, black foot, skin disease, lymphoma, and cancer in all parts of the body.

10.3.3. Asbestos

Asbestos is a group of naturally occurring fibrous and sharp needle-like crystalline silicate minerals widely distributed throughout the World. The industrial features that make asbestos commercially useful are the high tensile strength, flexibility of the fibrous, and resistance to fire, heat, abrasion, electrical and chemical changes.

The prolonged exposure and inhalation of asbestos fibbers cause serious human health hazard. The illnesses include malignant lung cancer, mesothelioma (rare form of malignant cancer) and asbestosis (a type of pneumoconiosis). The fibrotic changes which characterize the pneumoconiosis and asbestosis are the consequence of an inflammatory process set up by fibers retained in the lung. The European Union has banned all uses of asbestos and also the extraction, manufacture and processing of asbestos products.

10.3.4. Bauxite

Bauxite is the weathering product of aluminum bearing rocks and is used as the primary source of aluminum metal and refractory bricks. The mineral largely occurs in Australia, Brazil, France, Ghana, Hungary, Surinam and India. Several pulmonary disabilities (Shaver's disease) reported from the workers engaged in smelting/refining of bauxite. The health hazard is caused by the presence of free crystalline silica in bauxite ore.

10.3.5. Cinnabar

The mineral cinnabar is the primary source of mercury. Toxic effects of mercury include damage to the brain, kidney, and lungs.

10.3.6. Clay

Clay is composed of a large amount of free silica. Exposure and inhalation of clay dust is an occupational hazard in mechanized clay mines; hard rock crushing and ore dressing plants cause silicosis. Skin contact with wet clay causes skin drying and irritation.

10.3.7. Coal

Coal is a natural solid combustible material formed under sedimentary process of prehistoric plant life. Coal being extremely brittle generates fine dust during mining, mechanical transportation crushing, grinding and pulverization. Coal dust is easily venerable to explosion. The coal mine workers suffer from pneumoconiosis or black lungs disease due to long exposure to coal dust. The main hazards are coal mine fire all over the mining world.

10.3.8. Corundum

Corundum occurs as a mineral in mica schist, gneiss, marbles, low silica igneous and nepheline syenite intrusive. Corundum is primarily found in Zambia, Russia, Sri Lanka and India and is used as gemstone abrasive. Exposure to corundum dusts through inhalation, ingestion, skin and/or eye contact will cause irritation of eyes, skin, and respiratory system.

10.3.9. Feldspar

Feldspars represent a group of minerals composed of sodium, potassium, calcium, barium and aluminum silicate present in all common rocks. Chronic inhalation of feldspar dust causes silicosis due the presence of substantial amount of free silica.

10.3.10. Fluorite

Fluorite or fluorspar is a halide mineral composed of calcium fluoride (CaF_2). Fluorite often occurs as veins with metallic minerals. Fluorite is the principal source of fluorine and uses as flux in open hearth steel furnace. Elemental fluorine above 25 ppm is highly toxic and causes significant irritation and damage to gastric-intestinal system, eyes, respiratory tract, lungs, liver and kidneys. Moist hydrogen fluoride or hydrofluoric acid will make permanent damage to human skin.

10.3.11. Galena and Cerussite

Galena and cerussite are the principal sources of lead. The minerals occur individually and in association with zinc and copper deposits. Lead is intensely toxic to many organs and tissues including the heart, bones, intestines, kidneys, reproduction and nervous system.

10.3.12. Graphite

The mineral graphite is an allotrope of carbon (C). Graphite occurs mostly in metamorphic rocks as a result of reduction of sedimentary carbon compounds during metamorphism. Minerals associated with graphite include quartz, mica, and garnet. Graphite deposit contains silica and silicates. Long exposure and inhalation of carbon and associated dust during mining, processing and manufacturing of graphite products causes chronic type of pneumoconiosis. Graphite is hazardous to skin and eye contact may result mild irritation.

10.3.13. Gypsum

Gypsum is a soft dehydrated calcium sulfate ($CaSO_4$). The workers engaged in gypsum mining and processing are exposed to high atmospheric concentration of gypsum dust, furnace gases, smokes, high temperatures and risk of burns.

10.3.14. Mica

Mica group of silicate minerals occur as basal cleavage with tendency to generate flakes and powder. Chronic inhalation of mica and associated dust causes irritation of the respiratory track, silicosis and nodular fibrotic pneumoconiosis. Prolonged inhalation of vermiculite variety of mica containing asbestos causes asbestosis and lung cancer.

10.3.15. Pyrite

Pyrite (FeS_2) is ubiquitously distributed in almost all kinds of rocks in different proportion. Pyrite generates sulfuric acid when cones in contact with surface rain or underground water. This acidic water drains out and damages the soil, surface and underground water. Acid mine drainage is extremely harmful to the human, animals, agriculture and forest lands.

10.3.16. Radon Gas

Radon is a colorless, tasteless and odorless radioactive noble gas that occurs and accumulates naturally as the decay product of radium. It is one of the densest substances that remains in gas form under normal conditions and is considered to be a chronic and dangerous health hazard due to its radioactivity. Long and high exposure in radon gas environment causes chronic cancer to the miners that often results in fatal end.

10.3.17. Silica

The inhalation of dust containing free and combined silica damages the respiratory system and causes silicosis, a serious and potential fatal fibrotic lung disease.

10.3.18. Talc

Talc is an extremely white soft mineral composed of hydrated magnesium silicate and widely used as talcum powder. It occurs as foliated to fibrous masses. Talc containing silica and asbestos is harmful to human body. Chronic exposures to talc–silica–asbestos dust damage the lungs cancer, silicosis and asbestosis.

10.3.19. Wollastonite

Wollastonite is a calcium inosilicate mineral ($CaSiO_3$) $\pm$ Fe, Mg and Mn. It is usually white and soft. Wollastonite dust is often sensitive and causes irritation to skin, eye and respiratory system.

10.4. HAZARDS OF ROCKS

The natural hazards posed by rocks are rockfalls, rock movement, and rock climbing causing death of human and animals. Long exposure at workplace for rock processing causes lung diseases and permanent damage to respiratory system. Rock climbing is a fun and exciting sports for many. But it may end with many unexpected dangers including death by fall from high altitude, hitting with debris from falling rocks, muscle strain and pain, anxiety while climbing at high level and panic attacks.

10.4.1. Granite

The granite is the most abundant rock that constitutes the upper crust of the continental areas. The granite is coarse- to fine-grained igneous rock and in general composed of quartz (20–40%), feldspars (50–80%), and remaining with mica, amphibole. Granite is largely used as building and decorative stones and tiles and other construction purposes. Silicosis is the major health hazard for workers engaged in mining and processing of granite.

10.4.2. Limestone

Limestone is exclusively a sedimentary rock composed mainly of calcium carbonate (calcite) with high content of impurities like silica (siliceous), dolomite (magnesium carbonate) and clay (argillaceous) in varied proportion. The common health hazards in limestone quarries and processing are the exposure to airborne calcium, magnesium and free silica dust causing pulmonary changes, pharyngitis (throat inflammation), bronchitis (inflammation of mucous membranes) and emphysema (long-term lung disease).

10.4.3. Sandstone

Sandstone is a siliciclastic sedimentary rock consisting primarily of quartz sand that are often poorly cemented. The primary risks in mining, processing and construction of sandstone are silicosis and diseases in respiratory system.

10.4.4. Slate

Slate is an ultra-fine-grained foliated homogeneous metamorphic rock derived from an original argillaceous shale-type sediments composed of clay or volcanic ash through low-grade regional metamorphism. Silicosis and pneumoconiosis are the common health hazards for miners working in mining, processing and construction with slate. Chronic bronchitis and emphysema are frequently reported, particularly in extraction workers.

10.4.5. Rock-Fall

"Rock-fall" is defined as the quantities of rocks and debris falling freely from higher altitude or a significantly vertical or near-vertical rock exposure from the cliff face of the mountain. The rock-fall is caused by sliding, toppling, or free falling of detached fragment of rocks (big blocks or debris of varied sizes), that fall along a vertical or subvertical cliff and move down slope by bouncing, flying or rolling on talus or debris slopes. Rock-fall is a natural phenomenon and frequently occurs in the high hills and mountain ranges around the world. Rock-fall is accelerated by earthquake, steep slope, crushed and loose rock formation (Fig. 10.7), scanty forest and high rainfall. The common hazards caused by rock-fall include blockage of highways and railways in the mountainous terrain, closing major transportation routes for days at a time, blocking water channel effecting flash flood, injury and even killing of people and animals.

FIGURE 10.7 In June 2013, a multi-day cloudburst, centred on Himalayan Mountain belt of Uttarakhand, caused devastating flash floods, landslides, and rock-fall and rock movement. The worst natural calamity resulted loss of properties like housing, temples, cattle, human death toll exceeding 6000 and stranded millions of locals, tourists and pilgrims due to road blockage. *Source: Press Trust of India Ltd.*

10.4.6. Balancing Rocks

The "balancing rocks" are landform features predominantly of igneous rocks found in many parts of Australia, India, Zimbabwe, and Harare. It forms naturally by geological featuring of large rock mass or boulder of substantial size resting firmly on bed rock, other rocks or glacial till in a perfectly balanced state without any support (Fig. 10.8). The balancing rocks can form by (1) transportation and deposition of large glacial boulder by glaciers to a resting place, (2) large, detached rock fragment (perched boulder) transported and deposited by a glacier, and (3) an

FIGURE 10.8 Balancing rock, giant natural granite of Mahabalipuram, Tamil Nadu, India firmly resting on granite basement in perfect balanced state without any support.

erosion remnant of persisting rock formation after extensive wind, water and/or chemical wearing (Fig. 10.8). Balancing rocks are amazing natural art form, but if misbalanced by earthquake in the area, the gigantic ball will roll downward with tremendous uncontrolled force. It will make devastating damage to life and properties stand in the way.

10.4.7. Rock Fault

The "faults" are displacement of rock formation along a plane of discontinuity caused by various dynamic processes like change in gravity, stress, shearing, thrusting and earthquakes. There are two common types of fault namely (1) dip-slip fault (reverse and normal) and (2) strike-slip fault depending on movement along dip and strike, respectively. Faults can also be broadly classified into two main areas of "active" and "passive". Faulting is a natural process occurred in the geological past.

An active fault is one that had movement reported in the past and likely to have movement sometime in the future. Active faulting is regarded to be of geological hazards and related to earthquake as a cause. Effects of movement on an active fault include strong ground motion, surface faulting, tectonic deformation, landslides (Fig. 10.9), rock-falls, liquefaction, tsunamis and transmitting frequent seismic waves in the fault zone. Quaternary faults are often active faults that have been recognized at the surface and which have evidence of movement in the past.

FIGURE 10.9 Normal gravity fault, common in the Alps Mountain, sinking a vast mass of rock in "recent" time of geological scale. The displacement (background) changed the landform on which new habitation of human society and forest grows and continue (foreground).

10.5. HAZARDS OF EXPLORATION

The mineral exploration program can broadly be subdivided into four stages: reconnaissance, large area prospecting, prospecting and detail/ongoing exploration. The salient features include surface mapping, airborne and ground geophysical survey, geochemical study by collection of soil, rock and water samples, excavations (pits, sumps and trenches) and drilling to various extent and magnitude. The possible hazards are negligible during mineral exploration as it involves very minor excavation of the Earth surface and no acquisition of surface right. An appropriate compensation and rehabilitation is undertaken to satisfy the local inhabitants. A focus on community engagement process by facilitating employment opportunities to local community is important. The exploration program should include support, services, training and welfare to the community as a whole and youth in particular. This relation development model during exploration stages will pay dividends for future mining and related operations. This is the ideal time for development of fellow feeling and confidence building easily with the local administration and community. Compilation and evaluation of existing and new data on satellite images, topography, geological maps, sample locations, geochemistry (presence of mercury and other toxic metals), mineral occurrences, and quality of air, water, vegetation and forests will be of great value for creating the environmental base line. It will guide for future effort to reduce loss of life and property by lessening the impact of disasters.

10.6. HAZARDS OF MINING

Hazards during and after mining of a mineral deposit and rock are many with varied magnitude, seriousness, and/or sensitivity. However, each and all hazardous impacts can be mitigated technically and with good will.

10.6.1. Baseline Monitoring

A miner shall leave the area at mine closure in better and habitable form than he acquired it. Baseline monitoring is a significant component of monitoring programs for successful mining project. It commences at the reconnaissance phase and incorporates in feasibility studies. The base line information identifies the possible physical mining impact areas including economic and social issues for taking attention during the operating stages and their management. The system is continuously updated with periodical assessments to evaluate the extent of mining-related impacts and recovery following control of the impact or rehabilitation.

10.6.2. Surface Land

Surface land is a finite natural resource. The necessity of land is ever increasing due to rapid population growth in the developing countries and per capita enhanced industrial growth. However, the total of surface land requirement during actual mining (including beneficiation) is small ($<1\ km^2$) in comparison to other industries. The minerals are mined at the sites where it exists. In general, mining activity occurs in remote places far away from cities. The possibility of land and soil degradation is expected at these remote locations only.

The types of impact on land, topography and soils, and possible viable environment management solutions are the following:

1. Loss of agricultural and forest land

There will be complete loss of agricultural and forest land in and around the open pit. Underground mining uses limited surface land for the entry system and infrastructure development. In either situation, adequate compensation is provided to the land owners by cash, employment and rehabilitation. New agricultural land is developed and afforestation is done under land-use planning by enough plantations in nearby areas. No mining is permitted in reserve forest area under normal circumstances.

2. Top-soil and subsoil degradation

Surface mining affects the top soil and subsoil by changing the natural soil characteristics, e.g. texture, grain size, moisture, pH, organic matter, nutrients, etc. It is desired that the soil horizons within the selected mining limits are clearly defined. Top-soil and subsoil are removed separately and stocked at an easily accessible stable land. These soils can selectively be relaid simultaneously over the reclaim degraded land for agriculture at the time of mine closure. The removed vegetation from the mining zone should be replanted at suitable areas.

3. Changes of drainage pattern by blocking water and flash flood

The effect of unplanned mine waste dumping will change the surface topography and the local drainage pattern. The waste dumps may act as a barrier to the natural flow of rain water resulting in water logging, flash floods and damage to agriculture and to local properties downstream. It will also affect the seasonal filling of local reservoir and recharging of the ground water around the area. The changes in the drainage pattern can be anticipated from the expected postmining surface contours. Action plan for the surface drainage pattern can be designed accordingly for total water management and erosion control.

4. Landslide

Surface mining on hill slopes, particularly in areas of heavy rainfall, is vulnerable to landslides causing loss of human life, property and deforestation. This can be controlled by geotechnically designed slope of the mine and adequate support system.

5. Unaesthetic landscape

Mining activity changes the land-use pattern and alters the surface topography by increased surface erosion and excavations. This will result into unaesthetic landscape without proper reclamation. Open-pit mines must be filled with mine waste, rain or flood water for fisheries, water sports, etc.

6. Land-use planning

The procedures of land use are planned before the actual mining starts. The mine area should be reclaimed to the best possible scenario at the time of mine closure. It is the responsibility of the mining company to take into account the cost of reclamation in the project cost. The reclaimed land should preferably be reverted back to the erstwhile land owners under a mutual agreement. If it does not work, the land can be developed for the local society based on the overall planning of the region. The mode of operation can be decided by representatives from the mining company, local inhabitants, local authorities and state planning department.

10.6.3. Mine Waste

Principle of mining is not maximizing production but it aims at zero waste generation with long-term sustainable development for nonrenewable wasting assets. However, development of waste rock can never be unavoidable and handling of large amount of waste is a real hazard in mining. The quantity of solid waste likely to generate during mine development and production will be between 4 and 10 times of ore for surface and 0.25 times of ore in case of underground mining.

The coarse lumpy waste generated due to surface or underground mining can be used for reclamation of unused land in and around the mining area. This reclaimed land can be made into offices, industry, community buildings, amusement parks and playground. It can also be used as solid-waste fill of open-pit mined-out voids.

10.6.4. Mine Subsidence

"Subsidence" is the movement of ground, block or slope. It is caused by readjustment of overburden due to collapse and failure of underground operating mine excavation (Fig. 10.10), unfilled and unsupported abandoned stopes and excessive water withdrawal. It can be natural or manmade. Surface subsidence is common over shallow underground mines. The hazards due to sudden subsidence of ground include damage to human, material, topography, infrastructure and even mine inundation and development of mine fire.

The mine subsidence movement can be predicted by instrumentation, monitoring and

FIGURE 10.10 Surface subsidence over an operating underground zinc–lead open stope without any loss of man or material. The impact of the subsidence caused collapsing of crown pillars generating rich ore to draw without drilling and blasting costs.

analysis of possible impacts. The modification of underground extraction planning may help in minimum possible subsidence impact. Subsidence can be prevented by adequate support system (rib and sill pillars, steel and wood), cable and rock bolting, plugging of cracks, and backfilling by sand, cement-mixed tailing and waste rock.

10.6.5. Mine Fire

The "mine fire" (Fig. 10.11) is a common phenomenon all over the mining world. This is especially true in many of the coal seams and sometimes in high sulfides (pyrite)-rich deposits. Coal mine fire occurs due to presence of more methane gas, instantaneous oxidation property of coal when exposed to open spaces and generation of excessive heat. The intensity of fire depends on the exposed area, moisture content, rate of air flow and availability of oxygen in the surrounding area. The nature of fire may be confined to surface out crop, mine dump, open-pit benches and exclusively underground or even spread to surface. Fire in sulfides ore and concentrate is due to high pyrite bearing

FIGURE 10.11 An example of coal mine fire and recording of temperature by IR gun (left bottom) at Jharia coalfields, India. *Source: Dr A. Bhattacharya.*

dry stock pile exposed to open environment for long time under sun and atmospheric heat.

Mine fire poses serious hazards and causes impact with loss of economic, social and ecological nature. The losses are burning and locking of valuable coal reserves, polluting the air filled with excessive carbon monoxide, carbon dioxide and nitrogen, raising the surface temperature causing inconvenience to the people leaving nearby, damage of land, surface properties and vegetation, and lowering the ground water table. The common diseases that affect the local inhabitants are tuberculosis, asthma and related lungs disorder.

The nature of fire can be delineated precisely showing fire location, boundary, intensity and direction of movement. The change of temperature and gas can be recorded and measured by surface instrumentation or by airborne thermal scanner. Surface thermal IR measurements are more commonly used. The temperature anomaly is measured by a hand-held IR gun at the affected area on the surface or underground from various spots. The measurements are done in the predawn hours to minimize the effect of solar radiation. The prediction is done by simple contouring of temperature gradient or by applying different mathematical models. The depth and extent of fire can be determined by lowering probes into the fissures or along boreholes drilled in the affected areas. The temperature gradient is recorded by a digital recording unit connected by a long data transmission cable. This technique is less preferred due to the expensive drilling involvement and frequent damage of transmission cable. The drill holes also act as catalysts for additional air supply to the fire activity. The third technique is by airborne IR survey. The region is mapped by low-flying aircraft or helicopter fitted with an IR scanner. The airborne interpretation is refined by simultaneous collection of ground information on weather, soil moisture and vegetation. Once the mine fire is properly delineated, it can either be stopped or checked from further spreading. The possible remedies are the following:

1. Stripping or digging the fire out physically.
2. Injecting filling material like fly ash, water, mud, cement and sand to nonworking mines and voids through fissures, boreholes and other openings.
3. Isolating by large-scale trenching, fire-proof foam blanketing, impermeable layer of sand and debris, inert gas infusion, dry chemicals, and foams.
4. Plantation as much as possible to cool down temperature.
5. Fast action at the earliest to prevent spreading and change of fire position.

10.6.6. Airborne Contaminations

The "airborne hazards" can occur in solid, liquid and gaseous forms. The suspended particulate matter are small discrete masses of solids and liquids such as fine dust, smoke, fly ash, asbestos, lead, mercury, arsenic and other toxic metals. The gaseous pollutants are molecules of CO, SO_2, metal fumes, hydrocarbon vapor and acid mist, etc. The sources of air pollution are drilling (exploration and mining), blasting, crushing and grinding, ore and waste handling, workshops, vehicles, etc. Air pollution causes injury to eye, throat, breathing passage and lungs of the workers and the local inhabitants. Chemical pollutants are responsible for serious diseases like birth defects, brain and nerve damage, pneumoconiosis, tuberculosis and cancer.

The air pollution in the mining complexes can be controlled by the following:

1. Wet drilling applications in mine and grinding.
2. Dust suppression through mobile sprinkler along the haulage road and fixed sprinklers in the waste dumps and stockpiles.
3. Installation of dust extraction and dedusting facilities.
4. Chemical treatment at haul roads.
5. Selection of super-quality mine explosives.
6. Use of face mask.
7. Installation of dust/gas extraction system at crushers.
8. Ventilation fans and bag filters for cleaning exhaust gases from refinery.
9. Systematic stacking of waste and vegetation over inactive benches.
10. Afforestation/green belt development around the mine periphery.
11. Routine medical tests, monitoring and treatment of affected people.

10.6.7. Noise

Every person deserves "noise level" within acceptable standards in workplace and residential area. It is necessary to understand and measure the existing ambient noise level as part of the hazard assessment process at the early stage of project formulation. Measurements are conducted by automatic noise logger over adequate time period ideally during exploration. The recording is repeated prior to the mine being operational and while the mine is not operating to reflect the natural conditions. Excessive noise created by industrial machineries, transport vehicles and other associated sources increase beyond the acceptable level. Noise legislative framework is developed by most of the countries to combat problems caused by noise. The remedies are the following:

1. Community liaison and involvement in the decision-making process.
2. Periodical measurements and monitoring.
3. Control measure at manufacture level.
4. Change of blasting design and explosive control.
5. Evacuation of people from the blasting area.
6. Regulation of vehicular movements including night air traffic.
7. Acoustic barrier and green belt development.

8. Use of ear protection devices (earplugs and ear muffs) at workplace beyond 115 dB to reduce noise level exposure.
9. Location of residential and resettlement colonies away from noise generating sources.

10.6.8. Vibration

The major sources of vibration in the mineral industry are drilling and blasting in open-pit and underground operations, heavy machineries deployed for breaking and transporting ore and high-capacity crushing and grinding units at beneficiation plant. The increasing size and depth of open-pit and large-diameter long-hole blast in underground mines further aggravate the vibration. The other sources are movement of vehicles around the workplace, workshop, etc. Environment and mine safety authorities of several countries have laid down standards of acceptable vibration level to protect damages of existing structures and health hazards of workers based on their researches. The average ground particle velocity may not exceed 50 mm/s for soil, weathered and soft rocks. The limit for hard rocks is 70 mm/s. Any deviations in vibration level than standards may cause nervousness, irritability, and sleep interference. Routine ground monitoring equipment can identify sources and nature of vibration. The remedies are the following:

1. Modification measure at manufacture level.
2. Change in blasting design by hole-spacing, diameter and angle.
3. Avoid overcharging, use of delays and improved blasting techniques.
4. Use of superior quality explosive, explosive weight per delay and delay interval.
5. Control of fly rocks.
6. Green belt development.

10.6.9. Water Resources

Water is known as life indicator and is essential to sustain life. An adequate, safe and accessible supply must be available to all. The water can be classified into various groups depending on its source, use and quality. The primary sources of water are mostly from surface: oceans, rivers, streams, reservoirs and natural or manmade lakes. The other source is from subsurface aquifers that come to surface as springs. It can also be tapped by tube wells. In addition to the survival of human beings, animals and plants, it is also necessary for agriculture, industry and developmental activities. Water is rarely available in the purest form. It is usually polluted by various sources mainly through microbial, chemical and radiological aspects. Pollutants in the form of physical, chemical and biological waste make it unsuitable for uses. The physical pollutants are color, odor, taste, temperature, suspended solids and turbidity. The chemical pollutants are primarily related with geology and mining. The chemical contaminants are hardness, acidity/alkalinity, dissolved solids, metals (Fe, Pb, Cd, As and Hg), and nonmetals (fluorides, nitrates, phosphate, organic carbon, calcium, and magnesium). The microbial hazards cause infectious diseases by pathological bacteria, viruses, parasites (protozoa and helminthes), microorganism and coliform. The chances of dissolved pollutants like Fl, Pb, Cd, As and Hg are high in water bodies in the vicinity of mining and beneficiation industries. This is due to the presence of pollutant elements in the ore bearing host rocks and discharge of industrial effluents to the surface. Radiological hazards may derive from ionizing radiation emitted by radioactive chemicals in drinking water. Such hazards are rare and insignificant to public health. However, radiologic exposure from other sources cannot be ruled out. Mining operations frequently cause lowering of ground water table due to pumping of water to make mining safe.

Water, being a scarce and an essential material, needs to be used with long-term water management plan. The program must satisfy the industrial requirements as well as take care of domestic and agricultural needs of the surrounding villagers. The water balance exercise should cover study on requirements and availability of both quantity and quality for respective uses. The following water management program can be envisaged:

1. Identify all surface and subsurface sources of all types of water for adequate availability.
2. Introduce oil and grease trap and separator.
3. Construction of check dams, garland drains all around mine pit and waste dumps, soak pits, septic tanks, domestic sewage water and other water harvesting practices to arrest seasonal rain waters and any discharge of industrial effluent water for reuse in industry and plantation.
4. "Zero" discharge water management for mine pumps and recoup from tailing dam followed by sand bed filtering, treatment for pH and recycle mainly for industrial uses.
5. Low-density polyethylene lined for seepage control in and around the tailing dam and other mine water storage.
6. Minimize applications of fertilizers, herbicides, pesticides and other chemicals.
7. The dam and reservoir water (Fig. 1.4) is partially used for industry and domestic purpose of the township. A major portion is diverted to the surrounding villages for agriculture and drinking through a long-term water management master plan.

10.7. HAZARDS OF MINERAL BENEFICIATION

The lumpy run-of-mine ore is transported to beneficiation plant. The ore is crushed and grinded to very fine size for compete separation of ore and waste as well as ore and ore so that a particular mineral of interest can be separated to make concentrate. The concentration is achieved by virtue of the physical and chemical properties of individual mineral. The common practices are hand sorting, screening, gigging, tabling, gravity, magnetic, dense-media separation and froth flotation. There is likely to generate even up to 90% of fine waste (gangue minerals) out of the total ore at 10% recovered as concentrate.

The major hazard of beneficiation process is bulk generation of the fine waste in slurry form or tailings. The tailing is transferred through pipe lines to the tailing ponds for settling. The top of the tailing pond can be developed as green grassy park, playground, picnic spot or for other alternative uses. The tailing mixed with 5–10% cement can be directly diverted to the underground as void filling for ground support.

The second hazard is the use of various process chemicals in the froth flotation of metallic ore of uranium, copper, zinc, lead, and iron. The floatation chemicals are mainly isooctyl acid phosphate, sodium isopropyl xanthate and potassium amyl xanthate as conditioner and collector; methyl isobutyl carbinol (MIBC) as frothers; and sodium cyanide and copper sulfate as depressor. Cyanide is a useful industrial chemical and its key role in the mining industry is to extract gold. Acid leaching of low-grade Cu, Au, Ag, Pt ore and tailing pad is a common practice. The main leaching reagents are diluted hydrochloric, sulfuric and nitric acids. These hazard process chemicals, disposed to the tailing dam, are fast-acting poisons. The intake of these diluted chemicals over long time by gas inhalation, skin contact and through water, milk, vegetables and food pose toxic effect on human, animals, birds and insects. The chronic sublethal exposure, above the toxic threshold or repeated low doses, may cause significant irreversible adverse effects on the central nervous system.

10.8. HAZARDS OF SMELTING

The primary hazards are toxic gases and lethal chemical-enriched effluent water. Gas cleaning system and double conversion double absorption sulfuric plant is set up in smelter to minimize emission and prevent sulfur dioxide and other intoxicated gases to the environment through tall chimneys. Mercury removal plant aids to keep away from ingress of mercury in sulfuric acid and/or its entry into biocycle.

10.9. HAZARDS OF REFINING

The refinery waste is generally in the fluid form containing precious trace elements like Ag, Au, Co, Pt, Pd, etc. The value added metals are recovered by electrolytic metal/acid refinery process. The refinery discharge water contains large quantities of arsenic, antimony, bismuth, mercury and other hazardous elements. It must be neutralized and treated for effluent removal. The effluent water treatment plant is designed to remove heavy metals and other toxic components. The water discharged from various plants is collected in ponds, tanks and chambers. The water is often recycled for industrial purposes after neutralizing with lime following environmental compliance.

10.10. SUSTAINABLE MINERAL DEVELOPMENT

The principle of sustainable development promotes the thought of optimal resource utilization and leaving behind adequate resources for the future generations. The concept works on six keywords "Resource, Regenerate, Reduce, Reuse, Recycle and Replace". The sustainable mineral/mine development can be achieved by the following:

1. Science and technology
   a. Focus on pollution prevention, energy saving and health care.
   b. Clean technology that minimizes undesirable effluents, emissions and waste from products and process.
   c. Deficiency/excess of calcium, magnesium, potassium, iodine, zinc, selenium has to be optimized through grains, vegetables and fruits by using less chemical fertilizers and insecticides.

2. Fiscal measures
   a. Tax formula aims at minimizing damage to environment and ecological balance.
   b. Incentives to encourage reinvestment of income generated from mining in other mineral enterprises for sustainability.

3. Legislations
   a. Legislation is a universal means to enforce any policy.

4. Preservation of environment and forest
   a. Clean Water (Prevention and control of Pollution) Act, 1972.
   b. Clean Air (Prevention and control of Pollution) Act, 1970.
   c. The Environment (Protection) Act, 1986.
   d. The Forest Act, 1927.
   e. Wild Life Protection Act, 1972.
   f. Tribal's in Mining Projects.
   g. The Environment and Sustainable Development Act.

5. Regulated exploitation of mineral resources—sustainability and longer life
   a. Mineral Policy.
   b. Mineral Concession Rule.
   c. Mines Act.
   d. Mines Rules (Health and Safety).
   e. Mines and Mineral (Development and Regulation) Act.
   f. Mineral Conservation Act.
   g. Oil fields (Regulation & Development) Act, 1948.
   h. Coal Mines (Conservation & Safety) Act, 1952.

10.10.1. Mineral/Mining Sustainability

The mineral/mining sustainability is not merely about complying with the applicable regulations. Compliance is just the basic foundation of sustainability and more often it remains hidden from the eyes of most of the community and stakeholders. The visible issues are superstructure, track record of environmental care, biodiversity conservation, the socio-community development efforts, transparency and delivery of good governance. All these dimensions are relevant and integral to sustainable mineral/mining. The key management tasks in achieving sustainability in mineral/mining industry are the following:

1. Mineral and mining sustainability focuses around two themes: (a) concern about well being of future generations and (b) community development with humility.
2. Let us live with happiness for the present and leave enough for the future generations.
3. Mineral resources are limited, finite and nonrenewable. Once out of ground, lost forever.
4. Mineral exploration is a continuous process to augment the resource within certain limit.
5. Promoting environmental awareness within exploration and mining companies. Spread the message to the community people through programs. Share the concerns and commitments with them.
6. Educate and train employees and contractors. Adopt the method in practice.
7. Educate the local community people for economically sustainable program to achieve self-support in short period.
8. Early dialog for community development to establish trust and confidence. Encourage to work together. Build partnerships between different groups and organizations so that there is a sense of integrity, cooperation and transparency for shared focus to achieve mutually agreed common goal.
9. Developing community engagement plan involving employment with flexible work rosters, collaborative participation in decision making, services to the society, health care and medical advice, women education and child care, participation in community and spiritual festivals and handle with deep sense of humility.
10. Ensure sustainable postmine closure uses of land and all infrastructures toward creation of alternative employment.

FIGURE 10.12 "Little deeds of kindness, little words of love, make our earth an Eden, like the heaven above". *Source: Julia A. Carney (1845). Image source: Soumi.*

11. Full adaptation of compliance of national and international Impact Management Codes supported by independent audit.
12. Transparency and good governance much reflect in every plan and action.
13. Research, publication, knowledge-sharing seminars and participants at workshop.
14. Leave the area in much more environmentally beautiful, progressive and sustainable.
15. Let the future generations grow in an environment of love, affection, compassion, happiness, trust, genuineness and transparency ... (Fig. 10.12).

FURTHER READING

Books by Evans[17] and Chamley[6] are suggested for general reading of the subject. *A Guide to Leading Practice Sustainable Development in Mining* by David[28] as Principal Author, Government of Australia is an outstanding report with enormous case studies covering zinc, lead, copper, nickel, aluminum, diamond, uranium, coal and silica sand engaged in open pit and underground all over the World. Studies on hazards and development can be updated from Haldar.[24]

References

1. Bathurst RGC. Carbonate diagenesis and reservoir development: conservation, destruction and creation of pores. *Quart J Colo Sch Mines* 1986;**81**:1–25.
2. Blatt H, Tracy J. *Petrology: igneous, sedimentary and metamorphic*. New York: Freeman and Company; 1996. p. 529.
3. Boggs Jr S. *Petrology of sedimentary rocks*. Cambridge University Press; 2009. p. 600.
4. Bose MK. *Igneous petrology*. The World Press Private Limited; 1997. p. 568.
5. Bott MHP. *The interior of the earth*. London: Edward Arnold; 1982.
6. Chamley H. *Geosciences, environment and man*. Elsevier; 2003. p. 527.
7. Clarke FW. The data of geochemistry. *Bull U S Geol Surv* 1924;**779**:841.
8. Crnković B, Šarić Lj. *Građenje prirodnim kamenom*. RGN fakultet Sveučilišta u Zagrebu, Serija, Kamen; 1992 p. 184.
9. Dana ES. *A text book of mineralogy*. John Wiley & Sons, Inc; 1951. p. 851.
10. Boggs Jr S. *Petrology of sedimentary rocks*. Cambridge; 2009. p. 600.
11. Dott RH, Reynolds MJ. *Sourcebook for petroleum geology*, vol. 5. Am. Ass. Petrol. Geol. Mem; 1969. p. 471.
12. Dunham JB. Classification of carbonate rocks according to depositional texture. In: Ham WE, editor. *Classification of carbonate rocks*, vol. 1. Amer. Assoc. Petrol. Geol. Mem. 1962. p. 108–21.
13. Einsele G. *Sedimentary basins. Evolution, facies and sediment budget*. Springer; 1992. p. 628.
14. Embry AF, Klovan EJ. Absolute water depths limits of late devonian paleoecological zones. *Geol Rundsch* 1972;**61**(2):672–86.
15. Engelhardt WV. *Die Bildung von Sedimenten und Sediment-Gesteinen-Sediment–Petrologie, Teil III*. Stuttgart: Schweizerbart; 1973. p. 378.
16. Enoch P, Stawatsky LH. Pore networks in Holocene carbonate sediments. *J Sediment Petrol* 1981;**51**:961–85.
17. Evans AM. *Introduction to mineral exploration*. Blackwell Science; 1999. p. 396.
18. Füchtbauer H, Müller G. *Sedimente und Sedimentgesteine-Sediment-Petrologie II*. Stuttgart: Schweizerbart; 1970 p. 762.
19. Füchtbauer H. *Sedimente und Sedimengesteine*. 4th ed. Stuttgart: Schweizerbart; 1988. p. 1141.
20. Flügel E. *Microfacies of carbonate rocks: analysis, interpretation and application*. Springer; 2004. p. 974.
21. Gaines RV, Catherine H, Skinner W, Foord EE, Mason B, Rosenzweig A, et al. *Dana's new mineralogy, the system of mineralogy of James Dwight and Edward Salisbury Dana*. John Wiley & Sons; 1997. p. 1819.
22. Gilbert GK. Lake Bonneville-Mon. *U S Geol Surv* 1890;**1**:438.
23. Haldar SK. *Exploration modeling of base metal deposits*. Elsevier Publication; 2007. p. 227.
24. Haldar SK. *Mineral exploration: principles and applications*. Elsevier Publication; 2013. p. 374.
25. Haldar SK. Platinum–nickel–chromium: resource evaluation and future potential targets. In: *17th Convention of Indian geological congress and international conference;* 2011. p. 67–82.
26. Huges CJ. *Igneous petrology*. Elsevier; 1982. p. 551.
27. Klein C, Philpotts T. *Earth materials-introduction to mineralogy and petrology*. Cambridge University Press; 2012. p. 552.
28. Laurence D, Principal Author. *A guide to leading practice sustainable development in mining*. Department of Resources, Energy and Tourism, Government of Australia; 2011. http://www.ret.gov.au. p. 198.
29. Mason R. *Petrology of the metamorphic rocks*. London: George Allen & UNWIN; 1978. p. 254.
30. Mial AD. *Principles of sedimentary basin analysis*. 3rd, updated and enlarged ed. Springer; 2000. p. 616.
31. Moore CH. Carbonate diagenesis and porosity. *Dev Sedimentol* 1989;**46**:338.
32. Pettijohn FI, Potter PE, Siever R. *Sand and sandstone*. Springer; 1972. p. 615.
33. Pirsson LV. *Rocks and rock minerals*. John Wile & Sons, Inc; 1947. p. 349.
34. Popoff C. Computing reserves of mineral deposits; principles and conventional methods. *US Bur Mines Inf Circ* 1966;**8283**.
35. Potter PE, Maynard JB, Pryor WA. Sedimentology of shale, *Study guide and reference source*. Springer; 1980 p. 306.
36. Reading HG, editor. *Sedimentary environments and facies*. Blackwell Science Publ; 1986. p. 615.
37. Reeckman A, Friedman GM. *Exploration for carbonate petroleum reservoirs*. Wiley; 1981. p. 213.
38. Rösler H. *Lehrbuch der Mineralogie-WEB Deutschl*. Leipzig: Verl. für Grundstoffindustrie; 1987. p. 845.

39. Schmid R. Descriptive nomenclature and classification of pyroclastic deposits and fragments; recommendations of the IVGS subcommission on the systematic of igneous Rocks. *Geology* 1981;**9**:3–41.
40. Scholle PA. A color illustrated guide to carbonate rock constituents, textures, cements and porosities. *Mem Am Ass Petrol Geol* 1978;**27**:241.
41. Selley RC. *Applied sedimentology*. London–San Diego–New York–Boston: Academic Press; 1988. p. 446.
42. Slovenec D. *Sistematska mineralogija-Skripta Rudarsko-geološko-naftnog fakulteta*. Zagreb: Sveučilišta u Zagrebu; 1999. p. 229.
43. Slovenec D, Bermanec V. *Sistematska mineralogija-mineralogija silikata-Udžbenik*. Denona, Zagreb: Sveučilišta u Zagrebu; 2003. p. 359.
44. Society of Petroleum Engineers. *Guidelines for the evaluation of petroleum reserves and resources, a supplement to the SPE/WPC petroleum reserves definitions and the SPE/WPC/AAPG petroleum resources definitions*. USA; 2001 p. 141.
45. Stow DAV. Deep clastic seas. In: Reading HG, editor. *Sedimentary environments and facies*. Blackwell Sci; 1986. p. 399–444.
46. Tajder M, Herak M. *Petrologija i geologija*. Zagreb: Školska knjiga; 1966. p. 399.
47 Tišljar J. *Petrologija sedimentnih stijena-Rudarsko-geološko-naftni fakultet Zagreb*. Zagreb: Tehnička knjiga; 1987 p. 242.
48. Tišljar J. *Sedimentne stijene-Udžbenik*. Zagreb: Sveučilišta u Zagrebu, Školska knjiga; 1994. p. 422.
49. Tišljar J. *Petrologija s osnovama mineralogije-Udžbenik*. Zagreb: Sveučilišta u Zagrebu, Rudarsko-geološko–naftni fakultet; 1999. p. 211.
50. Tišljar J. *Sedimentologija karbonata i evaporita*. Zagreb: Institut za geološka istraživanja; 2001. p. 375.
51. Tišljar J. *Sedimentologija klastičnih i silicijskih taložina*. Zagreb: Institut za geološka istraživanja; 2004. p. 426.
52. Tišljar J, Velić I, Vlahović I. Facies diversity of the malmian platform carbonates in western croatia as a consequence of synsedimentary tectonics. *Géol Méditerr* 1995;**3–4**:173–6.
53. Tissot BP, Welte DH. *Petroleum formation and occurrence*. Springer; 1984. p. 699.
54. Tucker ME, Wright VP. *Carbonate sedimentology*. Blackwell Sci. Publ; 1990. p. 482.
55. Tucker ME. *Sedimentary petrology: an introduction to the origin of sedimentary rocks*. 3rd ed. Blackwell Science Ltd; 2003. p. 262.
56. Tyrrell GW. *Principles of petrology-an introduction to the science of rocks*. Redwood Burn Limited; 1978 p. 355.
57. UNFC. *United Nations Framework Classification for energy and minerals*. www.world-petroleum.org/publications/A-UNFC-FINAL.doc; 2004. p. 35.
58. USGS Bulletin 1450-A. *Principles of the mineral resource classification System of the U.S. Bureau of Mines and U.S. Geological Survey*; 1976.
59. USGS Circular 831. *Principles of a resource/reserve classification for minerals*; 1980.
60. Vrkljan M. *Mineralogija i petrologija-osnove i primjena-Udžbenik*. Zagreb: Sveučilišta u Zagrebu, Rudarsko–geološko–naftni fakultet; 2001. p. 207.
61. Vrkljan M, Babić V, Takšić J. *Mineralogija*. Zagreb: Školska knjiga; 1998. p. 413.
62. Wilson M. *Igneous petrogenesis*. Champan and Hall; 1995. p. 466.
63. Winter JD. *Principles of igneous and metamorphic petrology*. Prentice Hall; 2010. p. 702.
64. JORC. *Mineral resources and ore reserves*. www.jorc.org; 2004.

Index

Note: Page numbers followed by "b", "f" and "t" indicate boxes, figures and tables, respectively.

A

Abrasion, 125
Absorption, 125
Abu Simbel temples, 5f
Abyssal, 101
Accessory (minor) mineral ingredients, 98
ACD. *See* Aragonite compensation depth
"Acid" igneous rocks, 98–99
Acidic volcanic glass, 178–179
Actinolite asbestos, 67
Active volcanoes, 307–308, 308f
Adamellite, 107–108
Aegirine, 65–66
Aeolian processes, 129
Aeolianite, 129
Agate, 54
Agglomerates, 176t, 177–178
Aggradation, 130–131
Airborne contaminations, 319
Alabaster, 8f
Aletsch Glacier, 309f
Algal kerogens, 163
Alkali basalts, 119
Alkali feldspar, 75
Alkaline amphiboles, 67t
Allochemical diagenetic processes, 198
Allochromatic minerals, 49
Allodapic limestones, 257–258
Alluvial fans, 234–236
Alpine/Bleiberg deposits, 278
Amorphous minerals, 40–41
Amphibole, 15t–30t, 66
Amphibole schists, 225
Amphibolites, 227–228
Analcime, 79
Analogy base, petroleum reservoirs, 290–291
Andalusite, 8f
Andesite, 117
Anhydratization, 198–199
Anhydrite, 204. *See also* Gypsum; Sabkha anhydrite
Anisotropic minerals, 50
Anisotropy, 43
Annual cash flow diagram, 299–300, 300t
Anorthoclase, 77
Anorthosite, 31t–36t, 112–113, 112f
Anthophyllites, 66
Antigorite, 73
Apatite, 311
Aphanites, 102
Aplites, 120
Aragonite, 179–180
Aragonite compensation depth (ACD), 181–182, 181f
Arenaceous rocks. *See* Medium granular clastic sediments
Arenite sandstones, 155–157. *See also specific arenites*
Arfvedsonite, 68
Argentite, 15t–30t
Argillaceous sediments, 163–164
Argillites, 224
Arkosic arenites, 156–157
Arsenopyrite, 15t–30t
Asbestos, 311–313, 319
Asphalt, 163–164
Asthenosphere, 84
Atmosphere, 85b
Atterberg scale, 144, 145f
Augen gneisses, 220, 221f
Augite groups, 65
Aureoles, 41
Australia, zinc-lead-silver deposits, 266f
Autochthonous elements, 52
Avalanches, 128, 309

B

Bafflestone, 190f, 192
Balancing rocks, 314–315, 314f
Barite. *See* Baryte
Barren Island, 307–308, 308f
Barrier islands, 241
Barrovian metamorphism, 223
Baryte (barite), 10–11, 15t–30t, 57
Basaltic volcanism, 118
Basalts, 118
Baseline monitoring, 316
Basic igneous rocks. *See* Mafic igneous rocks
Batholiths, 100–101
Bauxites
  described, 174
  features and uses, 15t–30t
  hazards, 311
  karst, 174
  laterite, 174
  mineral resource, 14
  photo, 10f
  residual sediments, 172–173
Beach rock cements, 165
Bedded cherts, 211
Bedding
  external, 134
  forms from underwater slides/destruction of layers, 141–142
  internal, 134–137
  lower bedding plane structures, 140
  overview, 133–142
  upper bedding plane structures, 137–140
Belatan mine, 272f
Bentonites, 160
Beryl, 15t–30t, 62
Best linear unbiased estimator. *See* BLUE
Biochemical sedimentary rocks. *See* Chemical and biochemical sedimentary rocks
Bioclasts, 124, 186–187
Bioclasts limestone, 188
Biogenic silicon, 208–210

Bioherms, 189f, 190, 255, 258
Biological weathering, 128
Biosphere, 85b
Biostrome, 188, 189f, 190
Biotite, 70
Bioturbation, 141–142
Bismuthinite, 15t–30t
Bitumen, 163–164
Black augite, 65
"Black smokers" pipe-type deposits, 276
Black-pebble breccias, 148
Blasts, 216
Bleiberg/Alpine deposits, 278
BLUE (best linear unbiased estimator), 289–290. *See also* Kriging
Blurry currents, 128–129
Boehmite, 54
Bornite, 15t–30t
Boula-Nausahi chromite deposit, 112f, 274f, 275f, 303f
Bouma sequences, 245–247, 246f
Bounce marks, 140
Boundstone limestone, 190, 190f
Bowen, Norman L., 95
Bowen's reaction series, 95–96, 96f
Braggite, 15t–30t
Breccias, 149. *See also specific breccias*

C

Calcarenaceous sandstones, 159
Calcareous siltstone, 160–161
Calcareouse-evaporite complex, in Dalmatia, 206f
Calcite
  CCD, 181–182, 181f
  features and uses, 15t–30t
  group, 56t
  Mohs's Scale of Mineral Hardness, 48t
  photo, 9f
  secretion, 179–180
Calcite compensation depth (CCD), 181–182, 181f
Calcrete, 194
Caliche, 194
Carbonate debrites, 257–258
Carbonate hosted sulfide deposits, "Irish" type, 278
Carbonate lithic clasts, 142
Carbonate lithic detritus, 142
Carbonate platforms, 247–260
Carbonates, 55–57. *See also* Collector rocks of oil and gas; Limestone
  biogenic origin, 180
  described, 179
  extraction, 180
  pellets, 183
  sandy barrier islands, 250
  secretion in deeper water, 181–182
  secretion in shallow sea, 180–181
Carnallite, 57
Cassiterite, 15t–30t
Cataclasis, 145
Cataclastic sediments, 145, 146t
Cave limestone, 194–195, 194f
CCD. *See* Calcite compensation depth
Celadon, 70
Cement, 143
Cementation, 132–133
Cerussite, 11f, 312
Chain silicates, 63–68
Chalcedony, 207–209
Chalcocite, 15t–30t
Chalcopyrite, 10f, 226f, 272f
Chemical and biochemical sedimentary rocks. *See also* Dolomites; Evaporites; Limestone
  classification chart, 146t
  defined, 123
  overview, 179
Chemical composition, mineral/mineral deposit classification, 263, 264t
Chemical mineralogy, 40
Chemical weathering, 126–128, 126f
Cherts, 210–211, 210f
China clay, 173
Chlorite group, 72
Chlorite schists, 224
Chromite, 55
Chromite orebody, 272f
Cinnabar, 11f, 311
Clastic sediments and sedimentary rocks. *See also* Coarse-grained sediments; Collector rocks of oil and gas; Fine granular clastic sediments; Medium granular clastic sediments; Pyroclastic sediments
  diagenesis, 164–171
  genesis, 145–146
  groups, 145–146, 146t
  overview, 145
  residual sediments, 171–174
Clay
  described, 159–160
  hazard, 311
Clay shale, 159–160
Clayey sediments, diagenesis, 168–171
Claystone, 159–160
Clean Air Act, 322
Clean Water Act, 322
Cleavage, 49
Clinopyroxenes, 64, 64t
Coal, 263, 311–312
CoalMines (Conservation & Safety) Act, 322
Coarse-grained sediments (rudaceous), 146–153
  extraformational breccias, 148t, 149–152
  extraformational conglomerates, 152–153
Coastal marine environments, sand bodies, 239–242, 243f
Coated bioclasts, 186–187
Coated grains, 184
Cobaltite, 15t–30t, 311
Collapse breccias, 151–152
Collector rocks of oil and gas
  in carbonate rocks
    allodapic limestones, 257–258
    carbonate debrites, 257–258
    carbonate platforms, 247–257
    high-energy shallows, 248–250
    overview, 247
    peritidal carbonates, 250–253
    reef and perireef limestones- in carbonate platform, 254–257
    reef and perireef limestones- outside carbonate platform, 258–260
    restricted shoals, lagoons, inner shelf, 253–254
    turbidites, 257–258
  in clastic sedimentary rocks
    alluvial fans, 234–236
    debrites, 242–245
    deltas, 236–239
    overview, 234
    sand bodies in coastal marine environments, 239–242, 243f
    turbidity fans, 245–247

further reading, 260
introduction, 233–234
Collisional orogeny, 91
Compaction, 132
Concentric weathering, 118
Conglomerates, 146–147, 152. *See also* *specific conglomerates*
Contact metamorphism, 219, 220–223, 218t–219t
Contingent resources, petroleum, 297–298
Continuous reaction series, 96
Conventional/traditional classification, minerals, 293–294, 294f
Convolute bedding, 135
Cooling, of magma, 96–98, 97f
Cordierite, 62–63
Cordierite gneisses, 62–63
Core, 84
Corundum, 312
Covellite, 15t–30t
Cross-bedding, 135–136
Cross-section, estimation procedure, 285–287, 286f
Crude oil, 164
Crude oil and gas. *See* Collector rocks of oil and gas; Petroleum
Crust, 84
Crusty limestone, 194
Crystal clasts, 174, 178
Crystal forms, 45–46
Crystal lattices, 42f, 42–43, 48f
Crystal planes, 43–45
Crystal symmetry, 43–45
Crystal systems, 45–46
Crystal twinning, 45–46
Crystallization point, 40
Crystallographic axes, 43–45
Crystallography, 39
Crystals, 40–46
Cubic crystal lattices, 45
Cuprite, 15t–30t
Cutoff, 282
Cyanobacterial mats, 187, 187f
Cyclosilicates, 62–63

D

Dacite, 31t–36t, 116–117
Dalmatia, calcareouse-evaporite complex in, 206f
Debris, 146
Debris flows, 129
Debrite breccias, 151
Debrites, 129, 242–245
Decline Curve Analysis, 292
Deep seated deposits. *See* Large and deep seated deposits
Dehydration of gypsum to anhydrite, 205–207
Dellenite, 117
Delta front, 238
Delta plains, 238
Deltas, 236–239, 237f, 239f
Deltoid icosahedron, 60f
Denudation, 125
Deposition, 130–132
aggradation, 130–131
progradation, 131
retrogradation, 131
Depositional facies distribution, of Miocene sediments, 150f
Depth of occurrence, mineral/mineral deposit classification, 263–265, 268t
Descriptive mineralogy, 40
Desiccation cracks, 137–138, 138f
Desorption, 125
Destruction of layers/underwater slides, bedding, 141–142
Detailed exploration stage, 295, 298, 315–316. *See also* Reconnaissance phase
Detrite flows, 129
"Developed," traditional classification, 293
Diabase, 31t–36t, 110–111
Diagenesis. *See also* Limestone
clastic sediments and rocks, 164–171
evaporites, 205–207
limestone, 195–199
lithification, 132–133
Diallage, 65
Diamictite, 147
Diamond, 15t–30t, 48t, 52
Diapirism, 204, 204b, 207
Diatomaceous earth, 208–209, 209t
Diatomaceous mud, 208–209
Diatomite, 144, 208–209, 211
Dinosaur footprints, 139, 139f
Diorite, 108–109
Discontinuous reaction series, 96
Disten, 225–226
Dolerite, 110–111
Dolomite group, 56t
Dolomites
defined, 199–203
early diagenetic dolomitization, 201
late diagenetic dolomitization, 201–203
origin, 200–201
Dolomitization. *See also* Dolomites
limestone and, 199
in mixed zone of marine and freshwater, 202f
Dolostone, 31t–36t
Double-chain inosilicates, 66–68
Dripstone, 194–195, 194f
Dunham classification, 125, 190f, 191–192
Dunite, 114
Dykes, 101
Dynamic cutoff concept, 282
Dynamic metamorphism (kinetic metamorphism), 219, 220, 218t–219t

E

Early diagenetic dolomitization, 201
Earth
age, 88–89
origin, 86–89
structure, 83–85, 83f
Earthquakes. *See also* Faults
cause, 306
earthquake-prone countries, 307
epicenter, 306
focus, 306
Richter magnitude scale, 306, 307t
Sangetser Lake, 306f
seismometer measures, 306–307
tsunamis, 85b, 306, 315
Ecliptic plane, 88b
Eclogites, 230–232
Economic parameters. *See* Mineral economics
Edgewise breccias, 148
Emerson breccias, 151
Energy minerals, 263. *See also* Petroleum
Environment (Protection) Act, 322
Environment and Sustainable Development Act, 322
Environmental mineralogy, 40
Epicenter, 306
Epidote, 15t–30t, 61
Erosion, 125
Erosion channels, 140

Erosional caves and holes at Zion Canyon, Utah State, 212f
Estimation procedure, mineral resources/ore reserves, 284–292. *See* Large and deep seated deposits. *See also* Mineral resources/ore reserves; Petroleum reservoirs
  geostatistical method (kriging), 288–290, 290f
  small and medium size deposits, 285
  statistical method, 288
Evaporites, 203–207
  anhydrite excretion, 204
  classification of, 203–205
  defined, 203
  diagenesis of, 205–207
  gypsum excretion, 204
  halite excretion, 204
  mineral composition, 203–205
  origin of, 203–205
  petrology of, 205–207
Exploration hazards, 315–316
Exploration program, 315–316
External bedding, 134
Extinct volcanoes, 307–308
Extraformational breccias, 148t, 149–152
Extraformational conglomerates, 152–153
Extrusive (volcanic) igneous rocks, 102–104
  felsic, 99t, 116–117
  forms of, 102–103
  intermediate, 99t, 117–118
  mafic, 99t, 118–120
  mineral composition, 116–120
  textures, 103–104, 103f

F

Fault breccias, 151
Faults. *See also* Earthquakes
  active, 315
  Alps Mountain, 315f
  defined, 315
  dynamic metamorphism, 220
  hazards, 315, 315f
  plate tectonics, 89–91, 150–151
  transform, 89–90
Feasibility studies, 295, 298, 302, 316
Fecal pellets, 184, 184f, 187f
Feldspar group, 75–78
Feldspars
  features and uses, 15t–30t
  hazards, 312
  Mohs's Scale of Mineral Hardness, 48t
Feldspathic arenites, 157
Feldspathic graywacke, 158–159
Feldspathoids, 78
Felsic igneous rocks, 99t, 104b, 105–108, 116–117
Ferrohornblende, 67
Ferromagnesian minerals, 98
Field, petroleum reservoirs, 290
Fine granular clastic sediments (pelite), 159–164
  classification, 159–162, 160t
  clay and claystone, 160
  loess, 162
  marlstone, 162–163
  organic matter in argillaceous sediments, 163–164
  shale and mudstone, 161–162
  silt and siltstone, 160–161
Fire. *See* Forest fires; Mine fires
Flasar bedding, 137
Flazer cataclasite, 220, 218t–219t
Flint, 211
Floatstone, 190f, 192
Fluorite, 9f, 312
Fluorospar, 9f
Flute casts, 140
Fluvial processes, 128–129
Focus, earthquakes, 306
Footprints, dinosaur, 139, 139f
Forest Act, 1927, 322
Forest fires, 310, 310f
Fossil fuel, 164
Fossil records, 3–5
Freezing-thawing, 125
Freshwater limestone, 192–195
Froth flotation cells, 303f, 321

G

Gabbro, 109–110, 110f
Galena, 11f, 273f, 278f, 312
Garnets, 9f
Gas and oil. *See* Collector rocks of oil and gas; Petroleum
Gaussian probability distribution, 288, 289f
Genetic model
  Alpine/Bleiberg deposits, 278
  "black smokers" pipe-type deposits, 276
  "Irish" type of carbonate hosted sulfide deposits, 278
  magmatic deposits, 271–272, 275f
  "Manto-chimney replacement" deposits, 277–278
  metamorphic rocks, 274–275
  Mississippi valley-type deposits, 277, 277f
  overview, 266–279
  pennine deposits, 278
  placer deposits, 279
  residual deposits, 279
  SEDEX/stratiform deposits, 276, 277f
  sedimentary rocks, 272–273, 276f
  skarns, 223, 278–279
  VHMS, 275–276
  VMS, 275–276
Geographic distribution, mineral/mineral deposit classification, 263
Geological cross-section, estimation procedure, 285–287, 286f
Geostatistical method (kriging), 288–290, 290f
Geyserite, 208, 209t, 210
Gibbsite, 54
Gilbert deltas, 236–238, 237f
Glacial erosion, 125
Glacial processes, 130
Glaciers, 309, 309f
Glauconite, 71
Glaucophane, 68
Glaucophane schists, 224–225
Gneisses, 226
Gneissic structure, 214–215, 214f
Goethite, 54–55
Gold, 52
Graded bedding, 137
Grain support, 142, 143f
Grains
  packing of, sedimentary rock structure, 142–144, 143f
  sizes of, 144, 144t
Grainstone limestone, 190, 190f
Grand Canyon, 212f
Granites
  defined, 105
  features and uses, 31t–36t
  felsic intrusive igneous rocks, 105–108
  Half Dome, 105f
  hazards, 313

mineral constituents, 105–106
porphyritic, 105
Granitoids family, 105
Granoblastic texture, 216, 216f
Granoblasts, 216
Granodiorite, 108
Granulites, 230
Graphite, 312
Graphite schists, 226
Gravel, 146
Graywacke, 157–159
Great Pyramid, 4f
Green sandstones, 159
Green schists, 224
Groove marks, 140
Gutenberg, Bruno, 84–85
Gutenberg discontinuity, 84–85
Gypsum
defined, 57
dehydration of gypsum to anhydrite, 205–207
excretion, 204
features and uses, 15t–30t
hazards, 312
Mohs' Scale of Mineral Hardness, 48t
photo, 9f

H

Half Dome, 105f
Halides, 57
Halite, 15t–30t, 57
Harzburgite, 114
Hazards. *See also* Mineral hazards; Mining hazards; Natural hazards; Rock hazards; Sustainable mineral/mine development
defined, 306
exploration, 315–316
further reading, 323
mineral beneficiation, 321
refining, 322
smelting, 321–322
Hematite, 15t–30t, 54
Heulandite, 79
Hexagonal, 46
High-energy shallows, carbonates of, 248–250
Himalayan snow-capped peaks, 85f
Hindustan Zinc Limited, 304f
Hohentwiel, 307–308
Horizontal bedding, 134–135, 135f
Hornblende, 68
Hornfels, 220–223
Host rocks, 265–266, 271t. *See also* Mineral deposits
Humic kerogens, 163
Hummocky cross-bedding, 136
Hybrid sandstones, 159
Hydration
anhydrite to gypsum, 125, 206–207
defined, 126
Hydrosphere, 85b
Hydrothermal, 95
Hydrothermal metamorphism, 220–222
Hydroxides, 53–55, 53t
Hypabyssal, 101

I

Ice, 53
Idioblasts, 216
Idiochromatic minerals, 49
Igneous rocks. *See also* Extrusive igneous rocks; Intrusive igneous rocks
Bowen's reaction series, 95–96, 96f
chemical classification, 104b
classification, 85, 98–104
cooling of magma, 96–98, 97f
defined, 85
features and uses, 31t–36t
further reading, 120
lava properties, 94–95
layered igneous complex, 273f
magma properties, 94–95
mineral composition, 99t
uses and features, 31t–36t
veins, 120
Illite series, 70–71
Ilmenite, 111f
Impact casts, 140
Important mineral constituents, 98
"Indicated," traditional classification, 294
"Inferred," traditional classification, 294
Inner shelf, 241–242, 253–254
Inosilicates, 63f, 63–68
Insolation, 124–125
Intermediate igneous rocks
defined, 104b
extrusive, 99t, 117–118
intrusive, 99t, 108–109
Internal bedding, 134–137
Intertidal zone, 250–251
Intraclasts, 129, 154, 183, 184f
Intraformational breccias and conglomerates, 147–149, 148t
Intrusive (plutonic) igneous rocks, 100–102
felsic, 99t, 105–108
forms of, 100–101
intermediate, 99t, 108–109
mafic, 99t, 109–113
mineral composition, 104–115
shapes/structures, 102
textures, 101–102
ultrabasic, 113–115
ultramafic, 99t, 113–116
"Irish" type, of carbonate hosted sulfide deposits, 278
Iron meteorite, 81f
Isochemical diagenetic processes, 198
Isomorphism, 50–51
Isopach map, 291–292, 291f
Isostasy, 84, 85b
Isotropic minerals, 50
Isotropy, 43

J

Jadeite, 65
Jasper, 211
Jhamar Kotra rock-phosphate mine, 262–263, 272–273, 276f, 283–284, 303f

K

Kaolin, 15t–30t, 160, 173
Kaolinite, 127–128
Kaolin-serpentine group, 72–73
Karst bauxites, 174
Karst breccias, 151
Kayanite, 15t–30t
Kerogen, 163
Khetri Copper Mine, 9f, 10f, 12f, 13f, 219f, 226f, 229f, 272f
Kinetic metamorphism. *See* Dynamic metamorphism
Kohout convection, 195–196, 203
Komatiite, 31t–36t
Kriging (geostatistical method), 288–290, 290f
Kyanite, 61

L

Laccoliths, 101
Lagoons, 241–242, 253–254
Lamprophyre, 120

Landslide, 149
Lapilli tuffs, 175–177, 176t
Large and deep seated deposits, estimation procedure, 285–288
geological cross-section, 285–287, 286f
level plan method, 287–288, 288f
longitudinal vertical section, 287, 287f
Late diagenetic dolomitization, 201–203
Lateral/vertical continuity, of petroleum reservoirs, 290
Laterite, 31t–36t
Laterite bauxites, 174
Laterite soils, 173
Latite, 117
Laumontite, 79
Lava properties, igneous rocks, 94–95
Lead-zinc-silver deposits
Proterozoic Australia, 266f
Rajpurae-Daribae-Bethumni Belt, 267f
Leafy siltstone, 159–160
Lenticular bedding, 137
Lepidoblastic texture, 216
Lepidolite, 15t–30t
Leucite, 78
Leucocratic minerals, 98
Level plan method, estimation procedure, 287–288, 288f
Lherzolite, 114
Lightning
defined, 309–310
flash, 309–310, 310f
forest fires, 310
hazards, 309–310
Limestone. *See also* Marine limestones; Terrestrial limestone
aragonite
ACD, 181–182, 181f
excretion, 180
secretion, 179–180
bioclasts, 188
calcite
CCD, 181–182, 181f
secretion, 179–180
carbonates
biogenic origin, 180
extraction, 180
secretion in deeper water, 181–182
secretion in shallow sea, 180–181
classification, types, 189
composition, 179
crusty, 194
defined, 179–199
diagenesis, 195–199
diagenetic processes, 195–196, 195f
allochemical, 198
anhydratization, 198–199
deep-sea zone, 195–196
dolomitization, 199
evaporation zone, 196
greater depths of covering, 196–198
isochemical, 198
marine zone, 195
meteoric zone, 196
mixed meteoric and marine water, 196
shallow-sea zone, 195
silicification, 198
vadose zone, 196
features and uses, 31t–36t
foundation, physical-chemical-biological conditions, 179–182
freshwater, 192–195
lacustrine, 192
thinly laminated, 192–193
hazards, 313
reef-building organisms, 188
structural components, 182–188
coated bioclasts, 186–187
coated grains, 184
cyanobacterial mats, 187, 187f
fecal pellets, 184, 184f, 187f
intraclasts, 129, 154, 183, 184f
micrite, 182–183, 190f
noncarbonate authigenic minerals, 188
nonskeletal, 182–183, 183f
oncoids, 185–186, 186f
ooids, 184–185, 185f
oolites, 185
origin, 182
pellets, 183
peloids, 183–184
pisolites, 185
siliciclastic terrigenic, 188
skeletal-limestone, 183, 183f, 188
skeletal-shell organisms, 187–188, 189f
stromatolites, 183, 187, 187f, 190
Limonite, 55
Linoptilolite, 79
Lithic arenites, 156
Lithic graywacke, 158
Lithification, 132–133. *See also* Diagenesis
Lithoclasts, 174, 178, 257
Lithosphere, 84
Lizardite, 73
Loess, 159–160, 162
Loess dwarfs, 162
Lognormal probability distribution, 288, 289f
London Bridge, 3f
Longitudinal vertical section, estimation procedure, 287, 287f
Lopolith, 101
Lower bedding plane structures, 140

M

Mafic igneous rocks
defined, 104b
extrusive, 99t, 118–120
intrusive, 99t, 109–113
Magma
cooling of, 96–98, 97f
magmatic deposits, genetic model, 271–272, 275f
properties, 94–95
Magnesite, 15t–30t, 57, 56t
Magnetite, 54
Maithon Dam, 3f
Major mineral ingredients, 98
Mantle, 84
"Manto-chimney replacement" deposits, 277–278
Marble, 229, 230f, 231f
Marcasite, 15t–30t, 53
Marine limestones, 189–192
bafflestone, 190f, 192
boundstone limestone, 190, 190f
Dunham classification, 125, 190f, 191–192
floatstone, 190f, 192
grainstone limestone, 190, 190f
mudstone limestone, 190, 190f
packstone limestone, 190, 190f
rudstone, 190f, 192

wackestone limestone, 190, 190f, 191f
Marlstone, 162–163
Marmatite, 15t–30t
Material Balance Equation, 292
Matrix, 142–143
Matrix support breccias, 149f
Matrix-support systems, 142
"Measured," traditional classification, 294
Mechanical/physical weathering, 124–125
Medium and small deposits, estimation procedure, 285
Medium granular clastic sediments (arenaceous rocks)
arenite sandstones, 155–157
graywacke, 157–159
mixed/hybrid sandstones, 159
overview, 153
sandy sediments, composition/ distribution, 153–155
Mega-dunes, 129
Melanocratic minerals, 98
Melting point, 40
Mercury
cinnabar, 11f, 311
hazards, 315–316, 319, 321–322
Metallic minerals. *See also* Mineral resources/ore reserves classification system
diagnostic features/uses (list), 15t–30t
subclassification, 263
uses/diagnostic features (list), 15t–30t
Metamorphic aureole, 220–222, 265–266
Metamorphic rocks
classification, 86
defined, 86
features and uses, 31t–36t
further reading, 232
genetic model, 274–275
mineral composition of, 218t–219t
origin, 213–219
primary distribution, 218t–219t
structure, 213–219
textures, 216b
uses and features, 31t–36t
Metamorphism
Barrovian, 223
contact metamorphism, 219, 220–223, 218t–219t
defined, 213
dynamic metamorphism (kinetic metamorphism), 219, 220, 218t–219t
hydrothermal, 220–222
plutonic metamorphism, 230–232
pneumatolytic, 220–222
prograde, 217, 219f, 232
regional high-grade metamorphism, 223–229
regional low-grade metamorphism, 223–229
retrograde, 217, 219f
Meteorites, 81f, 82b
Mica
features and uses, 15t–30t
group, 70–71, 69t
hazards, 312
schists, 225
Migmatites, 232
Miller indices, 43–44
Millerite, 15t–30t
Mine fires, 318–319, 318f
Mine subsidence, 317–318, 317f
Mine waste, 317
Mineral beneficiation, hazards, 321
Mineral Concession Rule, 322
Mineral Conservation Act, 322
Mineral density, 49
Mineral deposits (mineral and rock deposits). *See also* Genetic model; Mineral/mineral deposits classification
defined, 262–263
further reading, 279
host rocks and, 265–266, 271t
Mineral economics
annual cash flow diagram, 299–300, 300t
cutoff and, 282
feasibility studies
baseline monitoring, 316
described, 295, 298, 302
investment analysis, 299–300, 299f
investment stages, 298
mineral resources/reserves and, 282
Order of Magnitude/Scoping Study, 300–301
overview, 298
prefeasibility study, 295, 302
Mineral exploration hazards, 315–316
Mineral exploration program
hazards, 315–316
stages, 315–316
Mineral hazards, 310–313. *See also* Asbestos
apatite, 311
bauxites, 311
cerussite, 312
clay, 311
galena, 312
graphite, 312
gypsum, 312
mica, 312
pyrite, 312–313
radon gas, 313
silica, 313
talc, 313
wollastonite, 313
Mineral oil and gas. *See* Petroleum
Mineral Policy, 322
Mineral resources/ore reserves. *See also* Estimation procedure; Mineral economics
defined, 282
economic parameters, 282
further reading, 304
introduction, 14
overview, 302–304
parameters, 282–284
cutoff, 282
minimum width, 283
sampling techniques, 292–293
sedimentary rocks as, 123–124
Mineral resources/ore reserves classification system
metallic/nonmetallic minerals classification, 293–296
conventional/traditional classification system, 293–294, 294f
JORC Classification Code, 293, 296, 296f
UNFC system, 293, 295–296
USGS/USBM classification scheme, 293–295
overview, 292–298
petroleum (oil and gas), 296–298, 297f
preparation by QPs, 293
worldwide, 293
Mineral sustainability. *See* Sustainable mineral/mine development

Mineral/mineral deposits classification
chemical composition, 263, 264t
depth of occurrence, 263–265, 268t
geographic distribution, 263
mode of occurrence, 265
nature of mineralization, 265
overview, 263–265
structural control, 265
Mineralogy
defined, 39–40
further reading, 79
introduction, 39–40
Minerals. *See also* Metallic minerals; Nonmetallic minerals
amorphous, 40–41
chemical properties, 47
defined, 261–262
diagnostic features/uses (list), 15t–30t
formation, 41
further reading, 37, 279
importance to society, 1–5
introduction, 5–12
number of, 262
physical properties, 47–50
rock forming
by chemistries, 52t
overview, 51–79
percents of, 52t
significance, 310–311
societal growth and, 310
uses/diagnostic features (list), 15t–30t
Mines Act, 322
Mines and Mineral (Development and Regulation) Act, 322
Mines Rules (Health and Safety), 322
Minimum width, 283
Mining hazards
airborne contaminations, 319
baseline monitoring, 316
introduction, 316–321
mine fire, 318–319, 318f
mine subsidence, 317–318, 317f
mine waste, 317
noise, 319–320
surface land, 316–317
vibration, 320
water resources
management plan, 320–321
mining hazards, 320–321
Mining sustainability. *See* Sustainable mineral/mine development
Minor (accessory) mineral ingredients, 98
Miocene sediments, depositional facies distribution of, 150f
Mirocline, 76–77
Mississippi Delta model, 238
Mississippi valley-type deposits, 277, 277f
Mixed kerogens, 163
Mixed sandstones, 159
Mode of occurrences, mineral/mineral deposit classification, 265
Mohorovicic, Andrija, 84
Mohorovicic discontinuity, 84
Mohs' Scale of Mineral Hardness, 48t
Molybdenite, 15t–30t
Monazite, 15t–30t
Monoclinic, 46
Monoclinic amphiboles, 67, 67t
Montmorillonite, 127, 160
Monzonite, 108
Moraines, 130
Mount Ashitaka, 307–308
Mount Buninyong, 307–308
Mount Huascarán, 309
Mud shale, 159–160
Mud-cracks, 137–138
Mudstone limestone, 190, 190f
Mudstones, 149
Muscovite, 70
Mylonites, 221f

N

Natrolite, 78
Natural gas, 164, 263. *See also* Petroleum
Natural hazards, 306–310. *See also* Avalanches; Earthquakes; Forest fires; Glaciers; Lightning; Volcanoes
Nature of mineralization, mineral/mineral deposit classification, 265
Nebular hypothesis, 88
Nematoblastic texture, 216
Nepheline, 78
Nesosilicates, 59, 59t
Niccolite, 15t–30t
Nickeline, 15t–30t
Noise, 319–320
Noncarbonate authigenic minerals, in limestone, 188
Noncollisional orogeny, 91
Nonmetallic minerals. *See also* Mineral resources/ore reserves classification system
diagnostic features/uses (list), 15t–30t
kinds, 263
uses/diagnostic features (list), 15t–30t
Norite, 111–112
Normal probability distribution, statistical analysis, 288, 289f
Novaculite, 209t, 211

O

Oceanic ridges, 89–90
Oil and gas. *See* Collector rocks of oil and gas; Petroleum
Oil fields (Regulation & Development) Act, 322
Oil shales, 161–163
Oil source rocks, 163
Oligomic conglomerates, 148t, 152
Olivine, 15t–30t, 59
Omphacite, 66
Oncoids, 185–186, 186f
Onion skin, 118
Ooids, 184–185, 185f
Oolites, 185
Opal, 41, 54
Opal-A, 207–208
Opal-CT, 207–209
Ophiolites, 91b, 115
Optical properties, of minerals, 50
Order of Magnitude/Scoping Study, 300–301
Ore, 283
Ore deposits, 283–284
defined, 283
important locations, 283–284
3D orebody wireframe model, 283f
Ore reserves. *See* Mineral resources/ore reserves
Orebodies
chromite, 272f
genetic model, 266–267
geographic distribution, 263
minimum width, 283

preferential geographical location, 283–284
stringers, 273f
Orogenetic movements, 91
Orthoclase, 75
Orthoconglomerates, 148t, 152–153
Orthometamorphite, 219
Orthopyroxenes, 42–43, 64t
Orthoquartzose conglomerate, 152–153
Orthorhombic, 46
Orthorhombic amphiboles, 66, 67t
Oxidation, 126
Oxides, 53–55, 53t

P

Packing of grains, sedimentary rocks, 142–144, 143f
Packstone limestone, 190, 190f
Pannonian Basin, 131, 151, 154, 234, 258–259, 259f
Paraconglomerates, 148t, 153
Paragonite, 70
Parametamorphite, 219
Pegmatites, 120
Pelite. *See* Fine granular clastic sediments
Pellets, 183
Peloids, 183–184
Pennine deposits, 278
Pentlandite, 15t–30t
Peridotite, 113–114
Peritidal carbonates, 250–253
Petroleum (crude oil and gas), 290–292. *See also* Collector rocks of oil and gas
  classification system, 296–298
  contingent resources, 297–298
  defined, 290
  energy minerals, 263
  prospective resources, 298
Petroleum reservoirs
  analogy base method, 290–291
  defined, 290
  described, 290
  field, 290
  performance analysis, 292
    Decline Curve Analysis, 292
    Material Balance Equation, 292
    Reservoir Simulation Models, 292
  traps, 290
  vertical/lateral continuity of, 290
  volumetric estimate, 291–292, 291f
Petrology
  of evaporites, 205–207
  further reading, 91
  introduction, 81–83
Petromict conglomerates, 152–153
Phillipsite, 79
Phlogopite, 70
Phonolite, 117–118
Phosphate sandstones, 159
Phosphates, 58
Phosphorites, 58
Phyllite, 224
Phyllonites, 220
Phyllosilicates, 68–73
Physical mineralogy, 40
Physical/mechanical weathering, 124–125
Pillar of Pompey, 108f
Pisoids, 182–184, 194
Pisolites, 185
Placer deposits, 279
Plagioclases, 110
Planar cross-bedding, 135
Planets, 88b
Plate tectonics, 89–91, 91b. *See also* Tectonic plates
Playa, 205
Plutonic gabbro, 31t–36t
Plutonic igneous rocks. *See* Intrusive igneous rocks
Plutonic metamorphism, 230–232
Pneuma, 95
Pneumatolysis, 42b
Pneumatolytic metamorphism, 220–222
Polarizing microscope, 50
Polymorphism, 50–51, 51t
Popocatépetl Volcano, 307–308
Porcelanite, 210–211
Porphyrite, 117
Porphyritic granite, 105
Porphyritic texture, 102
Porphyroblasts, 216
"Possible," traditional classification, 294
Postsedimentary diagenetic breccias, 152
Precipitation systems of sedimentary bodies. *See* Collector rocks
Prefeasibility study, 295, 302
Probability distribution, statistical analysis, 288, 289f
"Probable," traditional classification, 294
Prodelta, 238
Progradation, 131
Prograde metamorphism, 217, 219f, 232
Progressive metamorphism, 217, 219f, 232
Prospecting stage, 295, 298, 315–316. *See also* Order of Magnitude/ Scoping Study; Reconnaissance phase
Prospective resources, petroleum, 298
Proterozoic Australia, zinc-lead-silver deposits, 266f
Protolith, 213, 217
Protoplanet hypothesis, 87–88
"Proved," traditional classification, 294
Psilomelane, 15t–30t, 53t, 55
Pumice stones, 104, 116, 175, 175f
Pyrite
  defined, 53
  features and uses, 15t–30t
  hazards, 312–313
Pyroclastic breccias, 148t, 149, 152
Pyroclastic sediments. *See also* Clastic sediments and sedimentary rocks; Volcaniclastic sediments and rocks
  classification chart, 146t
  crystal clasts, 174, 178
  defined, 95, 146, 146t
  lithoclasts, 174, 178, 257
  pumice stones, 104, 116, 175, 175f
  scoria, 175
  tephra, 175, 177
  vitroclasts, 174–175, 178
Pyroclasts, 308–309
Pyrogenesis, 41, 42b
Pyrolusite, 15t–30t, 55
Pyrophyllite, 69
Pyroxenes, 15t–30t, 42
Pyroxenites, 115
Pyrrhotite, 12f, 272f

Q

Qualified persons (QPs), 293
Quartz
  arenites, 156

Quartz (*Continued*)
defined, 53–54
features and uses, 15t–30t
Mohs' Scale of Mineral Hardness, 48t
monzonite, 107–108
Quartz crystal
defined, 8f
stalagmite, 262f
Quartzite, 229
Quaternary faults, 315
Qutab Minar, 157f

R

Radiolarians, 208–210
Radiolarites, 144, 208–209, 209t
Radon gas, 313
Raindrop imprints, 139
Rajpura-Dariba-Bethumni Belt, zinc-lead-silver deposits, 267f
Reconnaissance phase
baseline monitoring, 316
hazards, 315–316
Order of Magnitude Study/Scoping Study, 300–301
in sequential approach, 298, 315–316
UNFC system, 295
Recycle keyword, sustainable development principle, 322
Red Mediterranean soil, 173–174
Reduce keyword, sustainable development principle, 322
Reef and perireef limestones, 254–260
Reef-building organisms, 188
Refining hazards, 322
Regenerate keyword, sustainable development principle, 322
Regional high-grade metamorphism, 223–229
Regional low-grade metamorphism, 223–229
Relict minerals, 217
Replace keyword, sustainable development principle, 322
Reservoir Simulation Models, 292. *See also* Petroleum reservoirs
Residual deposits, genetic model, 279
Residual sediments, 171–174
Residues, 145–146, 146t
Resource assessment. *See* Mineral resources/ore reserves
Resource keyword, sustainable development principle, 322
Restricted shoals, 253–254
Retrogradation, 131
Retrograde metamorphism, 217, 219f
Reuse keyword, sustainable development principle, 322
Rheology, 83b
Rhodochrosite, 12f
Rhodonite, 15t–30t
Rhombic dodecahedron, 60f
Rhyolite, 116
Richter magnitude scale, 306, 307t
Riebeckite, 68
Ring silicates, 62–63
Rinsed residues, 145, 146t
Ripple marks, 138–139, 138f
Rock deposits. *See* Mineral deposits
Rock faults. *See* Faults
Rock forming minerals. *See* Minerals
Rock hazards, 313–315
balancing rocks, 314–315, 314f
faults, 315, 315f
granite, 313
limestone, 313
overview, 313
sandstone, 313–314
slate, 314
Rockfall, 146, 146f, 314, 314f
Rockfall breccias, 150–151, 245
Rocks. *See also* Igneous rocks; Metamorphic rocks; Sedimentary rocks
classification of, 85–86
defined, 262
diagnostic features/uses (list), 31t–36t
further reading, 37
importance to society, 1–5
introduction, 12–14
stones compared to, 14
uses/diagnostic features (list), 31t–36t
Ruby, 54
Rudaceous. *See* Coarse-grained sediments
Rudstone, 190f, 192
Rutile, 15t–30t, 55

S

Sabkha anhydrite, 204–205, 207
Sampling techniques, 292–293
Sand bodies, in coastal marine environments, 239–242, 243f
Sandstones
arenite, 155–157
features and uses, 31t–36t
hazards, 313–314
mixed/hybrid, 159
Sandy barrier islands, 250
Sandy sediments
composition/distribution
arenaceous rocks, 153–155
diagenesis, 164–168
Sangetser Lake, 306f
Sanidine, 75
Sapphire, 54
Sapropelic, 163
Scheelite, 15t–30t
Schistose, 215
Schistosity, 215, 223–224
Schists, 224–226
Schists of metamorphism. *See* Regional high-grade metamorphism; Regional low-grade metamorphism
Scoping/Order of Magnitude Study, 300–301
Scoria, 175
Sea delta, 238–239, 239f
Secondary minerals, 98
SEDEX/stratiform deposits, 276, 277f
Sediment transport
aeolian processes, 129
fluvial processes, 128–129
glacial processes, 130
overview, 128
Sedimentary rock formation. *See also* Weathering
deposition, 130–132
lithification, 132–133
overview, 124
Sedimentary rocks. *See also* Clastic sediments and sedimentary rocks; Siliceous sediments and rocks; Volcaniclastic sediments and rocks
classification, 85, 144
defined, 85
features and uses, 31t–36t
formation, 122–123
function, 122–124
further reading, 212
genetic model, 272–273, 276f

as mineral resources, 123–124
significance, 122–124
transformation, 122–124
uses and features, 31t–36t
Sedimentary rocks- structure and texture. *See also* Bedding
depositional system, 212f
overview, 133–144
packing of grains, 142–144, 143f
Seismometer, 306–307
Sela Pass, 2f
Semivariogram, 288, 289f
Sericite matrix, 168
Sericite schist, 224
Serpentine, 73
Serpentinite, 31t–36t, 73
Shale, 31t–36t
Shine, of minerals, 49–50
Shoals, restricted, 253–254
Siderite, 56, 56t
Silica, hazards, 313
Silicates, 58–79
Siliceous sediments and rocks, 207–212
of biogenic origin, 208–210, 209t
chalcedony, 207–209
classification of, 207–208
of diagenetic origin, 209t, 210–212
diatomaceous earth, 208–209, 209t
diatomaceous mud, 208–209
diatomite, 144, 208–209, 211
mineral composition, 207–208
opal-A, 207–208
opal-CT, 207–209
origin of, 207–208
radiolarians, 208–210
radiolarites, 144, 208–209, 209t
spicule muds, 208–209, 209t
Siliciclastic detritus, 142
Siliciclastic grains, 142
Siliciclastic terrigenic components, of limestone, 188
Silicification, 198
Silicon sinter, 210
Sillar-tuffs, 178
Sillimanite, 225
Sills, 101
Silt, 160–161
Siltstone, 160–161
Silver-lead-zinc deposits
Proterozoic Australia, 266f
Rajpura-Dariba-Bethumni Belt, 267f
Skarns, 223, 278–279
Skeletal-limestone, 183, 183f, 188
Skeletal-shell organisms, 187–188, 189f
Skutterudite, 13f
Slate, 224, 314
Slaty rocks, 215
Slump, 141, 141f
Slump breccia, 149
Small and medium size deposits, estimation procedure, 285
Smectite group, 71–72
Smelting hazards, 321–322
Smithsonite, 15t–30t, 56t
Solar System, 88b
Sorosilicates, 61–62
Sperrylite, 15t–30t
Sphalerite, 13f, 277f
Spicule muds, 208–209, 209t
Spilite, 120
Spinel, 55
Spodumene, 66
Stalactites, 194–195, 194f
Stalagmite quartz crystal, 262f
Stalagmites, 194–195, 194f
Stannite, 15t–30t
Stars, 88b
Static cutoff, 282
Statistical method, estimation procedure, 288
Staurolite, 15t–30t, 61
Stibnite, 13f
Stone, rock and, 14
Stopes, 282, 293–294, 294b, 317, 317f
Stoping, 294b
Stormy breccias, 148
Stratiform deposits. *See* SEDEX/stratiform deposits
Stringers, 273f
Stromatolites, 183, 187, 187f, 190
Strontianite, 56t
Structural control, mineral/mineral deposit classification, 265
Subarkoses, 156–157
Subduction zone, 90
Sublithic arenites, 156
Submarine volcanic eruptions, volcaniclastic sediments, 176f, 177, 177f
Submarine volcanism, 94, 118–119, 175, 177, 177f, 275–276
Subtidal zone, 250
Sulfates, 57
Sulfides, 52–53
Sulfur, 15t–30t, 52
Supratidal zone, 251
Surface land
agricultural land loss, 316
drainage pattern changes, 316
forest land loss, 316
landslides, 317
land-use planning, 317
mining hazards, 316–317
top-soil/subsoil degradation, 316
unaesthetic landscape, 317
Sustainable mineral/mine development. *See also* Hazards
achievement methods
fiscal measures, 322
legislation, 322
preservation of environment and forest, 322
regulated exploitation of mineral resources, 322
science and technology, 322
further reading, 323
key management tasks, 322–323
keywords, 322
overview, 315, 322–323
visible issues, 322–323
Syenite, 109
Sylvanite, 15t–30t, 57
Sylvite, 15t–30t, 57

T

Taj Mahal, 231f
Talc
defined, 68–69
features and uses, 15t–30t
hazards, 313
Mohs' Scale of Mineral Hardness, 48t
schists, 225
Talc-pyrophyllite group, 68–69
Tectonic breccias, 151
Tectonic plates, 84, 85b, 223, 245, 307
Tectosilicates, 73–79, 74t
Tephra, 175, 177
Terra rossa (red soil), 173–174
Terrestrial limestone, 192–195
cave limestone, 194–195, 194f
crusty limestone, 194
dripstone, 194–195, 194f
stalactites, 194–195, 194f
travertine, 193–194, 194f
types, 193
Tetragonal, 46
Tetrahedron, 58f

Textures. *See specific textures*
Thawing-freezing, 125
Tholeiitic basalts, 118–119
3D orebody wireframe model, 283f
Tidal flats, 195, 240
Till, 147
Titanite, 61
Tonalite, 108
Topaz, 15t–30t, 48t
Tourmaline, 10f
Tower/London Bridge, 3f
Trachyte, 117
Traditional/conventional classification, minerals, 293–294, 294f
Transform faults, 89–90
Traps, petroleum reservoirs, 290
Travertine, 193–194, 194f
Tribal's in Mining Projects, 322
Triclinic, 46
Tripoli, 210–211
Troctolite, 109–110
Trough cross-bedding, 136
Tsunamis, 85b, 306, 315
Tuffite material, 177–178
Tuffs
  alteration of, 178–179
  defined, 178
  lapilli, 175–177, 176t
  sillar-tuffs, 178
  welded, 178
Turbidites, 257–258
Turbidity currents, 128–129
Turbidity fans, 245–247

U

Ultrabasic intrusive igneous rocks, 113–115
Ultramafic intrusive igneous rocks, 99t, 113–115
Underwater slides/destruction of layers, bedding, 141–142
Upper bedding plane structures, 137–140
Uraninite, 15t–30t

V

Variable cutoff concept, 282
Veins igneous rocks, 120
Vermiculite group, 71
Vertical longitudinal section, estimation procedure, 287, 287f
Vertical/lateral continuity, of petroleum reservoirs, 290
Vesuvianite, 62
VHMS. *See* Volcanic-hosted massive sulfide
Vibration, 320
Victoria Memorial Hall, 231f
Vitroclasts, 174–175, 178
VMS. *See* Volcanogenic massive sulfide
Volcanic basalt, 31t–36t
Volcanic bombs, 176t, 177–178
Volcanic breccia, 177
Volcanic glass, acidic, 178–179
Volcanic igneous rocks. *See* Extrusive igneous rocks
Volcanic-hosted massive sulfide (VHMS), 275–276
Volcaniclastic sediments and rocks. *See also* Pyroclastic sediments; Sedimentary rocks
  agglomerates, 176t, 177–178
  classification, 176t
  composition, 177–178
  defined, 123, 174–177
  formation/origin
    from pyroclastic flows, 175
    pyroclastic material from air, 175–177
    from turbulent flow of low density and high speed, 175, 177
  pumice stones, 104, 116, 175, 175f
  pyroclastic breccias, 148t, 149, 152
  scoria, 175
  volcanic bombs, 176t, 177–178
Volcanism. *See also* Submarine volcanism
  hazards, 319–320
  pyrogenesis, 41, 42b
Volcanoes, 307–308, 308f
Volcanogenic massive sulfide (VMS), 275–276
Volumetric estimate, petroleum reservoirs, 291–292, 291f
Vortex casts, 140

W

Wackes. *See* Graywacke
Wackestone limestone, 190, 190f
Water resources
  management plan, 320–321
  mining hazards, 320–321
Wave-formed ripple marks, 139
Wavy bedding, 137
Weathering, 124–128
  biological, 128
  chemical, 126–128, 126f
  concentric, 118
  physical or mechanical, 124–125
Weiss parameter, 43
Welded tuffs, 178
Wentworth scale, 144, 145f
Wild Life Protection Act, 322
Witherite, 56t
Wolframite, 13f
Wollastonite, 313

X

Xenoblasts, 216
Xenoliths, 100

Z

Zeolites group, 78–79
Zincite, 15t–30t
Zinc-lead-silver deposits, 266f–267f
Zircon, 15t–30t, 61
Zoisite, 61–62

Printed and bound by CPI Group (UK) Ltd, Croydon, CR0 4YY
03/10/2024
01040321-0003